From Clouds to the Brain

Series Editor
Jean-Claude Dupont

From Clouds to the Brain

The Movement of Electricity in Medical Science

Céline Cherici

iSTE

WILEY

First published 2020 in Great Britain and the United States by ISTE Ltd and John Wiley & Sons, Inc.

ISTE Ltd
27-37 St George's Road
London SW19 4EU
UK

www.iste.co.uk

John Wiley & Sons, Inc.
111 River Street
Hoboken, NJ 07030
USA

www.wiley.com

Library of Congress Control Number: 2020938719

British Library Cataloguing-in-Publication Data
A CIP record for this book is available from the British Library
ISBN 978-1-78630-595-4

MIX
Paper from responsible sources
FSC® C013604

Contents

Foreword

Céline Cherici's book explores the place of electricity in the history of medical science over a period ranging from the 18th to the 21st Centuries, from its discovery to the recent use of deep brain stimulation. Some contextual data helps us to grasp the full importance and scope of this undertaking.

The current importance of chemical and molecular representations, which have become consubstantial with biology, means it is easy that we forget that, from a historical point of view, it is physics that has constantly provided neurophysiology with explanatory models.

Galvani's use of artificial electricity as a stimulant led him, by analogy, to conceive of muscle fibers as small Leyden jars. Thus, the mysterious nature of nervous fluid, which was therefore electric, a model which was vary different from the humoral model described by the Encyclopedists, was unraveled. The fruitfulness of the polemic between the proponents of metallic and animal electrics is well known in the history of physics, since the voltaic pile battery triggered the development of electrophysics, electromagnetism and electrochemistry, the development of which, in turn, went hand in hand with the invention of two instruments of considerable heuristic value in 19th Century Life Sciences: the galvanometer and the impolarizable electrode. These instruments led to the development of the great German electrophysiology, that is to say, the classical works of du Bois-Reymond, Helmholtz, Hermann, Bernstein, etc., which made it possible to propose various hypotheses on the nature of the action and rest currents. These currents were often considered to be the result of purely physical phenomena occurring in living tissue. In accordance with the dominant physical and technical models, nerve fiber was likened to a metal wire, a circuit, a magnet and, finally, a battery – a living battery the functioning of which physicists sought to understand by taking inspiration from the model of the electric battery. The origin of the battery's electromotive force, how the nerve impulse was conducted and the changes produced in the fibers by this impulse were also investigated, leading to imported notions such as "polarized structures" or "local circuits". In France, it was this tradition, that of the physical explanation of the functioning of living organisms, that was taken up by Louis Lapicque when he likened nerve fiber to a radio transmitter.

This electrical power was also supported by electrodiagnosis and electrotherapy. Studying the effect of electricity on the living was done in many ways, depending on the intention. The problem of elucidating the nature of nerve impulses, excitability and underlying processes seemed to be distinct from the problem of investigating the possible use of electricity in medicine. For a physician, knowledge of the laws of excitability, or even that of simple correlations or modifications of excitability according to the pathology making electrodiagnosis possible, as well as the revelation of the therapeutic effects of electricity, could be quite sufficient. In this sense, the electrophysiological studies carried out in the laboratory on animals seemed to have little relevance to medicine. Under the influence of positivism, empiricism and pragmatism, a myth of 19th Century medical thought was forged, that of the independence of the experimental and the clinical.

In fact, the history of electrodiagnosis attests to the close and constant links between the laboratory and the clinic from the very beginning. One of its founders, the clinician Duchenne de Boulogne, who gave his name to various myopathies, conceived it as a true experiment in the examination of patients. In the 19th Century, electrophysiological instrumentation was common to both the laboratory and the hospital. The characters who have marked the history of electrodiagnosis were often both laboratory technicians and clinicians. Duchenne used the induction coil (faradic current), with which it was possible to excite nerves and muscles through the skin at certain points. He established the topography of these motor points, skin regions where the electrodes had to be placed to obtain the muscle jolt with the least possible intensity (bipolar excitation). He thus inaugurated electrical semiology: either the muscles no longer responded to induction excitations (faradic hypoexcitability; Duchenne reaction), or the excitability was normal. As for anatomopathology, it was the electrodiagnosis inaugurated by Duchenne that contributed to the creation of a nosological group, that of degenerative diseases, and, more generally, of an electrical semiology of muscular and neuromuscular diseases.

In France, chairs of medical physics were developed and journals were created for the new specialty, such as the *Archives d'électricité médicale* (1893) or the *Annales d'électrobiologie* (1898). Following Duchenne, prestigious names, often both doctors and physicists, devoted themselves to medical electricity and wrote treatises in which electrology held a central place, such as Jean Bergognié, founder of French radiobiology and creator of the first anti-cancer centers, Emmanuel Doumer, who studied the use of electricity in surgery, or Jacques d'Arsonval, a student of Claude Bernard, a Brown-Sequard collaborator who worked with diathermic currents, electrocoagulation (electric scalpel) and electrotherapy with high-frequency currents (darsonvalization, 1913). Forms of electrodiagnosis were developed that were not based on neuromuscular excitability (electrical resistance of the human body, voltaic vertigo), preceding an electrodiagnosis no longer based on stimulation, but simply on detection (EEG, electromyography). It should be noted that while English-speaking countries were engaged in electromyography (Adrian and Bronk), the Netherlands in electrocardiography (Einthoven, Nobel Prize 1924), Germany in electroencephalography (Berger), France's focus in the thirties remained on chronaximetry with Louis Lapicque, and thus demonstrated a considerable delay between the two wars.

Electrotherapy was another cornerstone of electricity in the medical sciences. In the 18th Century, ignorance of the exact nature of nervous fluid did not prevent the empirical use of electricity for therapeutic purposes by great names such as Nollet, Jalabert, Aldini, or Marat. During the 19th Century, electrotherapy instrumentation developed considerably in the field of galvanization, faradization, galvano-faradization, franklinization and hertzian and other darsonvalization, up to the point of the electroshock of patients.

It is necessary to underline how poor and disappointing neurochemistry remained for a long time, in comparison with this wealth of medical electricity. Moreover, some physiologists still had the diffuse idea, inherited from the 19th Century, that it was not necessary to precisely identify the charge carriers, that is, to penetrate to the deepest level of the phenomena, in order to explain cerebral or nervous functioning. Electricity was still "the essence of life". The dominant paradigm and culture of electricity explained, for example, the skepticism with which the first real experimental argument in favor of chemical neurotransmission at the ends of the autonomic nervous system was greeted, the experiment of Otto Loewi (1921), and then, from the 1930s onwards, the results of Henry Dale's school concerning chemical transmission at the lymph node and neuro-muscular scales. In 1936, when Loewi and Dale were awarded the Nobel Prize, the accumulation of divergent data led to major difficulties in chemical theory, giving way to the development of elaborate electrical designs following Lapicque's ancient chronaxial theory, such as that of John Eccles. While, at the beginning of the 1950s, a consensus had been established among pharmacologists in view of the considerable therapeutic perspectives offered by chemical theory, it remained limited to the peripheral nervous system, especially since explanations of the characteristics of the most elementary central activity, reflex activity, had been proposed, avoiding any departure from a strictly electrical determinism using neuronal circuits. For these reasons, and for reasons of anatomical complexity, the penetration of chemical theory at the central level was slow. It was finally made possible by techniques (microiontophoresis), by the renewal of the neurochemical context (neuroendocrinology), and, above all, by the appearance of psychotropic drugs, despite the subsequent discovery of electrical synapses. Chemical theory offered a considerable range of interpretations of the mode of action of psychotropic drugs, from which one sought to extrapolate the pathogenesis of neurological (Parkinson's, epilepsy) or even psychiatric (depression, psychoses) diseases.

It is the importance of this culture of electricity that Céline Cherici develops in her book. The judicious choices of the long period and the resonance of different disciplinary fields (physics, physiology, medicine, in particular, neurology and psychiatry) around this theme of electricity have allowed her to propose research at the crossroads of the history of techniques and the history of biology and medicine. Céline Cherici demonstrates the imprint left by electricity in culture and life sciences, focusing her analyses on the links between electricity and the nervous system ("from the clouds to the brain") and the medical appropriation of electricity. The aim is to describe the establishment of this culture, and to analyze, beyond the origin of ideas and facts, the beginnings (Canguilhem), the epistemical (Foucault, [FOU 06]) and the phenomenotechnical (Bachelard) bases that made this establishment possible. She thus shows how the questions of materiality and the location of the soul and faculties are closely linked, in the 18th Century, to the promotion of electricity as a tool for the treatment of convulsive illnesses.

One can only agree with this idea of a long and profound influence of electrical culture if we remember the resistance to brain chemistry reported above. During the 20th Century, the electric brain model, supported by electroencephalographic data, still predominated, despite the experimental evidence to the contrary. Similarly, when Cherici studies deep brain stimulation applied to the field of psychiatry, which is also an exploratory technique, it is to show its importance in the development of a normal and pathological model of brain function. By paying attention to the instrumental context of the discoveries, Céline Cherici underlines the extent to which Gaston Bachelard's invitation to understand science as an "empirical inventive thought" applies to medicine. Medical practices require instruments, which they transform as needed. Conversely, medical instruments transform practices, and model representations of diseases, directing them towards certain theoretical options concerning body functions. The history of nervous diseases covered and medical electricity described by Céline Cherici is a masterful illustration of this. From these subtle interactions between the instrument and the concept results a true invention of nervous diseases and a construction of brain models, which a history of medicine had to consider. There are, however, few works that address the question in such a way and to such an extent, François Zanetti's recent book being limited to the France of the Enlightenment (2017). Cherici's work thus undoubtedly fills an editorial gap. It also testifies to the vitality of an approach, that of a history of methods and concepts, in the tradition of French epistemology, which the author knows how to extend by insisting on its historical foundation, but also by renewing it, opening it happily to the history of techniques, anthropology and cultural history.

Jean-Claude DUPONT
University of Picardy Jules Verne
28 May 2020

Introduction

From the clouds to the brain: this is the journey, both historical and epistemological in nature, that this book sets out to retrace. The focus of this book is the history of medical electricity, and it is this journey that is explored, accompanied by a look at electricity's medical effects on the body, up to and including the center of the human brain. Although this history of explorations and brain simulation may seem recent, we can broaden the scope of this investigation further and take into account the philosophical, scientific and technical roots of medical electricity in the 18th Century. I am part of an important secondary literature, often written in English, tracing the major historical stages of electricity in its links with the body, the living and neuroscience. Indeed, with the exception of François Zanetti's excellent French-language book [ZAN 17], published in 2017, the summaries on the history of this force are, for the most part, in English. The research on the effects and applications of electricity on the body, the brain and living things has been collected and correlated, in order to show the connections between these fields. The works of Iwan Rhys Morus, published between 1998 and 2011 [RHY 98, 99, 11, 09a, 10, 02, 04, 09b], have also been analyzed to show how the electrical sciences have permeated society and science from the end of the 18th Century onwards, thus opening up a dimension of cultural history. On the other hand, the treatises on the links between the applications of electricity and the birth of neuroscience, notably by Stanley Finger [FIN 94, 99, 11, 12] as well as the works on the exploration of electric life by Marco Piccolino and Marco Bresadola [PIC 03, 13] have been studied at length. The historical period covered by this book is between 1740 and 2010. In addition, epistemological analysis has been tightened around the correlations between electricity, the brain and its nerve ramifications. Thus, we find the representations of an *electric culture* [RHY 11, p. 9], applied to the body in its physical and moral dimensions, from the second half of the 18th Century. Far from being reduced to a cycle of failures and errors, this period shows the emergence of an electrical tool applicable, not only to human ills, but also to the exploration of the mechanisms of a living being, a category which, of course, includes the human species. As early as 1746, static electricity machines were built, the body was a member of the family of conducting bodies and medicine, marked by physics, became electric. The peak of this movement was reached during the controversies between animal and metal electricity, around 1790.

But how do we retrace that story? How do we differentiate the origins of the knowledge of electricity from its beginnings as knowledge itself? While electricity refers to the Greek term ἤλεκτρον (*êlektron*), which means Baltic amber, it does not mean that knowledge was being built at that time. However, Thalès de Milet, in the 7th Century BCE, recorded the fact that amber, if rubbed, had the ability to attract light objects and to produce, though not systematically, sparks. Moreover, Hippocrates, Plato and Galien described the remarkable properties of electric rays, so frequent in the Mediterranean. Galien used them on living patients in the treatment of rheumatic afflictions and headaches. In addition, amber, a physical electricity present in nature, was also noted. Sribonius used electric shocks [SRI 55] to treat a wide variety of diseases, including headaches and various kinds of paralysis. Around 1600, William Gilbert (1544–1603) recognized that the property of attracting light bodies was common to certain minerals and stones. Otto von Guericke (1602–1686) made one of the first electrical machines, around 1660, and compared the phenomenon caused to the attraction of the Earth on animate and inanimate bodies.

So, when do we talk about the beginnings of electricity? Do we have to trace them back to Greco-Roman antiquity? To 17th Century mechanics?

> For what was electricity when Thales of Miletus discovered it? And what became of it for a long series of centuries, in the hands of Pliny, Strabo, Dioscorides and Plutarch? It was, during this long interval, only a seed stuck in the ground, waiting for happier hands to bring it out [...]. [ALD 04, ij, author's translation]

Its beginnings were initiated by the explorations of the forces of nature through 18th Century physics, which became systematic and also corresponded to a vast questioning of humanity's place in Nature and its links with the laws at work there. In the same way, it was necessary that a particular epistemology enabled the questioning of the localization of the soul in the brain, the materiality and innateness of the faculties, making them free to develop, in order to found medical electricity as a tool of care for the illnesses of the psychological sphere. This discrepancy between the moment of origins and that of beginnings also made it possible to understand the immediate appropriation of electricity in the medical field. Indeed, as early as 1746, when the Leyden jar experiment by Musschenbroek and his assistant proved dangerous and painful, this first capacitor immediately catalyzed the hopes of a new medicine which was technical, interventionist, economical and beyond all metaphysical considerations.

The history of medical electricity, beyond its periodization, is based on a questioning of the concepts at work in it, such as human nature, natural laws and the study of forces. It also requires an in-depth study of the techniques that are constantly revising its applications, making them more precise and more reliable, as well as the theme of contexts, which appear to be so many different fields of experimentation and the setting up of new protocols. In addition to representing a relatively long period, the period from 1740 to 2010 required more work on the primary bibliography. For example, the *Bibliographie francophone des ouvrages et articles relatifs à l'électricité et au magnétisme publiés avant 1820* [BLO 00] has no fewer than 2,000 titles. This is why the theme is centered around the links between electricity, medico-philosophical questions on the naturalization of faculties and the brain as the place where these issues are anchored. It is an epistemological journey to which we are invited by the different chapters of this book.

Research, more than progress, around electric power immediately marks a strong imagination where humanity takes precedence over nature and over itself. First mixed with experiments on magnetism and mesmerism, electricity is part of the context of investigations and experiments on the energies at work in the universe. For example, Mesmer (1734–1815), whose medical thesis was on the influence of the planets on the evolution of humanity, developed the idea that living beings are linked together by a universal magnetic force. This force, present in the macrocosm, could, according to him, have a major influence on health and balance. He thus posed as a practitioner capable of rebalancing the flow of animal magnetism in the body. During public sessions, he used magnets to restore the flow of magnetic fluid in subjects suffering from disorders as varied as hysteria and blindness. While his concept of animal magnetism did not survive the report by the commission of the *Académie Française des Sciences* (French Academy of Sciences), requested in 1784 by Louis XVI to evaluate his practices, the idea that there were links between the laws governing the universe and the mechanisms of the body permeated research on electricity. The roots of this conception also appealed to the neo-hippocratism that developed in the 18th Century. The advent of electricity, in the field of physiology and therapy, marked a never-ending intertwining of exploration and care. Its entry as a physiological configuration, conceived in terms of organic fluid, was a sign of a break between a medicine still tinged with metaphysics and a medicine of the Enlightenment, intended to be rationalizing. Its developments during the 18th Century were marked by the naturalization of animal spirits, the shift from the notion of fluid to that of energy, the entry into a secularized medical era opening up a materialistic perspective of the psychological and physical nature of the human being. In any case, these are the representations delivered by the research of Jean Antoine Nollet (1700–1770), Benjamin Franklin (1706–1790), Allamand, Sans and Ledru (1731–1807) from the second half of the 18th Century. One of the paradoxes of this history of the appropriation of electric force by medicine and, more broadly, by physiology and the experimental biological sciences, is that it is, above all, made up of errors and failures, punctuated by the resurgence of hopes carried by electricity.

These developments are structured around six chapters. The first chapter proposes tackling the concept of electric imaginary born of the hopes raised by the new techniques generated from the 18th Century onwards. It is inseparable from the analyses of the different periods in the history of medical electricity[1]. Masars de Cazeles, considered the designer of care practiced by electric friction, recalled the metaphors of a divine, animist electricity whose applications have been integrated and developed within a medicine that has become experimental:

> However, if I were allowed to reason according to the authority of my own people, I would dare to say that the fable of Prometheus stealing the Celestial fire from the wheel of fire of the Sun to animate our clay is, perhaps, only an allegory of the effects of Electricity, formerly glimpsed, little known in the aftermath, brought to light by modern Physicists, & made more interesting by the way in which they now fix the attention of Doctors. [MAS 80, p. 15, author's translation]

The second chapter addresses the intellectual, scientific and experimental path from physics to electrifying physicists: the theme of studies on the laws of electricity will be addressed in order to show that it is physicists who seek to decipher the mechanisms of electricity, who are primarily interested in its effects on the body as well as its therapeutic potential[2]. On the one hand, the philosophical stakes for the inscription of humanity in nature will be taken further, but so will the dependence that this link creates with physics, between electrical therapies and machines. Did medical engineering arise in the 18th Century? Thus, between 1745 and 1765, electricity appeared, in the visual sense of the term, as an instrument of movement, initiator of involuntary mobility. It was in the context of the link, born accidentally following the experiments with the Leyden jar, between movement and electric power, that the first actors of a physics that was becoming medicalized applied it to paralysis, while continuing to explore its mechanisms in nature [NOL 46, ENG 56, SAN 72]. In the third chapter, we will discuss the initial electrical turmoil marked by

---

1 See Appendix 1, in which chronological tables are provided to give the reader a guide to the major stages of this history.

2 See Appendix 2, in which extracts from the tables of contents of physicists, inventors or demonstrators, Nollet, Franklin, Jallabert and Morin, have been selected to highlight their research combining physical knowledge with considerations of the body.

failures that demonstrated controversial applications of this care to tackle nervous and mental illnesses and show how this force began its descent onto the brain. Indeed, after a few years of increased mistrust due, on the one hand, to often fatal electrocutions and, on the other hand, to the ineffectiveness, however distressing, of these treatments, medical electricity moved, between 1770 and 1800, towards the treatment of nervous, convulsive and mental illnesses [LED 83, GAL 91, PET 02–03]. It is also divided, in this same period, between cures of static electricity and medical galvanism. The fourth chapter is thus devoted to the breakthrough generated by Galvani's (1737–1798) discovery of animal electricity. Between medicine and physiology, perspectives on the living were marked by electrocentrism. At the end of the 18th Century, biology appropriated electricity to make it inherent to matter. This stage marked the definitive appropriation by physiology and medicine of this physical energy, as well as the beginning of Galvani's research on electric neuro-fluid. Galvanism, which traveled beyond the Italian borders while Europe was suffering from the political consequences of the French Revolution, opened an extremely heuristic program, both for electrophysiology and for future resuscitative medicine. Thus, bodies came to life like automatons, becoming fields of exploration for the delineation and knowledge of the dying process, the central nervous system and its ramifications throughout the body. The successor to Descartes' (1596–1650) concept of the animal machine, galvanism intended to explore the nervous mechanisms of living beings, as well as medical in its treatment of hysterical and, more broadly, magnetic phenomena. Chapter 5 then discusses the specialization and development of the different branches of biomedical electricity. Between laboratory explorations and clinical applications, electrical medicine, by confronting diseases with vast symptomatologies, contributes to differentiating the fields of psychiatry and neurology. At the same time, the activities of the nervous system are quantified, measured, recorded, objectified and made visible in the form of signs, plots or diagrams. Electrophysiology met Volta's desire to involve measurement and mathematics. Electroclinical and electrophysiological explorations developed between 1900 and 1950 complement each other. While electrotherapy is equipped with machines and techniques that hope to leave their mark, with regard to the problem of the reversibility of psychoses, particularly in the emergency context of the two world wars, electrophysiology measures, models and describes the impacts of electricity in the body. Finally, Chapter 6 discusses the first applications of electrical neurostimulation therapies, and tries to show that the field of mental illnesses was a favorite one as early as 1950. The aim will be to delineate two aspects of the history of brain electricity and its therapies: a long history beginning in the late 18th Century, completed by a shorter history taking its roots in the second half of the 20th Century. Between 1980 and 2010, brain stimulation techniques, deep or external, following research on brain implantation, which considered the field of mental illness as a therapeutic target, stand out in their renewed applications in the field of psychiatry.

Thus, not only can we speak of therapeutic electricity before Galvani's discovery of animal electricity, but it is also a question of making one of the paths of medical electricity in the brain sciences, future neurosciences, and within society intelligible. This is where a long investigation begins: it was necessary to differentiate the stages, the phases of enthusiasm and decline, each period marked by the improvement of techniques and advances in knowledge on the different ways of applying currents (galvanization, faradization, etc.), as well as on the human brain and the ills it can be affected by. Marked by its polymorphism, electricity in the medical field requires a broad epistemological study, both at the level of its temporality and that of the knowledge explored. In an epistemological tradition inherited from Canguilhem (1904–1955), "Philosophy is a reflection for which all unknown material is good, and we would gladly say, for whom all good material must be unknown" [CAN 78, p.8].

Thus, everything is material for thought: the success of a theory, but also its failures and errors. The epistemologist, always in search of lines of convergence and divergence, must approach all the states of the scientific discipline under discussion, respecting both its singularity and its continuity. This continuity, in the case of medical electricity, is marked by a large number of technical, societal and scientific breakthroughs that punctuate the waves of successive crazes and discredit that hinder its development. It is built within an experimental design, conceived in terms of trial and error, marked by failures as much as by fantasized or misunderstood successes. The historical, scientific and philosophical interactions between the concepts of machines, techniques and the brain have necessitated historical back-and-forth, to the benefit of the problematization of the subject. This work takes place in the context of an open epistemology[3].

How can we analyze the failures of an electrical method that has been constantly changing since the 18th Century? How can we understand the links between physics, medicine and current electrical therapies, whose psychiatric applications are multiplying? Does going back to the roots of the applications of medical electricity on the human brain allow us to understand its past and present implications?

3 Cornelius Borck speaks of "open epistemology" [BOR 18a, p. 264].

1

The Birth of an Electrical Culture: From Frankenstein to Hyde

The notion of the *electrical culture* [RHY 11, p. 9], developed by Rhys Morus in his book *Shocking Bodies; Life, Death & Electricity in Victorian England,* came about at the beginning of the 19th Century, coming to the fore through the discussion of two issues in which were mixed scientific aspects and the imagining of a force that seemed to possess all powers, such as:

– "re" creating life: indeed, experiments on the bodies of convicts were adjacent to the theme of electricity as the driving force of life. Aldini and Cumming, by re-animating corpses, dramatized demonstrations, thus spreading the links between galvanism and vital properties;

– control of behaviors: in the middle of the 19th Century, electrical medicine broke with the dualistic paradigm of the electrified automaton to locate, in the brain, the areas that would allow the control of behaviors through these therapies. This movement followed a more general shift from moral issues to psychiatric disorders.

From *Frankenstein* [SHE 18] to the main character in the novel *The Strange Case of Dr. Jekyll and Mr. Hyde* [STE 86], two periods, foundational for medical thought, are articulated. They both contributed to making electricity an intelligible instrument of exploration and treatment, and participated in a strong popular imagining about the possibilities opened up by the application of electrical techniques in medicine. This cultural and medical imagining weaved a context in which these applications took place. The first period took place from Galvani's experiments through to those by Doctor Ure; then after 1840, a second period marked the passage from a dualistic medicine to a holistic medicine, where consciousness was embodied in convolutions. What seemed to correspond to an objectification of the applications of medical electricity, referred to the construction of a culture in which electricity represented a fantasized scientific positivism.

The use of electricity in the "resuscitation" of those who had drowned and the apparently dead was first proposed in 1778 by Charles Kite (1768–1811) to the Royal Humane Society in London. An active member [ALZ 05] of this learned society, he wrote *An Essay on the Recovery of the Apparently Dead* [KIT 88] for which he received a medal. In this essay, he distinguished suspended animation from irreversible death and described the importance of collecting the necessary information on each victim of drowning to assess a possible return to life. In his presentation, he stressed that a body that no longer reacts to electrical shocks should be considered dead. Electricity, in addition to revealing the properties of matter, was imagined, early on, as an instrument to explore the boundaries between life and death. In addition to having an important impact on the definition and description of the dying process, it was conceived and massively disseminated in scientific, literary and popular circles as a means of bringing people back to life.

1.1. "Re"creating life?

There was only a short leap from experiments where animals were electrified, then revived, then electrified again through to the formulation that *electricity is life*. The underlying idea was that as long as a limb could be electrified, it still had signs of life. To what extent could medical electricity become an instrument of resurrection? Many of the episodes in the history of medical electricity revolve around questions of life and death, and as early as 1740, human control over the boundaries between these two states and medical power emerged. The implication of electricity in the resurrection process was deduced from experiments to bring the dead back to life. This concept prefigured the advent of future resuscitation techniques and is part of the context of an interventionist and dualist medicine, the body being like an automaton that can be animated in the manner of a machine. Experiments where electricity killed, where it allowed animals to be revived, preceded research on the bodies of convicts, which did not fail to question the links between the body and consciousness. Thus, Pierre Bertholon, in his treatise *De l'électricité du corps humain dans l'état de santé et de maladie* [BER 80], demonstrated animal experiments relating to the effects of electricity conceived as a vital fluid. In particular, he spoke of the experiments by physicist Daniel Bernouilli (1700–1782), "This illustrious geometrician brought drowned birds back to life, using only electric sparks as a means of restoring them to life" [BER 80, p. 54, author's translation]. The experiments of which he spoke were also quoted in a treatise published in 1738 [BER 38]. In 1780, electricity was already considered a vital remedy, especially against asphyxia:

> 'Then I shot him,' he says 'a few sparks from the tip of his nose, which made him stand up on his legs to complete his healing, I gave him a couple of fairly light jerks. All this work didn't last six minutes when with the third shake the animal ran away, [...]'. [BER 80, p. 55, author's translation][1]

Thus life was first "given back" to the animals, the subjects of experiments, to understand the links between physiology, asphyxiation phenomena and electricity. After 1791, medical galvanism was considered as a stimulant to revive muscular actions:

> The *Ecole de Médecine de Paris* (Paris Medical School) tried to subject asphyxiated animals to Galvanic action; in its research it set out to determine the action of this stimulant on the muscular organs. It has mainly experimented with rabbits and small guinea pigs. The state of susceptibility of the nervous and muscular organs presented particular phenomena, depending on the difference in the causes of asphyxia. [CAS 03, p. 34, author's translation]

The concept of death at the end of the 18th Century encompassed reversible states of unconsciousness. Here we have an important point to understand the role that electricity played in the medical imaginary. The definition of death had not yet been decided, this force was about to play the role of an objective element to differentiate between living and non-living states. Moreover, if as long as the body was excitable, there was life, then it became a primary ingredient in the idea of the creation of life by Man:

> Can I name one more experiment where electricity brought a dead dog back to life? I say dead; for they have taken away part of his brain: & in this state, they put him on the cake, & they electrify him: he comes back to life, breathing, strong, gets up on his legs as if to run away. One stops electrifying it, it falls back into the inertia & the numbness of death; one starts electrifying again, & the movement starts again. [BIA 77, p. 36 quoted in ROZ 77, vol.9, p. 429, author's translation]

1 Bertholon alludes, in this passage, to the research of Nicolas, M. "Experiments made on some animals fallen into asphixia caused by coal vapors", in *Observations sur la physique, sur l'histoire naturelle et sur les arts*, [NIC 86, p.231, author's translation].

The epistemological status of animal electricity in the 19th Century was a symbol of life. It was made into a spectacle during the electrification of the bodies of those executed, who found themselves animated, without coming back to life, if we think of it in terms of consciousness. Like automatons, they were shaken by disordered movements that imitated those of the living. Medicine, marked by Cartesian dualism and 18th Century materialism, was able to experience the limits and properties of life on a Man who had become a machine. A symbol of atheism, revolution and reductionism, the experiments of the first third of the 19th Century contributed to the construction of a culture of physical, medical and sociological electricity. The bearer of hope, electricity was like the fire stolen by Prometheus to be given to humanity, and symbolized a materialistic progress where humans could gain access to knowledge and control over it. The notion of the electrical body, including its relationship with the soul, was constructed during the 19th Century through the study of the links between physics and the body. As a legacy of the 17th Century, the analogies of mechanics with human and animal physiology developed. Alongside the applications of electricity, the imagining of the mechanized body, obeying the laws of physics, was developing. While the way in which electricity connected the soul and the body remained a subject of speculation and questioning, the body became the site of investigations into the limits of life and the beginnings of death. How do gain control over these limits? Which organs help maintain life? How much room is there for the brain? The fact that the body could react to electrical simulations, that the heart starts beating again, was not enough to bring it back to life. The issues of the brain's role in understanding human singularity were central to the applications of this exploratory electricity. In this way, organs acquired a very strong symbolic value that can still be found to this day. Aldini, Galvani's nephew and colleague, spread galvanism beyond Italy's borders, notably by electrifying the bodies of the tortured. The analogies between the galvanic cell and the organization of nerves and muscles, which seem to form organic circuits designed to conduct electricity, reinforced the idea that the body has a mechanics that can be known and mastered by the medical sciences. As early as 1791, electricity was considered the most important function of animal economics, especially for Joseph Priestley [PRI 67, 71] (1733–1804), for whom it revealed the nature of things. How can we understand the expression "culture of electricity"? If you look at it from a physical point of view, it's hard to pinpoint. But if we consider from its very beginnings, the dimensions of spectacle and supernatural powers that surround its inscription in society, it becomes enlightening. Society was faced with a new technology, used as early as the first third of the 18th Century, as a trick and form of entertainment. Gray's 1730 flying boy experiment is emblematic of these beginnings:

> All metals, wood, reed or hemp, are conductors [...] but also: soap bubbles, water, an umbrella, a slice of beef, or a young boy! [GRA 31–32, p. 35, author's translation]

Figure 1.1. *In 1730, Gray experimented with the "flying boy", putting the human body on the list of conducting bodies*

The spectacular dimension of these experiments, in addition to aiding the differentiation between insulating and conducting bodies and including the human body among the latter, went beyond the scientific field and profoundly marked the popular imagination:

> He did the first experiment on a child aged 8 to 10, suspended on two silk cords, in a horizontal position. Then putting the tube close to the child's feet; his head, his hair, his face became electric; the same thing happened to his feet, when the tube was brought close to his head. [MAN 52, p. 10, author's translation]

The imagination was all the more marked by the fact that, following Musschenbroek's accident, accidents due to electric shocks all too often proved fatal. Electricity, the powerful power of nature, could not be easily tamed. Self-electrification, which was spreading in academic and cultural circles, conveyed an image that was sometimes unflattering. Alongside this frightening depiction of the uses of this force, scientists were conducting experiments confined to artistic fields [BOZ 54, p. 28] and were spreading a more positive image of them:

> Electric shocks had become well known, so it was disguised in a thousand different forms. Everyone was eager, big & small, learned & ignorant, hastened to experience such a singular phenomenon on themselves. Thirty, forty, one hundred people at a time took pleasure in feeling the same blow & in shouting just one cry. [MAN 52, pp. 30–31, author's translation]

Anonymous novels are devoted to this energy, while Franklin imagined the electric spit:

> On this principle, Mr. Franklin has imagined an electric wheel that turns with extraordinary force, & which, by means of a small wooden arrow raised perpendicularly, is able to roast a large bird in front of the fire, which is then threaded onto it. That's what he called the electric spit. [MAN 52, p. 184]

Medical electricity owes its success less to the credulity of the sick than to its air of progress, and to its promises for the mastery of human finiteness which were spreading throughout all the countries of Europe. Thus, the bodies became electrified by becoming the meeting place of Volta's metallic electricity and Galvani's animal electricity. They provided a spectacle during electrifications, notably in 1802 during the demonstrations by Rossi and Vassali:

> After I had explained to Professor Rossi on July 15, 1802, the effects I had obtained on those who had been tortured, he told me that on that same day there was an unfortunate man condemned to be beheaded; but the impossibility of combining a series of experiments in such a short time made him go to the hospital alone, where he saw, for the first time, the results I have mentioned. [ALD 04, p. 90, author's translation]

As early as August 1797, some of Aldini's experiments on tortured people were reproduced at the Academy of Turin, while in the perspective of applying galvanism to the knowledge and mastery of the living, he explored, following Kite, the idea that galvanism could be an agent of resurrection. This research therefore formed part of the activities of the *Royal Humane Society,* which since 1774 had been investigating the possibilities opened up by new techniques for resurrecting these victims. One of the medical, but also philosophical, challenges was to understand the process of dying. What was to die? Could the steps be reversed? [BAR 06]. As Zanetti summarized:

> If capital execution, carried out under the control of physiology, allows the precise analysis of the different stages of the passage from life to death, is there no hope of going the other way? The decapitations and the hangings of London and Glasgow are only the prelude to a medicine of reanimation, which throughout the 19th Century was concerned with the freshest cadavers, multiplying the discussions on the definition of the thresholds of death and its reversibility. [ZAN 17, p. 39, author's translation]

In this context of exploring the reversibility of death, after much research on electrified animals, scientists were turning to the possibilities offered by judicial executions. Most experiments were designed to use visual observations and emphasize two analogies: the first between life and motion, the second between electricity and vital motion:

> One of the earliest experiments on criminals condemned to death took place in Germany in 1791. In the presence of physicians and students assembled at the site of an execution by decapitation, the investigator began by demonstrating that exposed parts of the torso's neck muscles quiver when touched with a probe. Deeper contact caused muscle contractions strong enough to arch the back and to abduct the arms that had been folded with fingers interdigitated. A light touch of the probe on the cut end of the spinal cord in the neck likewise evoked facial muscle twitches, especially around the lips, and occasional retraction of eyelids. Deeper probing again caused massive contraction of all facial and tongue muscles. Such grotesque grimaces forced some shuddering observers to leave. The results led to the conclusion that consciousness probably persisted after decapitation. [KEV 85, p. 219]

Sœmmerring's experiments are discussed here. Indeed, he experimented, following Galvani's experiments [DIC 22, v. 6, p. 294], on the properties of bodies in a *post-mortem* context. In fact, as early as 1791, electricity was considered capable of revealing the conditions of the passage from life to death. An all-powerful medical imagining was at work:

> It was not enough for science to have made itself master of the fire of the sky by means of lightning rods; to have learned to reproduce at will most of the circumstances of the terrible phenomenon, to have found in the battery a device from which the electric fluid escapes in a continuous burst which the hand of man provokes and stops, activates and slows down, directs and uses in a thousand ways; to have, by the combination of the electric fluid with the magnetic fluid, given rise to the mechanical and physiological agent whose effects we have reported so varied and so powerful; to have, in a word, applied electric force to the accomplishment of so many wonders which without it would have remained forever chimeras whose thought the most ardent imagination would hardly have dared to conceive; [...]. [MAN 63, pp. 131–132, author's translation]

After starting his experiments on animals and the dead at the same time in Italy, Aldini came to repeat his experiments at the Veterinary School of Alfort. For example, he connected the head of an ox placed on a table to an electric current. Its eyes opened and rolled in their sockets, its ears quivered, suggesting that the animal felt anger. They contributed to the transition from the warm-blooded animal model to the human model. At the beginning of the 19th Century, scholars were engaged in comparative thanatology to understand the effects of galvanism on the vital forces:

> I repeated on the corpse of a beheaded criminal the observations I had made on the head and torso of an ox. I established an arc from the spinal cord to the muscles: a prepared frog was part of this arc. I always got strong contractions without the help of the battery, without the slightest influence of metals. I have observed proportionately the same result on naturally dead men. [ALD 04, pp. 9–10, author's translation]

Then he experimented on:

> [...] the head of a dog, passing the current of a strong battery: this single contact excited truly frightening convulsions. The mouth opened, the teeth clattered, the eyes rolled in their sockets; and if reason did not stop the struck imagining, one would almost believe that the animal had returned to suffering and life. [ALD 04, pp. 9–10, author's translation]

These descriptions, worthy of horror novels, contributed to the imagining of the mad scientist who creates life from a subject presumed dead. In November 1803, in Mainz, the leader of a group of bandits, named Schinderbannes, was beheaded, along with 19 of his accomplices. The town's doctors hastened to recover the bodies in order to submit them to the galvanic experiment. Nevertheless, the delays in the arrival of the bodies did not allow them to experiment on more than four torture victims [FIG 67, v. 1, p. 650]. They derived the following physiological principles from it:

> That the muscular contractions which were obtained by means of the Voltaic pile on recently killed individuals reproduced mechanically, in a most perfect manner, the movements performed during life; That the action of the battery was all the more sensitive, the more precisely the electric current followed the direction of the nerves; That the muscles subjected during life to the influence of the will obeyed, better than those which are independent of it, the electric agent. [MAN 63, p. 192, author's translation]

At least two remarks can be made: the expression "mechanically reproduced the movements of life" shows the aspect of a medicine marked by the human automaton model. Furthermore, the idea that the muscles subject to consciousness were those that "obeyed" electricity best was not insignificant. This point, in apparent contradiction to the first, underlines the medico-philosophical significance of this research: do feelings and willpower persist for some time after beheading? [TIL 15]

On January 17, 1803, in London, in front of members of the Royal College of Surgery, Aldini experimented on the body of Georges Forster, a criminal hanged for the murders of his wife and children.

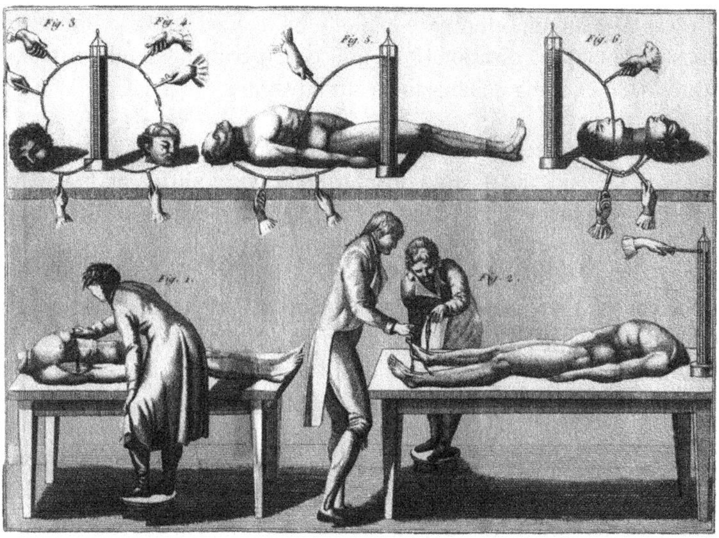

Figure 1.2. *Aldini tests the muscular reactions of a human head and then of the whole body [ALD 04, slide 4, fig. 1-6]*

By using the Voltaic pile, it led to waves of contractions and convulsions, marking in a first series of experiments, the face of the grinning murderer:

> The head was first subjected to the action of galvanism, by means of a pile of 100 silver and zinc plates: two metal wires, one from the base and the other from the top of the pile, came to the inside of the two ears, which were moistened with salt water. I first saw strong contractions in all the face muscles, which were contorted so irregularly that they imitated the most awful faces. The action of the eyelids was very marked, although less sensitive in the human head than in the ox's head. [ALD 04, p. 70, author's translation]

After animating his face, the scientist plugged a cable into the ear and another directly into the rectum. Forster's body then began to move frantically, in a disarticulated manner. You can imagine the impression made on the assembly. His experiments were driven by a genuine scientific curiosity on Aldini's part and could not be reduced to a mere spectacle. The latter, on the other hand, played an important role in the dissemination of knowledge. For example, on the subject of sensitivity or insensitivity of the brain, he carried out animal electrophysiology to show that with: "[...] an iron plate, or by touching them with silver nitrate: then the live animals feel the most pain, as when they are inflamed" [ALD 04, p. 87, author's translation]. Moreover, these *post-mortem* experiments on whole corpses enabled a deeper understanding of the respective places of the heart and the brain in the dying process. It was also an opportunity to explore the technical possibilities of restarting the heartbeat beyond the cessation of the pulse. Experiments on the galvanization of the heart provoked exciting debates in electrophysiology in the early 19th Century:

> This muscle[2] which, according to Haller's principles, is the first to receive life and the last to lose it, follows a different law when subjected to the action of galvanism. [ALD 04, pp. 99–100, author's translation]

Aldini explored the influence of galvanism on the heart while experimenting with direct galvanization of the isolated brain. The challenge was to determine which of the two organs could be considered as a physiological center in the dying process and therefore whether galvanization could have the most important action:

> Then Dr. Mondini, with all his skill, tried to separate in the brain the medullary substance, the corpus callosum, the striated bodies, the layers of the optic nerves, and the cerebellum. All these parts were successively brought into an arc, and the results of the experiments previously carried out on the bodies of other criminals were confirmed with full success. [ALD 04, p. 82, author's translation]

Volta considered the heart to be an insensitive organ, whereas Humboldt and Grapengiesser claimed, on the contrary, that it reacted strongly to galvanic stimulus. Thus, within the framework of controversial experiments, a field of medical observations opened up on the action of galvanism on the cardiac organ, the cessation of which was the first symptom, at least until 1940, of the process of dying. The polemics on its status played a dual role: on the one hand, the experiments on its state, after the passage to immobility and unconsciousness, enabling the passage from a resurrection envisaged medically to the techniques of eanimation; on the other hand, they played a fundamental role in the electrical

2 The muscle referred to in this quote from Aldini is the heart.

imagining which developed around questions on life, death and the porosity of these two states:

> What was my surprise, in carrying out this kind of research, to recognize that the heart, despite all the assertions made to date about its insensitivity to galvanism, is of all the organs, the one that retains its excitability the longest under the influence of this agent, and therefore occupies the first place in relation to the duration of galvanic excitability, [...]. [NYS 02, p. 8, author's translation]

Thus, Nysten (1771–1818)[3] conducted galvanic experiments on dogs freshly poisoned with opium. He used a vertical Volta device, consisting of 38 zinc discs, 3 coins and 38 cloths soaked in a solution of ammonia muriate:

> As a result, this substance does not annihilate the galvanic excitability; but I did not pay enough attention to this kind of experiment to dare to ensure that it does not diminish the energy of this property [...]. [NYS 02, p. 14, author's translation]

In order to test the excitability of the heart in a *post-mortem* situation, he tried to recover[4] the bodies of convicts just after their execution. Thus he obtained the body of a 27-year-old man, considered to be of a hot-headed temperament due to his having committed a criminal act and having just been beheaded:

> I opened his thoracic cage [...]; the muscles that I irritated with my scalpel while making this opening, such as the sterno-humeral (large pectoral), the costo coracoidian (small pectoral), the sterno-pubic (large right of the abdomen), contracted strongly. I freed the heart from its pericardium: the sinus of the pulmonary veins (the left atrium) and the aortic ventricle (the left ventricle), mechanically

3 A Franco-Belgian physiologist, Nysten, together with Bichat, conducted pioneering electro-medical experiments, particularly in connection with the determination of the stages of rigor mortis, described in 1811 and known as "Nysten's law". This research cannot be detached from experiments on the galvanization of the heart, nor can it be detached from investigations into the length of time the tissue retains potential excitability.

4 To understand the atmosphere and the race of these physiologists and experimenters to obtain these bodies, it is exciting to read Nysten's story of his adventure to arrive just at the moment when the tortured man gave his last breath. Thus he saw the announcement of execution, went urgently to the director of a medical school who could receive such a subject and assisted him in motivating the request: "I testify to him the desire that I have to try on the heart of man the experiments that I had already made on the hearts of several animal species. I add that we are going to torture a criminal, [...] I obtain authorization by virtue of which the body of the person to be killed was placed at my disposal, after his beheading, [...]. No sooner had I got there than I saw the fatal knife fall" [NYS 02, p. 18, author's translation].

irritated, remained perfectly immobile; but the sinus of the vena cava (the right atrium) and the pulmonary ventricle (the right ventricle), showed obvious contractions. [NYS 02, pp. 17–18, author's translation]

Let's take an interest in the temporality of these experiments: Nysten started at 2.45 pm, and by 6.30 pm he was still stimulating the body. He pointed out that after four hours of galvanic applications, the only organ still sensitive to the action of this agent was the heart, since the sinuses of the pulmonary veins and the vena cava, essential parts of it, continued to contract. From 6.45 pm onwards, the contractions became completely imperceptible. While the body had been stimulated for four hours, death could no longer be considered as an instantaneous state but as a process taking place in stages. Moreover, the reaction of each organ to galvanism and the duration of their sensitivity were here the subject of a precise temporal determination, which went in the direction of a rationalization of the action of galvanism on the body. Such an approach can be seen in the tables shown in Figure 1.3.

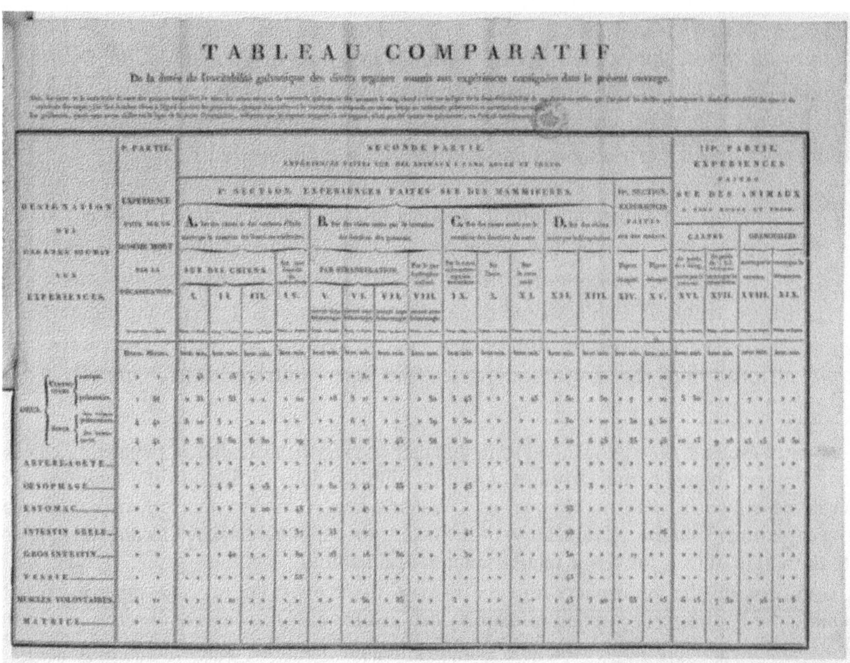
TABLEAU COMPARATIF

Figure 1.3. *Comparative table of the duration of galvanic excitability of the various organs subjected to the experiments recorded in this book [NYS 02]*

Using an electrophysiological approach, Nysten compared the excitability times of human organs to those of animals. What did he conclude from his experiments? The long influence of comparative anatomy made itself felt in his conclusions: he advocated a unity of the laws of Nature according to which there is no reason why galvanization should have different effects on humans or warm-blooded animals. Yet the context of the death, the state of the body, human or animal, were all conditions that influenced these effects. He therefore highlighted the importance of the experimental protocol, which must include everything that is relevant to the information on the experimental model chosen. Indeed, on a man who died a violent, natural death or disease, or on an animal, stimulation influenced differently the energy of the heart, its durability and its sensitivity. The experimental protocol could be refined, specified in relation to the specificities of the organs' reactions to stimulation. For example, the high frequency of stimuli caused an inhibition of activity that need not be confused with permanent cessation. These were all mechanisms that the galvanist needed to take into account. Nysten, by maintaining a central status in the process of dying, paved the way to experiments of restarting its activity beyond the limits of life. Between the heart or the brain, which organ determined the state of death? The brain, in the perspective of the naturalization of the intellectual faculties, was becoming increasingly important. So he argued for the place of the primary organ with the heart. Nysten was involved in these quarrels:

> But I have not only determined the contractions of the heart longer than the physicists of Turin, I have also proved that this organ is of all, the one which retains the longest faculty of contracting under galvanic influence, and I have thereby retained for it the title of ultimum moriens[5] which it was about to lose. [NYS 02, p. 35, author's translation]

The main conclusion being that galvanic action, in addition to maintaining the excitability of the heart, brings it back when it is ready to die:

> The sinus of the pulmonary veins was apparently insensitive to any excitation, either mechanical or galvanic; but its excitability, so to speak extinguished, was revived by galvanic action, to the point that it then contracted not only by this action, but also by that of mechanical agents [...]. [NYS 02, p. 36, author's translation]

Bichat, in the second part of his book *Physiological Researches On Life and Death* [BIC 99–00], recorded his work on galvanism, discussing the influence of

5 *Ultimum moriens* is Latin for the last thing to die. In cardiac anatomy it describes the right chamber of the heart, considered as still susceptible to contraction after the rest of the heart and as the last part to stop.

brain death on heart death in the context of the *ultimum moriens* debates. He reported on five experiments in which he tended to show that the heart does not immediately stop working when brain function is interrupted. He saw galvanism as a technique to shed light on the relationship between the heart and the brain:

> There is another kind of experiment similar to these, which can still shed light on the relationship between the heart and the brain: that of galvanism. I will not overlook this means of proving that the first of these organs is still currently dependent on the second. [BIC 99–00, p. 393, author's translation]

Bichat chose to experiment on the animal model of the frog and followed a classical experimental protocol, made of zinc and lead metal frames and organic materials:

> I have equipped several times in a frog, on the one hand his brain with lead, on the other hand his heart and lower limb muscles with a long zinc blade which touches the first one by its upper extremity, and the second one by its lower extremity. [BIC 99–00, p. 394, author's translation]

Although he established communication between the muscles and the brain, no acceleration was noticeable in the heart while it was still beating, and no movement occurred after it stopped. His physiological research focused not only on the passage from life to death, its different stages and the possibility of modifying its parameters through stimulation, but also on the interactions between organs. Following the idea that every body is subject to the harmonious functioning of the systems of the animal economy such as the nervous, cerebral or blood systems, Bichat explored the idea of hierarchy to determine which organ lives or functions the longest between the heart and the brain. In addition, the *Société Médicale d'Émulation de Paris*, which he founded in 1797, rewarded the text written by Malacarne (1744–1816) [MAL 03, 98, 99] in 1802 on the physiological landscape formed by the different physiological systems. It is a fundamental theme for the understanding of brain and heart physiology. Does the brain directly influence the heart? Bichat also investigated whether there is an irritating movement of the heart that can be differentiated from cardiac movement:

> 1st to detach the heart from the chest; 2nd to place it in contact, with two different metals, by two points on its surface, or with portions of flesh armed with metals; 3rd to make the armatures communicate by a third metal. [BIC 99–00, p. 396, author's translation]

Given the proximity of the physiology of warm-blooded animals and that of humans, he experimented on the former. However, the brain-heart circuit does not give specific movements:

> These different considerations, are a manifest proof that the communicating branches of the ganglions, should no more be considered as a continued nerve, than the branches, which pass from each of the cervical, lumbar and sacral nerves, to those which are immediately above and below them. In fact, notwithstanding these communications, each pair of the latter mentioned nerves is regarded as a separate pair. [BIC 99–00, p. 397, author's translation]

This lack of movement of the heart in relation to the brain would indicate that the two organs die independently[6]. Bichat, in order to avoid movements due to the effect of galvanic irritation and artifacts due to experimental protocols, first placed his circuit and established communication in a second step:

> 2dly in dogs and guinea pigs, I have repeatedly applied the metals, first to the brain and the heart, then to the trunk of the spinal marrow, and the heart; then to the par vagum and the heart. The communication being made, was followed by no apparent result. 3dly, on making the communication between the metals, when applied to the cardiac nerves and the heart there was no very sensible motion. [BIC 39, p. 261]

As long as he tried to provoke cardiac movements by connecting the heart to the central nervous system, it achieved little. It had to be detached, insulated and brought into direct contact with metals:

> Besides, in admitting even these different results, I do not see how it is possible to refuse acknowledging, that with respect to the stimulus of galvanism, there is a wide difference between the susceptibility of the muscles of the animal life, and those of the organic life. Again, supposing that the galvanic phenomena were the same in both sorts of muscles, the fact would prove nothing more […]. [BIC 39, pp. 261–262]

---

6 Historically, this debate is part of older quarrels about the respective places of the heart and the brain. Moreover, it accompanies the progressive cerebralization of faculties. "Such a relationship is not self-evident and has not been established without difficulty. Aristotle and his school located the center of sensations in the heart, and gave the brain, a cold and humid organ, the only function of tempering the body's internal heat. The idea that the brain is the center of intelligence is found in a cursive and often metaphorical way among some Presocratics, Hippocrates and Plato. But it was Galen (131–200) who, refuting Aristotle, provided the first systematic demonstration of this, based on convergent anatomical and pathological arguments" [LAP 70, p. 599, author's translation].

This experiment showed that cardiac movements reacted to a localized and direct galvanism, independently of a direct action of the brain. Bichat differentiated between the effects of galvanization on animal muscles, subject to the action of the brain, and on those of organic life, which depended only very occasionally on the will. So there was no direct action of the brain on the heart, nor any technical means of triggering it. The two organs did not cease to function simultaneously. The knowledge of the life of the organs but also the definition of death was one of the stakes of this research. The description of brain death, which was not completed until the middle of the 20th Century, depended on research on resuscitation and on the possibility of identifying all the stages that marked the boundaries between life and its cessation.

Aldini also noticed that the cardiac function appeared to be subdued in some of the suppressed patients, while it was easily reactivated in all muscles in others. He used three procedures to verify the resumption of cardiac movement *post-mortem*:

– by supplying the spinal cord with a lead cylinder inserted into the canal of the cervical vertebrae and then bringing one end of a silver arc over the surface of the heart and the other to the spinal cord framework. The heart, which in the individual subjected to galvanism still enjoyed great vitality, immediately presented very visible and quite strong contractions;

– by supplying the nerves of the wave pair and the large sympathetic nerves without the help of a battery;

– by means of Volta's devices and making use in general of a battery composed of 50 silver discs and as much zinc with the cards soaked in a solution of muriate of soda.

He noted in several decapitated people, very strong cardiac contractions and concluded that the tip of this organ was, of all its parts, the most mobile and the most sensitive to galvanic influence. The contractions produced by the last of these three processes were not only stronger but also longer lasting. In the same perspective of verifying or corroborating scientifically the irritability of the heart and thanks to electrophysiology applied to the body as a whole, he was able to test the different organs separately and together and to specify the duration of excitability during which they could still imitate the movements of life:

> 1. That if the mechanical irritation of the pin and scalpel excited, from the beginning, visible contractions in the intestines, heart and diaphragm; the same contractions were much stronger with the battery. 2. That when neither the heart nor the diaphragm was no longer irritated by the scalpel, the scalpel still excited contractions in the muscles of the extremities. 3. That after the intestines, the heart

> first lost galvanic susceptibility; then the diaphragm, and finally the muscles of the extremities. [ALD 04, p. 93, author's translation]

Thus, Aldini's experiments on the actions of the heart muscle formed some of these whole-body demonstrations. The medical imaginary, deeply embedded in scientific explorations, could hardly be separated from them. While Aldini chose tortured bodies to experiment on, the reasons were not only related to the history of the links between the bodies of those condemned to death and the history of vesalian anatomy, but also, and perhaps above all, to the possibility of having subjects whose vital forces had not been altered by disease and in which the springs of the fibers were not destroyed. In addition to galvanizing excitable points on the body, Aldini also connected isolated brains to raise their reflex actions:

> I sawed the skull to determine the action of the battery on the different parts of the brain, in the same order as they were presented by the anatomical dissection. All these parts obeyed the force of galvanism; but the corpus callosum and the cerebellum gave a more lively action. [ALD 04, pp. 62–63, author's translation]

Excerpts from Aldini's texts allow us to highlight an underlying questioning of the links between facial expressions caused by galvanization and possible thought content retained even after the cessation of visible life. The following passage retranscribes both this questioning, already found in the revolutionary period on the subject of decapitated heads just after the killing when the eyes or face are still affected by movement, and a scientific theme on the links between muscles and the mechanics of emotions that were discussed in more detail by Duchenne de Boulogne (1806–1875):

> So I placed the two heads of the torture victims horizontally on a table (sl. 4, fig. 6.), so that the two sections communicated with each other by animal moisture alone. It was wonderful, and even frightening, to see these two heads at the same time making horrible faces at each other, so that some of the spectators who did not expect such results, were truly frightened. [...] Having removed the upper part of the skull by a dissection parallel to the base of the brain, I incized the meninges, and made an arc from one of the ears to the medullary substance: first I saw sharp convulsions in the facial muscles. [...] I separated the lobes of the brain, and applied the arc to the corpus callosum and to the ears, and then to the lips: and there was a violent shock over all the head, and over all the muscles of the face. Some of the spectators even believed that the corpus callosum had been affected by a convulsion of its own [...]. [ALD 04, pp. 73–74, author's translation]

Thus, by applying localized galvanization directly to selected nerves and not to the spinal cord, Aldini used the expression *experimental myology* [ALD 04, p. 100], the only one capable of making the fixed and mobile points of muscles and the true term of their action sensitive to the eye. Did Aldini really seek to bring the dead back to life? Nothing is less certain. In any case, it appears that he did not seek to bring criminals back to life, but accident victims, drowning victims or suicides caught in an apparent and perhaps reversible death. Indeed, the industrialization of the 19th Century, accompanied by the growth of cities, created the study of new neuroses, a psychiatricization of mores but also an increasing suicide rate, as Brierre de Boismont points out:

> It is impossible not to notice, when going through them, that the study of suicide touches upon the great questions of our time, such as pauperism, work, salary, family, property, the future of craftsmen, the future of society perhaps, etc. All these subjects and many others find in the etiology of many teachings, at the same time as they reveal the depth of an evil that has claimed no less than 300,000 victims in France since the beginning of this century. [BRI 56, viii–ix, author's translation]

Thus, Aldini's perspective was that of a public health problem affecting several countries and of emergency medicine coming to the aid of unfortunate people who, as soon as they died, may have already regretted their action:

> Nevertheless, it cannot be opposed that the help of galvanism should be given, together with any other, to these unfortunate people who, in despair, have sought their destruction by strangulation or other means. Such accidents are unfortunately all too frequent in large cities; and galvanism deserves all the more confidence because its application, in all these cases, suffers no delay. [ALD 04, p. 141, author's translation]

Helping people in distress, defining death, understanding it in the depth of its physiological mechanisms were the goals pursued by Galvani's nephew who promoted galvanism as an instrument of medical philanthropy. In fact, if galvanization contributed to differentiate, beyond the absence of movement and consciousness, the living from the dead, then it could help to save people who were not dying:

> A host of facts have shown us many times that people have raced to their graves before death struck them irrevocably. [ALD 04, p. 143, author's translation]

It was thus conceived as a diagnostic tool to make visible the conditions conducive to life:

> Before we finish talking about the useful applications of Galvanism, we must indicate it as a means that could be used to prevent the premature burial of people who had fallen into lethargy. To judge whether death was real or apparent, it was sufficient to insert the tips of a Galvanic exciter into the muscle parts. [CAS 03, p. 38, author's translation]

While Bichat positioned himself to provide a physiological description of the process of dying[7], Aldini placed his studies in a societal perspective by linking them to the development of a resuscitation medicine[8]. The demonstration made in London on Forster is detailed in the newspaper *The Times* from January 22, 1803. It made a strong and lasting impression on the minds of scientists but also on all those present. As we have seen, Aldini did not seek immortality or miraculous resurrections. This discrepancy, between the imaginary linked to the representations given of his research and the scientific aspects, was the catalyst for Mary Shelley's novel featuring a scientist facing his contradictions. Indeed, the idea that electricity can be the source of a resurrection is more a cultural representation than a medical theory:

> If, after the extinction of general life, a remnant of life remains in the corpse, this remnant still shows electrical phenomena: once the nerves and muscles are dead, the electrification no longer produces any movement in them, and there is no electrical discharge that is able to revive a corpse. Electricity is therefore not the principle of life; it is only a form in which the principle manifests itself, a form of activity which the organism possesses in common with inorganic bodies, but to which it nevertheless imprints a particular modification. [PAL 47, p. 60, author's translation]

Aldini had a much more humble goal: to demonstrate that galvanism could be a useful tool in a variety of resuscitation procedures for people who had died from asphyxiation or drowning. His stay in London was financed by the *Royal Humane Society*, for which it had been a concern for several decades. Thus, his work is considered on the borderline of the transition from resurrection to resuscitation:

---

7 Aware of the ambiguities of the state of proven death, Aldini underlined the fact that an individual can be paralyzed and alive or, on the contrary, moving and dead: "[...] that a man whose paralyzed limb refuses to undergo muscular contractions can very well be alive; whereas muscular contractions can easily be obtained by galvanism, or any other stimulant, in a subject who is truly deprived of life." [ALD 04, pp. 222-226, author's translation].

8 These preoccupations to differentiate between the state of life and that of death were developed during the 19th Century [DES 51; CAR 15].

> I invite them today to employ in similar cases the action of galvanism in the manner I have proposed. It is good to multiply the means to relieve our fellow human beings, especially in circumstances where old medicine offers us very few resources. In the meantime, I think it would be useful to do some tests on asphyxiated animals in different ways. These tests could be valuable, and provide a lot of perspective to saving the lives of men. [ALD 04, p. 98, author's translation]

The fact that the experiments were practiced on individuals who had died without having suffered from illness brings to light the process described by Aldini for returning consciousness and movement. This process, far from being imbued with magic, required an organism whose unaltered organization could make a return to functionality:

> Galvanism must not be placed among these chimerical agents; its action is real and well observed, its devices and their construction are not hidden, their strength is known and felt by everyone. [ALD 04, p. 139, author's translation]

While it was possible to bring subjects back to life, it was not in any biological condition, which already underlined a demarcation between the theme of resurrection and that of reanimation. Indeed, only subjects who were not dead could be brought back:

> I think, therefore, that the application of this highly-active stimulus should be limited to cases where the animal suspension is affected at a single point, which still gives hope for the recovery of life. If the heart, if the circulation, if the lungs, if the nervous system is inactive, provided that the organism still exists, and that the vital functions are not suspended for long, galvanism can be administered. [ALD 04, p. 141, author's translation]

Following this, the idea of the profound connection between galvanism and vital phenomena traveled beyond the spheres of physics and medicine. His experiments in reviving the hearts of bodies marked human power over life itself. This spectacular and public dimension played a decisive role in the fact that electricity could impress society and be seen as the cure for all ills. A visible sign of this was the high presence of electroshocked bodies in dissecting rooms or anatomical theaters. This development placed the electric body in its singularity among bodies and technical tools:

> Making the electric body a viable cultural construct was to make it blend into its environment. The body gained its significance in the same way as any other technological artefacts that inhabited the same space. It mattered therefore for the sense of Aldini's or Ure's versions of the electrical body that they were articulated in a dissecting room. [RHY 02, p. 114]

The human body, by becoming an electric body, became the *post-mortem* electro-experimental field of invisible energy. It represented the possibility for galvanists to experiment on the body as if it were alive. While the animal model could be experimented with *in vivo*, the human body was reified as a scientific object when it was inert. Galvanism opened up the possibility of experimenting on a body that had been temporarily and mechanically put back into motion. Electricity made it possible to make the link between the dissected body and the animated body and was part of the fantasized vision that society had of science and its actors. In a reprint from 1839, Bichat related an experiment on the setting in motion of pieces of human bodies and how they regained mobility and precision of movement:

> The last galvanic experiment was made by transmitting the electric fluid from the spinal marrow to the cubital nerve near the elbow; the fingers moved quickly like those of a performer on a violin; one of the assistants who endeavored to keep the hand shut, found that it opened in spite of his efforts. A wire was applied to a slight incision made at the end of the first finger; the hand had been previously shut; the finger was instantly extended, and, after a convulsive agitation of the arm, the dead man seemed to point his finger at the spectators, some of whom thought that he had come to life. [BIC 39]

The imaginary supernatural powers attributed to galvanism transcended public boundaries and were recounted even in certain scientific treatises. Thus, in the early 19th Century, Petetin described it as follows:

> The recent discovery of a new fluid is occupying all the savants. Its appearance has been accompanied by prodigies. It was not far from resurrecting the dead; at least it has the well-established property of reproducing movement in them: it is already considered to be the precious fluid which animates all parts of the human body; and this insight gives hope of giving man long youth and prolonging his life. [PET 02–03, pp. 2–3, author's translation]

Experiments on human automatons continued during the 19th Century, with some paradigmatic cases of fascination with the vital spark returned to these bodies. The soul of the criminals who inhabited them was often perceived as the soul of life. Expressions such as "*demonic smile*" and "*crazy eyes*" flourished in the accounts of these demonstrations. The experiment by Scottish chemist Andrew Ure (1778–1857), carried out on November 4, 1818 [URE 19], on the corpse of a torture victim, which he subjected to a voltaic pile 10 minutes after he had been taken down from the gallows, is a significant episode in this story. He performed this demonstration in an anatomy theater at the University of Glasgow where students mingled with curious anatomists and doctors. From vesalian dissection to the galvanization of lifeless bodies, there was an epistemological separation between an inert, resolutely *post-mortem* anatomy and an animated electrical anatomy. The obstacle that was overcome between the 16th and the 19th Centuries was that of animating human bodies. This moment when the body started moving again corresponded to a technical feat of the galvanist who, by combining the new technologies of his time, became a symbol of the almighty scientist. Ure exposed the spinal cord by ablation and made incisions to uncover the sciatic, ulnar and diaphragmatic nerves. By electrifying these different nerves with two metal rods charged by a 270-plate voltaic pile, contractions were caused in the torso and limbs:

> The success was truly extraordinary. Full and labored breathing began at once; the chest was raised and lowered; the abdomen felt movements corresponding to those of the diaphragm [...] Third experiment. The supraorbital nerve was exposed at the point on the forehead where it exits from the supraciliary hole above the eyebrow; one of the ends of the apparatus was brought into contact with this nerve, and the other with the heel. The strangest grimaces appeared on the face. [...] Rage, horror, despair, anguish, atrocious smiles were painted in turn on the murderer's face, with a hideous expression that no brush could render. Several of the curious ones ran away shivering, and one of them lost consciousness. [BAR 06, pp. 38–39, author's translation]

It is said that the leg was thrown so violently that a helper was almost knocked over. Thus the corpse rose, seeming to look at the spectators and showing signs of breathing movements:

> By placing the electrified corpse's gesticulations in the context of the stylized gestures of Regency stage and art, Ure was also reminding his audience, as Aldini had done with his experiments, that electricity seemed to hold out the very real possibility of restoring the dead to life. [RHY 02, p. 98]

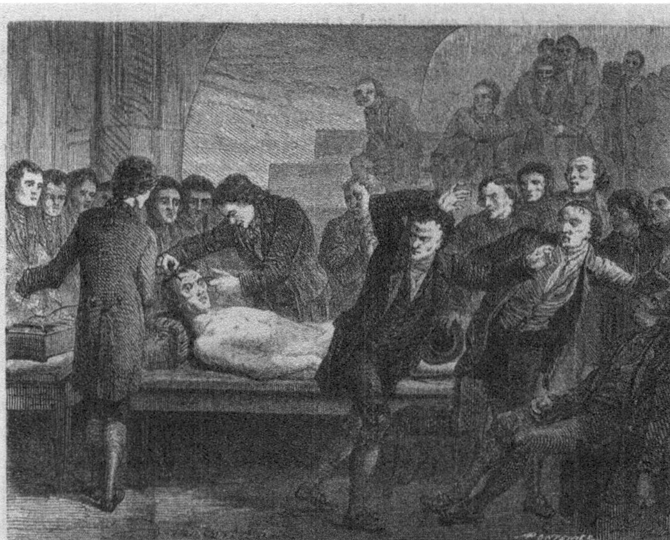

Fig. 333. — Le docteur Ure galvanisant le corps de l'assassin Clydsdale.

Figure 1.4. *In this 1867 illustration, a crowd of scientists watch in horror as Andrew Ure shakes and shocks Matthew Clydesdale's lifeless body with electricity*

In 1819, Thomas Weem, executed for murder, was galvanized by James Cumming (1777–1861), a chemist at Cambridge and considered a pioneer of electrical instrumentation [STO 76, p. 29]:

> The execution of Thomas Weems for murder on 6 August 1819 has become very famous in criminal histories. His condemned body was subjected to a number of quasi-scientific experiments to explore the nature of electricity, resuscitation and brain death, all associated with Mary Shelley's *Frankenstein* (1818). [HUR 16, p. 241]

Cumming tried to make the body breathe by exciting the vagus nerves from the brain to the heart, lung, and digestive organs. His idea, according to the identity of the nervous fluid with the electric fluid, was to revive the organism's regulating nervous fluid by means of galvanization. From Aldini to Cumming the concept of "experimental resurrection of inanimate flesh" was developed [BAR 06, p. 38]. Between these galvanic experiments, from the beginning of the 19th Century, and contemporary reanimation, the potential of the control of life mechanisms by medicine was taking shape:

> Friday 6 August 1819 – The body, after being suspended for an hour... was immediately conveyed in a cart to the Chemical Lecture of the Botanical Garden, where Professor Cumming had prepared a powerful galvanic battery (which formerly belonged to Professor Tennet) with the intention of repeating some of the experiments lately described by Dr. Ure of Glasgow in the Journal of the Royal Institution.[9]

The Cambridge Journal kept a full account of the experimental stages of this research and helped to disseminate it in society, where it asked a fundamental question: what is dying?:

> The beheaded and hanged men of London and Glasgow were only the prelude to a medicine of reanimation, which throughout the 19th Century was concerned with the freshest cadavers, multiplying discussions on the definition of the thresholds of death and its reversibility. [BAR 06, p. 41, author's translation]

Between the belief in the persistence of a consciousness that some believed they could see through the distortions of galvanized criminals' features and the fact of managing death for a definitive and irremediable moment, the refusal of human finiteness emerged. These experiments were the culmination of medical utopias and the imagination of a potential victory over death. They tell the story of galvanism in terms of the connection between its applications and the mechanisms of life:

> Sturgeon's Lectures[10] constructed a strikingly novel genealogy for galvanism. Rather than tracing the development of experiments and ideas from the Volta's pile through Humphry Davy's heroic experiences, he pointed to a different history. His heroes were Galvani and Aldini (....). The history of galvanism, on the one hand, was the history of the discovery and demonstration of electricity's connections with and role in the animal economy. [RHY 98, p. 129]

In terms of physiology, knowledge acquired on an animated anatomy and for future resuscitation techniques, they were extremely heuristic:

> If these (still?) fantastical practices seem anecdotal, they nevertheless reveal the preponderance of the utopia of victory over death and over

9 "Execution of Weems", *Cambridge Chronicle and Journal*, August 13, 1819.

10 William Sturgeon is known to have introduced the first electromagnet in 1825. In 1832, he introduced the first electric motor. In 1836, he published the journal *Les Annales de l'Électricité* and invented the galvanometer.

> the body corrupted by life, of which we have brought to light during a few moments. They also point out to us that under their strange, curious and disturbing aspect, the experiments of Andrew Ure or Giovanni Aldini are always on the horizon of the technical utopias of our time. [BAR 06, p. 43, author's translation]

The idea of the reversibility of death referred to the notion of stages in this process on which medicine could still intervene. This idea was based on the fact that the vital principles present in the body do not stop working immediately. The heart plays an important role. Seen as a pump, a mechanical part, it is considered as being able to be re-launched and capable, by its action on the blood circulation, to start again to excite the centers of innervation, those of the general motricity, and thus to ensure the redistribution in all the points of the organism, of an always present energy:

> However speculative such assumptions may be, they nonetheless provide a precise measure of what death means: they establish the exact meaning of this physiological transformation. Just as life as a whole was nothing but a result and harmony, its disappearance cannot be the ruin of any principle, and the death of this whole is reduced to an accident which restores to their freedom – a fatal freedom, since it must this time bring about a real annihilation – the partial energies whose association was necessary for the constitution of the individual. The death of the whole is thus only the breaking of the united pact that creates individuality: there is no life that perishes, for real life is concentrated entirely within the organic element. [BER 05, p. 18, author's translation]

The spirit of the early 19th Century permeates the novel *Frankenstein,* or *The Modern Prometheus* [SHE 18], which Mary Wollstonecraft Shelley (1797–1851) published in 1818; the book quickly became a classic in the field of Gothic horror literature:

> Mary recalled that, after some days of '*blank incapability*', the night of 16 June, she had a '*waking dream*' that was at the origin of her own story: '*I saw the hideous phantasm of a man stretched out; and then, on the working of some powerful engine, show signs of life, and stir with an uneasy-half-vital motion.*' This description shares many similarities with what happened in London on January 17, 1803 when Aldini showed that Galvanic stimulation of the brain seemed able '*to give an appearance of re-animation*'. In the second chapter of Mary's

> novel, the echo of the electrical experiments on dead bodies became even stronger, as Dr. Frankenstein specifically refers to electricity [...]. [CAM 18, p. 28]

Mary Shelley was the daughter of the British philosopher, political theorist and novelist William Godwin (1756–1836) [GOD 20]. He was fascinated by the recent discoveries of galvanism, which he felt was an agent of materialism. These considerations led him to conclude the materiality of the soul and the non-necessity of a god. Moreover, inspired by the spirit of Enlightenment and the French Revolution, he proposed societal reforms designed according to the data of reason. His notoriety, as a defender of the most innovative ideas, failed to have him accused of materialism and atheism in the same way as his friend, Thelwall (1764–1834). He is considered part of the reformists and utopians projecting a society based on an egalitarian system, against whom the economist Malthus (1766–1834) wrote. Thelwall [THE 02] used electricity to argue for the material basis of life and human rights as a natural consequence of the laws of nature. While electricity was a sign of scientific progress, it became a paradigm for political progress:

> Machines appeared to provide a concrete way of articulating and making sense of new relationships between natural and political economies, between human labour and the natural forces increasingly being harnessed to power industrial progress. From this perspective, the human body itself could be regarded as a machine, embodying the newly articulated doctrine of the conservation of energy in just the same way as did an electric battery or a steam engine. [RHY 99, p. 249]

This inscription of the history of electricity in the post-revolutionary context makes it possible to understand to what extent the craze that it provoked symbolized the advent of a society marked by the progress of technology. The passage in Mary Shelley's novel in which the scientist, Victor Frankenstein, explains how he glimpsed the links between light and the possibility of animating matter offers several interpretations. The most likely is the link between natural electricity such as lightning and the vital spark that animates matter *in vivo:*

> […] from the midst of this darkness, a sudden light broke in upon me – a light so brilliant and wondrous, yet so simple, that while I became dizzy with the immensity of the prospect which it illustrated, I was surprised that among so many men of genius who had directed their inquiries towards the same science, that I alone should be reserved to discover so astonishing a secret. Remember, I am not recording the vision of a madman. The sun does not more certainly shine in the heavens than that which I now affirm is true. Some miracle might

> have produced it, yet the stages of the discovery were distinct and probable. After days and nights of incredible labour and fatigue, I succeeded in discovering the cause of generation and life; nay, more, I became myself capable of bestowing animation upon lifeless matter. [SHE 18, pp. 75–76][11]

Shelley's text is not without ambivalence on the technical aspects. As mentioned above, her father was a supporter of galvanism. Her husband, Percy Bysshe Shelley (1792–1822), took part in lively debates, notably with Dr. William Lawrence (1783–1867), on the physical origins of life and the possibility of sparks bringing the dead back to life. The novel *Frankenstein* represents the fascination for the possibility of artificially creating life, but also, and perhaps above all, a surpassing of the dualism present in the fact of animating human automatons, insofar as the creature is not a machine but possesses a consciousness embodied in its matter. Mary Shelley, in a holism and materialism in which spirit is generated by matter, linked the organic parts from which the creature is formed, coming from different criminals, and their influence on consciousness. From the project of distinguishing the stages that range from life to death, the powerful myth of making life or giving it to organic matter in its modern, technical and progressive version, represented in the *Frankenstein* epic where materialism is next to the technical power of Man, was born:

> From the suggestion of these experiments came a wide range of Gothic literature, especially in Germany, England and America, of which the most famous expression, but also the most open to multiple interpretations, was Mary Shelley's Frankenstein. [FRE 14, p. 254, author's translation]

Thus the exploration of the possibilities of reanimating matter had clearly taken hold of many experimenters in medical electricity. The voltaic pile, symbol of major technical progress, made the hand of the modern Prometheus powerful. Works of fiction on these issues were not uncommon and some can be considered gems of the genre. Here is an excerpt from a play, published in 1854, symbolizing the fascination and fear of galvanism but also the human ambiguity of its position in nature. The scene takes place in Bologna, in 1797:

> (Galvani alone, sitting by the table.) Bringing the dead back to life! Recalling the divine breath in an inanimate body! O thought of breaking the brain! A thought that contains a hundred times more pride than it took to lose the first man! Giving life to the dead! But it is to want to correct the work of God; to want to be God himself! And

11 In his film *Frankenstein* (1994), the director Coppola shows the doctor plunging the patched body into a tank filled with electric fish. This other interpretation of the novel refers to the work on these animals, and also to a natural electricity, here organic.

> yet, it is certain that since the earth has been turning, there has been a misunderstanding between the Creator and his noblest creature: that one should die when one has reached the limits of extreme old age, when the springs of the organism are worn out, one understands it: this is the universal law; but in the prime of life, in the flower of youth, to die! To die altogether is a nonsense that God did not commit. [AND 54, p. 3, author's translation]

In the rest of the story, a murder and a galvanic resurrection are staged during which a female character stands at one end of a semi-organic circuit with a Leyden jar and transmits the electric fluid to her lover's lifeless body, under the instructions of the doctor [AND 54, p. 123].

Another renowned author, who was not insensitive to the electric imaginary and the Promethean culture that emerged from it, is Edgar Allan Poe (1809–1849). An assiduous reader of the *Medico-Chirurgical Journal;* or, *London Medical and Surgical Review* edited by Dr. James Johnson (1777–1845) and published between 1820 and 1847, and of the famous medical journal *The Lancet,* he dispersed in his tales and essays episodes of galvanism or dead people coming back to life, not without a certain irony.

In his 1844 short story "The Premature Burial", he captures the societal anguish of being buried alive and emphasizes the galvanic apparatus as a diagnostic tool for death:

> The mention of the galvanic battery, nevertheless, recalls to my memory a well known and very extraordinary case in point, where its action proved the means of restoring to animation a young attorney of London, who had been interred for two days. This occurred in 1831 and created, at the time, a very profound sensation where it was made the subject of converse. The patient, Mr. Edward Stapleton, had died, apparently of typhus fever, accompanied with some anomalous symptoms which had excited the curiosity of his medical attendants [...]. An incision of some extent had been actually made in the abdomen, when the fresh and undecayed appearance of the subject suggested an application of the battery. One experiment succeeded another, and the customary effects supervened, with nothing to characterize them in any respect, except, upon one or two occasions, a more than ordinary degree of life-likeness in the convulsive action. It grew late. The day was about to dawn; and it was thought expedient, at length, to proceed at once to the dissection. One student, however, was especially desirous of testing a theory of his own, and insisted

> upon applying the battery to one of the pectoral muscles. A rough gash was made, and a wire hastily brought in contact, when the patient, with a hurried but quite unconvulsive movement, arose from the table, stepped into the middle of the floor, gazed about him uneasily for a few seconds, and then spoke. What he said was unintelligible, but words were uttered; the syllabification was distinct. Having spoken, he fell heavily to the floor. [POE 44, pp. 757–758]

The theatricalization underlines the stupor of the subject who awakened fully conscious, differentiating him from the automaton to which mobility is infused in the experiments of Aldini or Ure. The body was clearly not brought back to life since it did not bear the *post-mortem* alterations. So he was alive but unconscious. Beyond the literary aspects, galvanization was rationally applied to a body that could not be considered dead. The point here is to differentiate between galvanic culture and the imaginary symbolized by the novel *Frankenstein*. Another of Poe's texts will help us to make this distinction. In *Some Words With a Mummy* [POE 45], he recounts, not without humor, the case, much closer to the myth than to a societal application of galvanism, of a mummy three or four thousand years old. His nerves exposed and feeling the first effects of the application of a galvanic cell caused movements very close to Aldini's descriptions of the tortured corpses. Absorbing contractions of the palpebral muscles and lower limbs, the mummy ended up giving a strong kick to a scientist. The fictitious dimension made it possible to go further than the description of automatic movements. This text refers to the popular fear of the dead returning from the dead, the novelty of which lies in the fact that the mummy's return was under the effects of the technique and not under supernatural conditions. The comparison of the two texts shows very divergent cultural and literary instrumentalizations of galvanization. While in the first, Poe relates the fictional conditions of reanimation; in the second, under the guise of mocking traditional fears, he takes the experience of the tortured to the absurd by taking as his object a mummy several millennia old. Poe's reference to electricity corresponds to a long reflection on the unity of the laws of physics in the universe, culminating in *Eureka* [POE 48], an essay dedicated to Alexander von Humboldt. Body and soul of the universe, electricity allows us to understand nature in its totality, material and spiritual, to which we must relate the phenomena of vitality and consciousness:

> The cosmos was a vast electrical machine that could be understood and manipulated in much the same way as they understood and manipulated the electrical machines and artefacts with which they plied their trade. Human bodies were part of the electrical universe too. [RHY 02, p. 102]

From the research on clouds, whose meteorological disturbances generate thunderstorms, a culture of the powers of electricity was born. From the sky to the body of the torpedo fish, electricity seems to flow everywhere. In this materialistic mechanics, it found its political and cultural foundation in the revolutionary period and developed to the core of organic fibers. From inorganic to matter, it permeated all 19th Century research. Several medical branches developed from the knowledge of electricity and the way it intervenes in vital phenomena: diagnosis, resuscitation, electrophysiology and electrotherapy are all fields born from the physics of this energy. The history of electricity is not only played out in medical circles, it permeates societies shaken by political events. In addition to allowing the development of medical and physiological experimentation, it poses the exploration of the boundaries of life and death as a philosophical, medical and societal issue. The concepts of the electrical body, electrical culture or galvanic culture are the results of this intertwining of electricity and society. Within a materialistic philosophy, it promotes the awareness that not everything stops with voluntary movement and that there is a certain permanence of life in death, especially through chemical and physical processes. From the culture of electricity to the culture of galvanism, the notion of scientific scales of the living developed to understand the sequence of non-life, life and the living as a whole. Moreover, the idea of a galvanist culture refers to a certain type of representation of the individual in the 19th Century: an individual, living in ever-larger cities whose mores must meet the standards of the society in which they evolve. The result of this societal expansion of galvanism is that electricity, in addition to stimulating a body that has become electric, also heals its consciousness. The historical value of medical electricity is not reduced to the technical prowess it achieves, but responds to the epistemology of each context, by being inscribed and renewed in it. From success to failure, at each stage it answers medical-philosophical questions about human beings and matter, about an organism whose functions generate consciousness:

> One reason why electrical engineers saw electrotherapeutics as an area where their own expertise was relevant was because they recognized the electrical body as being made up of the same kind of components, organized in the same kind of system as that on which they plied their trade. [RHY 02, p. 106]

The creation of the electric chair is rooted in the large number of deaths by electrocution as well as in research into the medical boundaries between life and death and is the result of the acute awareness of the power taken by this energy. In North America, it thus became a killing instrument in the early 1890s. This invention was not insignificant in relation to the fact that electricity was developing as the guarantor of a certain societal order. While at the beginning of the 19th Century it was a question of bringing people back to life, electricity could then be used to punish criminals for their crime. But the cultural impact of electricity did not stop at the physiological understanding of states of life and death. From 1770, medical electricity was applied to nervous and psychic illnesses, particularly in the work of Ledru [LED 83]. This research into the fundamentals of electrotherapy for nervous and psychic diseases will be developed further. From 1840 onwards, galvanic electricity became a symbol of control and standardization of the subject. From the myth of Frankenstein, we move into the context of Dr. Jekyll unwillingly transforming himself into Hyde. From this beast that sleeps in each one of us, electricity symbolizes the cage:

> And it chanced that the direction of my scientific studies, which led wholly toward the mystic and the transcendental, re-acted and shed a strong light on this consciousness of the perennial war among my members. With every day, and from both sides of my intelligence, the moral and the intellectual, I thus drew steadily nearer to that truth, by whose partial discovery I have been doomed to such a dreadful shipwreck: that man is not truly one, but truly two. [STE 86, p. 75]

This shift in the applications of these techniques cannot be detached from the organicist and materialist movement that integrates the mental faculties within the brain structures.

1.2. Changing and regulating behavior

Galvani had already spoken about *neuro-electric fluid* [GAL 53, p. 64] and attributed the genesis of animal electricity to the brain. Correlate to this the fact that the nervous structure was thought of in terms of wires for the nerves and voltaic pile for the brain organ. At the same time, the work of Ledru and abbé Sans on the links between artificial electricity and nervous diseases put forward the latter to be treated by the applications of electric fluid. Thus, at the time when electrotherapy was developing, the 19th Century was marked by an imaginary control by electricity of mores and behaviors considered harmful. If humans were beings in which matter generated consciousness, they seemed to be inhabited by a beast that sometimes pushed them to adopt harmful societal behaviors. On the literary side, we find this idea in Stevenson's novel (1850–1894) or in *The Beast in Man* where Zola (1840–1902) [ZOL 85] stages the archetype of the mad murderer. This idea also developed in the imagining of galvanic doctors who, realizing the complexity of the links between the brain and consciousness, conceived these therapies as the guarantors of a certain moral security. Electricity then became a tool for the standardization of the individual, inherited from the research of Delgado or Heath in the mid-20th Century:

> The key to understanding – and disciplining – the body in this respect was standardization. Just as other items of electrical apparatus and equipment needed to be standardized to function effectively within the new networks of power, so did the electrical body. [RHY 02, p. 102]

There was a division between the idea of reducing the vital functions of a body to that of controlling the subject and his behaviors. A conceptual division between a medicine marked by the Cartesian dualism of the animal machine, then by the materialism of the 18th Century, and a holistic medicine where faculties and consciousness were integrated into the human machine. Yet these two approaches complemented each other. From the beginning of the 19th Century, neuro-anatomical research was looking for Galvani's neuro-electric fluid within the human brain, in order to record it and correlate it with the expression of mental faculties. In 1808, Malacarne was between a *post-mortem* approach of the brain organ to the idea of acting on its electricity. The analogy of the electrical machine to organic structures extended to the nervous system:

> Are not brains, nervous ganglions, and nerves, which are evidently the seat of vital action, in the identities we call animal, real electrical machines; similar in principle, as they are similar in substance and in structure, to the electrical discharging apparatus of the gymnotus and torpedo, which consist of large brain-like ganglions connected with the spinal cord? [MAC 31, p. 94 in RHY 98, p. 131]

Thus, Malacarne published a text entitled "*Conoscendo dalla organizzazione del cervelletto in ispezie, e forse anche da più attento esame del cervello e dalla midolla spinale che queste viscere formano qualche cosa di somigliante alla colonna galvanica del Volta*" [MAL 08, pp. 122–130]. Probably the first localizer of the faculties within the cerebral organ, his research was marked among the scholars of the 19th Century [CHE 16]. Bayle quotes him in his *Encyclopédie des sciences médicales* about his notorious influence on Reil's reflections who:

> Reasoning from Malacarne's observations of the proportion between the development of intellectual faculties and the number of superimposed blades of the cerebellum, he argues that this organ is formed by an aggregation of small galvanic cells. [BAY 25, pp. 798–799, author's translation]

In 1808, he developed a project in which he designed a series of electrophysiological experiments in which brain slices were connected together like a galvanic column, in order to find the cerebral source of the human machine's animating fluid. He thus presented the experimental steps necessary to verify the brain organ as a source of animal electricity. The physiological theory according to which the brain would produce, in the same way as a voltaic pile, electricity allowing the body to move, was based on a double metaphor: the first was mechanistic and compared the functioning of the brain to that of a machine; while the second was based on the identity of the shape of the galvanic column with the morphological organization of the brain structures. Animal electricity was conceived as the cause of the anatomopsychological nature of man. Beyond the source of the movement, this project tended to seek in galvanism, a "[...] phenomenon as important for animal functions" [MAL 08, p. 126, author's translation], a similar functional principle for the expression of intellectual faculties and nervous mechanisms. Brain function was thus brought back to form: "[...] by beginning to subject to it those parts of the brain whose structure is visibly closest to the galvanic column, i.e. the cerebellum, and then moving on to the brain itself [...]" [MAL 08, p. 128, author's translation].

The heuristic fertility of this analogy between the electricity-generating machine and the nervous productions of the brain lay in the resemblance of the structures, from which was induced an identity of the mechanisms:

> Let us assume six hundred and add the three hundred slices of the cerebellum that we observe and compare this machine with the galvanic column formed by nine hundred discs, also assuming some analogy in the activity of exercising galvanism. Shouldn't we expect these powerful phenomena we admire in individuals to be produced by a hitherto unknown prerogative of the nervous and cerebral system? [MAL 08, p. 125, author's translation]

This research was linked, on the one hand, to neuro-anatomy, since no experiment could be undertaken without prior anatomical knowledge of the central nervous system, and on the other hand, to the experimental dimension of the recordings that Malacarne planned to make. The analogy of the cerebral structures with the galvanic column is traditionally attributed to Luigi Rolando (1773–1831), who from 1809 [ROL 09] stimulated the different parts of the column with current and thus caused violent convulsive phenomena in animal cerebella. In 1810, von Paula Gruithuisen (1774–1852) [GRU 10] described the cortical substance as an inexhaustible source of nerve power and stressed that it was a secretory organ. Reil imagined the cerebellum as a kind of voltaic pile, based on its histological aspect and the idea of a natural circuit. In 1840, Baillarger (1809–1890) [BAI 40] described the cell layers of the cerebral cortex and recommended that with its six alternating sheets of white and gray substances, it would be most similar to a battery and therefore suspected of making animal-electric fluid. The analogy of brain structures with a battery provided a morphophysiological model of a natural circuit conducting electricity from one point in the body to another:

> In the cerebellum, there are a large number of separate slices superimposed on each other and joined together by a kind of conductor like the two elements of a galvanic cell. It is also noticeable that the nerves are formed of uninterrupted threads from the brain or spinal cord to their destination, and that these threads are usually wrapped in a fatty material that completely isolates them from each other and from neighboring parts; this gives these nerve conductors much resemblance to the silk-covered metal wires so often used to conduct electricity from one place to another without loss of power. [PAL 47, p. 39, author's translation]

Although animal fluid and electric fluid were often compared, there was still some ambiguity: their different applications did not require the same equipment. Animal fluid referred to the organic secretion of an energy similar to electricity. However, imagining a brain conceived as the driving force of the human machine, in a holistic perspective, included the production of normal and pathological thought, ideas and mental content. Considered to be responsible for behaviors considered as deviant, disturbances of the electrical cerebral organization needed to be medically and technically taken care of. This research was the catalyst the historical and conceptual articulation of the dominant influence of physics on the life sciences and of a medical discourse that attempted to understand the particularity of cerebral mechanisms in their organic and moral dimensions. Thus, galvanism referred to an imagination of the power of the human brain and contributed to basic cerebral myths on its unknown powers, such as telepathy or telekinesis:

> When the Voltaic pile produces incredible phenomena, when all the bodies of nature act and react upon each other, when electricity perhaps presides over all physical and vital phenomena, when its powerful action is perhaps not alien to the reproduction and evolution of living bodies, can we say that nervous action, the nervous fluid, the animal electricity, the magnetic fluid, any word, emanating from the brain of man, whose two substances and their numerous and deep convolutions perhaps form an animated electric instrument, can it be affirmed, I say, that this nervous fluid, after having been powerfully directed at the fingertips, cannot go beyond the limit of the nails? Can it not ally, unite and correspond with another person's nervous system and impress them? [PIG 39, p. 41, author's translation]

The beginning of the 19th Century marked the construction of a culture of physical, biological and medical electricity, of which today's medicine still bears the traces. The links between the development of techniques for transmitting, stimulating and measuring electricity and progress in understanding the body were close. They were strengthened by work on the brain, with the conceptual background of the problem of how electricity maintains unity between mind, body and world. For galvanic doctors, there was no longer any doubt about the electrical nature of a mind-body connection. Thus was born the concept of an electric brain on which it was possible to intervene in order to modify, repair and/or contain the mind. As early as 1810, Matthew Yatman published a treatise entitled *Galvanism, proved to be a regular assistant branch of medicine; also, "Interesting Inquiries concerning this influence, with regard to Living Actions*" [YAT 10]. As a preamble, he describes animal fluid as the key to understanding all scales of life, from involuntary movements to conscious acts:

> The animal vital principle, formerly called 'The Nervous Fluid' is the connecting medium between mind and body; the source and regulating spring of animal sensation and expression, action and motion, both voluntary and involuntary; comprehending the circulation of the blood, respiration and all the other vital functions, or *living actions*. [YAT 10, p. 3]

This harmonious communication between all levels of the organism took place, under what he called, the electrical influence formed in the lungs by the action of respiration, producing oxidation and chemical changes in the blood. Indeed, he took up the theme, largely theorized after 1740, of the influence of natural electricity present everywhere and absorbed by the mechanics of the body. The consideration was that the brain plays a central role in this modeling of the body as it separates electricity from the blood and transmits it throughout the vascular and muscular system, producing action.

Electricity developed as an exploratory science of the central nervous system of control and improvement. This conceptual framework, linked to the current concept of increasing and improving capabilities, triggered a flurry of advertising promoting the merits of electrical accessories capable of improving mores by improving behavior. The fact that electrical treatments were seen as cures for virtually all diseases of the body and mind, from neurasthenia to epilepsy, was a result of this societal imaginary. The term culture, applied to electricity, took on its full meaning:

> Such ideas about the hierarchical ordering of the body and society also made sense in light of the connections formed in the early nineteenth century between nervous impulses and other imponderable forces. If nervous impulses acted like electricity, for example, it stood to reason that the nerves existed in a continually active state. Luigi Galvani's connection between electrical *and nervous* impulse and Johannes Müller's law of specific energies both emphasized the extent to which

> the body could be understood via analogy with machines powered by imponderable forces. [GRE 02, p. 82]

The history of electricity ranges from sensationalism to the moral control of the instincts that lie dormant in every person. From the representation of internal electricity as an indispensable ingredient of life to external electricity applied to moral disorders, we pass from the myth of Frankenstein to that of Hyde between 1803 and 1840. Carpenter (1813–1885) assumed that society was anxiety-provoking, disrupting the links between body and mind, leading to nervous diseases related to the eruption of uncontrollable behavior on the surface of the brain and nerves [CAR 46]. Critical of phrenological localizations, he developed a dynamic and functional brain model, ranging from the most automatic brain layers to those most likely to generate free will. His representations of an organic and hierarchical human nature were based on the evolutionary model developed by Jackson:

> On both comparative and anatomical issues he highlighted phrenology's flaws, and presented an alternative physiology of the mind rooted in anatomical considerations. The human mind was split not into faculties but into levels: automatic reflexes, which were the most basic; instinctual behaviour, of only limited psychological function; and consciousness, the highest level, which was anatomically located in the cerebral cortex. While all cerebral control of the body was mediated in the same way, through the reflex machinery Hall elucidated, there was a key difference between purely reflex and volitional actions, as the latter were dependent on the action of the will. The highest level, consciousness, interacting through the highest centres of the brain, was constituted and acted as a whole: this was the indivisible and non-material aspect of the human mind. [FIN 12b, p. 45]

Electricity was designed as the tool to keep behavioral overflows contained. It focused the hopes of a power that could be mastered by humanity and, acting upon it, would make it possible to discipline the behaviors generated by the most automatic and oldest layers of the brain. Carpenter, in the context of his studies on a physics of a humanity-inclusive world, also argued in favor of the idea that the different forces involved in inorganic processes were modifications of a single life force, itself correlated with the forces at work in matter. Electricity, magnetism, and other forces, including those of the body, would only be the different scales of one and the same energy, the difference then consisting of the devices by which they were manifested and measured. Thus, the mid-19th Century saw a resurgence of mesmerism. The discovery of electromagnetism had as cultural consequences the desire to unify these forces even in the knowledge of the human being on which the medical treatment of mental illnesses depended. This resurgence corresponded to the strong craze for the forces of nature, newly integrated into electrical therapies. In 1845, Thomas Courant, a follower of electric and magnetic medicine, founded the *Société Philantropico-Magnétique de Paris* as well as a newspaper called *L'union magnétique : journal de la Société philanthropico-magnétique de Paris* published between 1854 and 1869 [UNI 54–69]. His works, quoted many times by the press,[12] had a magical atmosphere close to mesmerism. The words of Crichton-Browne (1840–1938) reflect the universal role played by electricity since the 19th Century:

> We used to explain electricity in relation to matter; now we are trying to explain matter in electrical terms. Could not electricity now be understood through the much more subtle manifestation of a psychic energy that has perhaps always been beyond the reach of research in physics, but in which we live, move and have our being? [CRI 38, p. 187, author's translation]

After 1840, it became the guarantor of cerebral security by participating in the psychiatrization of moral disorders. Analogies with machines served as theoretical materials for understanding the electrical functioning of the nervous system. At the morpho-functional level, the nerves were considered like electrical wires and conducted instructions from the mind to the body, even if they were recalcitrant. It is in this context that Thomas Laycock's cerebral-moral considerations on electrical control [LAY 40] came into play:

> Thomas Laycock concurred that electricity could prove to be the therapy of choice in treating hysterical women. Electrotherapy could return to normality the bodies of those whom hysteria had transformed from 'the gentle, truthful and self-denying woman' into victims of 'insane cunning, destructiveness, infanticidal impulses, morbid appetites, etc.'. [RHY 02, p. 105]

Laycock was part of this monistic scientific atmosphere where electricity should enable us to understand all the laws of nature, from the phenomena of life to the great terrestrial and cosmic phenomena. Nevertheless, while he drew analogies

12 To give just a few examples of French language references of articles on Thomas Courant's therapies: *Le Siècle* from September 18, 1821, article by M. Pecatier; *La France Musicale* from January 25, 1852, article "Castil-Blaze"; *Le Constitutionnel* from June 20, 1852; *L'Assemblée Nationale de juin* 1853, article by Doctor Aussandon; *Le Siècle* from October 19, 1853, article by Doctor Aussandon; *Les Débats* from September 10, 1854, article by M. F. Barrière; *Le Siècle* from July 13 and January 25, 1855, articles by Léon Plée; *Le Pays* from September 16, 1855, article by M. Lecouturier; *Le Courrier médical* from April 30 and July 23, 1861, *Le Mouvement médical* from August 20, 1805 (Scientific observations); *Le Temps* from October 14, 1865, *les ceintures du docteur Courant sont recommandées contre le choléra*, by Doctor O. de Langenhagen, *Le Pays* from January 10, 1866, article by M. J.-F. Gall; finally, the English language *Daily Post,* London, August 2, 1864.

between the great forces of the universe and mental phenomena, he pointed out that the latter belong to us intrinsically while other phenomena remain external. What did this clarification made in 1863 mean? It appears that Laycock highlighted the fact that while electricity allows us to understand our environment, it also opens up the possibility of knowing ourselves to the depths of our mental schemas, which are themselves of an electrical nature. This electrical definition of our mental sphere complements his remarks on hysteria, a convulsive disease that was the target of medical electricity at the end of the 18th Century:

> And it is clear, too, that the primary or essential phenomena of electricity, chemical affinity, heat, light, and even gravity, are just as much beyond the reach of observation as those of mind [...]. There is an important difference; however, in favor of mental phenomena in this respect, in the fact that they are in immediate relation with our consciousness, whereas those of all other forces are only in a mediate relationship. [LAY 63, p. 169]

As early as 1840, several representations were at work in the description of the links between the nervous system, consciousness, hysteria and electricity: on the one hand, the woman was considered as the potential victim of her emotions; on the other hand and in a less obvious way, electricity was conceived as the coercive treatment of hysterical disturbances:

> The consequences of all this is, the young female returns from school to her home a hysterical, wayward capricious girl; imbecile in mind, habits and pursuits; prone to hysteric paroxysms upon any unusual mental excitement. [LAY 40, p. 142]

He argued for restoring communication between the nervous system and the body through galvanization. Gray matter was compared to a galvanic cell in which an electric current was generated, while white matter was like the electric wires of telegraphs that conducted current within the body. Hysteria was the catalyst for these concerns to the extent that:

> Nervous illness, hysteria is movement, excessive movement of body and mind. It has always been said that the hysterical person is unstable, capricious, irregular, that she gives in to 'fantasy'. Pathological because it is too representative ('more woman than other women,' it is often said), this hysterical woman 'changes ideas and feelings with inconceivable speed'. [BAC 12, p. 167, author's translation]

Indeed, women were described as being unable to control their bodies, which were different in their nervous system from that of men, and were capricious machines requiring technical guidance in order to function. His conceptions marked a change in the consideration of behavioral disorders which were seen as psychiatric and technical objects that could only be regulated by an electrifying doctor:

> Electricity not only provided a dynamical physical mechanism that explicated the link between mind and matter, morality and nature, but also served as a tool that allowed the doctor to intervene directly and correctly regulate the imbalances in the female physiological machine. [RHY 98, p. 245]

Laycock presented galvanic medicine as a solution to a social and moral problem relating to the normalization of women. While his arguments concerned the British context, we nevertheless find this medical imagining in the developments of war psychiatry or more generally in the history of electric shock therapy. Using the case of women, Laycock explored the possibility that electricity could be used to improve human behavior. The notion of improvement was conceived in relation to what society understood by this term. He argued for the electrical restoration of communication between mind and body by advocating localized electrification. In the specific case of hysteria, electric current needed to be applied to the sexual organs, without which there was little chance that the mental disorders it caused could be restored. Galvanism could act as a cleanser, resulting in the disappearance of cerebral pathological phenomena and the return of the mind to normality:

> However, the scientific physician enlarges the sphere of his inquiries, the good of man is his great object – the end of all his labours being to prevent moral and corporeal disease, to alleviate pain, to restore health. [LAY 40, p. viii]

For doctors such as Laycock or Millingen, hysteria, trauma or Victorian male hypochondria had in common that they were embodied in the whole body, including the moral and intellectual dimensions. From this incarnation of psychic evils in all the nervous structures, the concept of unconscious cerebration was born:

> One of the reasons 'unconscious cerebration' is more than a Victorian curiosity is that cognitive scientists have picked up this Victorian thread in theorising the 'adaptive unconscious' as opposed to the Freudian unconscious. [LEW 19, p. 77]

This concept is associated in Laycock's work with the concept of brain reflex [LEF 03, pp. 26–27], made visible, in particular, in phobic reactions translated into reflex and automatic fear. The links between these concepts, heuristics for the development of cognitive sciences and electrical treatment, have scientific but also cultural stakes. If all ailments are nervous-based, then the nervous system becomes the preferred place of treatment for electrical therapies. Beyond his assertions about the feminine nature, Laycock proposed a more profound conception of the exploration of life and the living, from the scale of consciousness to the molecular scale:

> They form the connecting link between the phenomena of consciousness, and the molecular changes in organic matter upon which the phenomena of heat, electricity, galvanism and magnetism depend. They point out a new path of experimental inquiry into the phenomena of life and thought and, if traced out in all their relations, cannot fail to change the whole aspect of mental philosophy. [LAY 40, p. 100]

In 1848, Millingen published *The Passions or, Mind and Matter* [MIL 48], a treatise in which he discussed the galvanization of mores and behavior. While he reinforced a very classical vision of hysteria, linked to the female sexual organs, he correlated this phenomenon to the weakness of "the energies of the brain or the sensorium of woman" [MIL 48, p. 44]. In a literary and romantic style, he questioned the essence of life:

> Galvanism, it is true, may produce actions similar to those of many of our functions; but who would dare to assert that life is the result of galvanism or electricity? [MIL 48, p. 121]

He interpreted brain movements, which consist of processing external stimuli, in terms of electrical speed. This point must be understood through the image of a machine brain at the controls of a machine body. Thomas William Nunn (1837–1909) published in 1853 a treatise entitled *Inflammation of the breast, and milk abscess* in which he extended the comparison of the cerebral organ to a galvanic machine, to the uterus, the breast and the ovaries. They would have, according to him, a morphological organization comparable to a reproductive galvanic cell completed by the female nervous system:

> The ovaria, uterus and mammae form, as it were, a reproductive pile, the circuit being completed by the nervous system. [NUN 53, p. 3]

The functional analogy of nerves with electric wires was taken up in a comparison of brain function with the electric telegraph, designed by Baron Schilling in St. Petersburg in 1833:

> Again by analogy, just as we have compared the constitution of a swamp to a vast galvanic apparatus, we can also liken the human body to a voltaic pile, since it is also formed by the contact of heterogeneous elements whose nerves and muscles are the conductors, and solids and fluids are both the generators and conductors of electricity. [PAL 47, p. 232, author's translation]

Electrical therapies cannot be separated from the invention of new technologies. In the same way that the telegraph helped to maintain order by allowing criminals to be reported more quickly; electricity guaranteed moral order by restoring electrical brain power immediately. While the nerves conduct instructions from the body to the mind, communication still has to work. The development of the telegraph, thus gave a model to the nervous functioning, conceived in terms of transfers and electrical communications:

> As I have already observed, these instruments of mental transmission, although they are consecutive in their operation, and may be considered sequent in their course, yet act in such a simultaneous manner, that sensations are submitted to the test of our judgment and reason with electric rapidity. [MIL 48, p. 137]

Ada Lovelace (1815–1852), daughter of the poet Byron (1788–1824), is a figure in the history of the brain as a machine. A pioneer in computer science [KIM 99] and creator of one of the first computer language programs, she showed an early interest in electricity and the brain-machine. As a patient of Laycock, she crossed paths[13] with Andrew Crosse (1784–1855) with whom she evoked the fact of making electrical experiments a tool to reach a new understanding of vital mechanisms and consciousness.

The societal and medical stakes of the application of electricity were multiple: from the knowledge of Man in his materiality to the possibility of intervening on his mental physiology, the range was wide and is still developing today. While the cerebralization of behaviors and faculties seems to result from a rational movement, it also stems from an interventionist imaginary and the desire of the human species to control itself. In this context, the imaginary of convulsive behavior joins the electric imaginary. Both of them have had a lasting impact on the history of medicine. While the 19th Century was described as the century of convulsions, its conceptions of human nature were based on the electrical conception of the subject and the secularization of diseases of the mind. Thus, Laycock, Marshall Hall (1790–1857) and William Carpenter wanted to demonstrate that mind-body intricacies were very complex, that much of what the mind did to the body took place on the surface

13 Andrew Crosse is known for his research on electrocrystallization.

of consciousness and thought, but also that electricity played a crucial role in this research with the technical perspective of manipulating the currents that continually work between these substances. Electrification ranged from the whole body to the brain, making visible the important notion of functional localization, involving a representation of the brain organ as the organic substrate of human instincts, faculties and behaviors. This key notion of cerebralization was concretized and prolonged in a process of internalization of psychological evils. Gradually, the electrical stimulation penetrated deeper into the brain to better reveal its organization and functioning:

> At the dawn of the 20th Century, the disturbing strangeness was displaced, no longer in a mysterious Other, but in oneself; in the darkness of one's own psyche. [BAC 12, p. 184, author's translation]

While between 1801 and 1840, electricity represented a counter-culture to atheism and materialism, capable of giving life back to the deceased, from 1840 onwards, it became the guarantor of the standardization of mores and a certain representation of happiness. Its developments thus marked the domination of Man over the evolution of his species.

Figure 1.5. *French advertisement dating from 1911 for the "Herculex" electric belt*

Electrical treatments were seen as universal remedies, or in any case were disseminated as such within public opinion. Everybody could compensate for the weaknesses of their animal fluid and re-establish good connections between their consciousness and their body. Devices were becoming more compact, easier to handle, resulting in a wave of companies producing *healthy* electrical items. These paramedical products, such as the electric belt, easily accounted for a quarter of advertisements in 1880 [LOE 99].

These devices referred to the fact that in addition to taming the world, bringing light and progress to it, electricity was able to discipline the body and mind. This medical movement, which had its roots in the second half of the 18th Century, could have died out in the face of the uncertain results initially brought about by electrical treatment. Because it corresponded to a time when society was looking for new, stable and rational points of reference to regulate the lives of individuals, its posterity in the history of neuroscience, understood in the broadest sense, is still relevant today. These applications of electricity to a body that had symbolically become a machine could be conceived as a step contrary to hypnotism, insofar as it was not a question of reaching consciousness by disconnecting its link with the body but of intervening directly on the cerebral circuits to regulate behavior. In the context of the development of electrotherapy rooms, we can speak of a naturalization of behavior. While convulsion referred to illnesses that are difficult to differentiate from each other, electricity appeared to be an instrument that could act both on the frozen condition and on the disordered movements, able to differentiate a psychological illness from an organic pathology. Thus, cataleptics, hysterics, ecstatics and epileptics resembled each other and merged together in the medical discourse, in that their lists of symptoms had in common that they did not present a visible organic disorder:

> Others, such as the supporters of the *École de la Salpêtrière* in the 1880s, made it a simple symptom combined with other neuroses: hysteria above all, but also ecstasy, epilepsy, apoplexy, death, chorea. Some speak of 'hysterical catalepsy', others of 'cataleptic ecstasy'; others still of 'hystero-catalepsy'. [BAC 12, p. 173, author's translation]

The excerpt from an article in the French newspaper *Le monde illustré*, dated August 14, 1887, describing the intense therapeutic activity in the electrotherapy department of Salpêtrière, uses the argument of the number of patients treated to assert the effectiveness of electrical treatments. This point underlines the fact that this treatment responded to a societal problem involving ailments about which little was known but which affected a large number of people:

> Thousands of patients have been treated in recent years at the Salpêtrière. At each consultation, the average number of patients

> varies between two hundred and fifty and three hundred. [...] It's called the electric bath. Under its influence, we can observe various physiological phenomena (heat, blood circulation, etc.), too technical to find their place here. Localized electrification is done by means of appropriate exciters. The main ailments that are treated at the Salpêtrière clinic belong to two classes; nervous diseases (hysteria, neuralgia, all kinds of paralysis) and nutritional diseases in which we understand dyspepsia, stomach dilatation, chlorosis, anemia, rheumatism, etc. The ever-increasing number of patients coming in for each consultation is the best proof of the effectiveness of this treatment. Already known, but not yet known enough, this new therapeutic method, which has already taken the largest extension, is destined for the brightest future.[14] (author's translation)

In addition, the development of psychoanalysis and Charcot's therapeutic hesitations moderated what could be conceived as a general craze for electrical interventionism on mental ills. Freud, who was first interested in Erb's work, ended up considering this treatment as chimerical:

> Whoever wants to make a living from the treatment of nervous patients must obviously be able to do something for them. My therapeutic arsenal contained only two weapons: electrotherapy and hypnosis, as sending to a hydrotherapy facility after a single consultation was not a sufficient source of gain. I referred to W. Erb's manual for electrotherapy, which gave detailed prescriptions for the treatment of all the symptoms of nervous diseases. Unfortunately, I soon had to admit that my docility in following these prescriptions was of no avail, that what I had taken to be the result of accurate observations was nothing but a phantasmagorical structure. [FRE 50, p. 40, author's translation]

Thus, it appears that the notions of the electrical body and then of electric consciousness followed one another during the 19th Century and nourished an already well-established culture of electricity. Difficult to separate from technical advances and exploratory and therapeutic applications, medical electricity permeated research on the integration of Man in nature, on the materiality of his mental mechanisms. From the objectification of this electrical nature, an imagining of the power, technique and capacity that electricity has to change us ourselves was born. Alongside the Neohippocratic movement was also the idea that individuals possess sensitivities to electricity on which their character traits depend or influence.

14 Excerpt from the newspaper *Le monde illustré*, dated August 14, 1887.

1.3. Possible electrical profiling?

The notions of hot-headed, irritable temperaments that punctuated the texts of electrifying and/or galvanizing doctors until the end of the 19th Century accounted for the observations that two subjects, dead in the same conditions and not subjected during their lifetime to the same temperament, will have organs that do not react in the same way to electrification or galvanization. Moreover, this electrical individualization also influenced the person during his or her lifetime: did the person have a more explosive temperament caused by an excess of animal electricity? Or on the contrary, a softer temperament?

In 1787, Petetin drafted a typology of personalities likely to have behavioral disorders. His analysis was based on the idea that everyone has an innate quantity of electric fluid circulating in their body, which causes predispositions to these ills:

> The violent & fleeting convulsions which characterize it, are announced in advance by the signs of a dominant electricity in the whole animal economy; such are the supernatural strength, inconceivable agility, the vivacity of ideas, combined with the greatest volubility in expression, a more lively heat spread over the torso, the head & the arms, while the lower extremities are usually devoid of them, a sometimes voracious appetite, the desire for cold & sour drinks, the fire of the eyes, insomnia, or turbulent sleep, all the passions of the soul, exalted. [PET, 87, p. 86, author's translation]

In a gendered tradition of understanding hysterical symptoms, he described passions in women that "...accumulate too much fire principle in the brain, & dispose them to frequent convulsive outbursts" [PET, 87, p. 91, author's translation].

These points developed by Petetin are interesting. Indeed, when Henri Gastaut (1915–1995) tried to outline a typology of electroencephalographic tracings [ADR 54] in connection with archetypal personalities, we find the way in which these links between the nervous system and electricity developed.

Then in 1803, Cassius explained on the one hand the direct effects of galvanism on subjects and on the other hand the link between temperament, in the psychological sense of the term, and the ability to be stimulated by galvanism. The notion of temperament refers to a trait of character as well as to a link with the supposedly singular strength of the nervous fluid specific to each individual:

> Among the influences that Galvanism exerts on the animal economy are the acid and sometimes alkaline taste, the production of lightning, the pain one feels, the involuntary movements one makes when

> touching the galvanic devices, the convulsive state into which it plunges those with the very irritable nervous type. [CAS 03, pp. 22–23, author's translation]

Experiments were made to discover, or not, the singularities of the nerves of people suffering from "electrical diseases":

> When an epileptic died, I had a surgeon cut some of his nerves, and I also got similar ones from another corpse which had not been subject to nervous diseases during his lifetime. These nerves being well dried out I rubbed them in the dark & I saw a lot of electric light, between the rubbing & the nerves of the one who had been subject to spasmodic movements; I saw very little of it in the nerves drawn from the second body. The nerves of a person prone to spasms are therefore more electric than those in whom one has never, ever noticed similar convulsions as the experience I have made, & that I have repeated several times with the same success, having kept these nerves for more than two years. [PAL 47, p. 225, author's translation]

One can wonder about this imaginary electrical singularity of an organism suffering from nervous disorders. It must be brought closer to the notion of a body designed as a natural electrical machine. If the sensitivity and excitability of the nerves depended on a personality marked by a more or less nervous temperament, the medical applications of galvanism were fairly quickly focused on nervous disorders. This fact influenced its reception in the medical world:

> No doubt people of excessive nervous susceptibility are usually well served by a bath or simple galvanic current, so M. Labeaume's method must be useful to patients with an eminently nervous constitution. [LAB 28, p. 25, author's translation]

Also experimenting on human chains[15], Humboldt quickly noticed the variability of conductivity from one person to another, which could be almost zero in some people. This observation completed the question of a subject's electrical temperament; this temperament influencing both body and consciousness.

15 Aldini also experimented with human conduction chains: "Let four or more people, holding hands moistened with a solution of muriate of soda, form a long animal chain; let the first person hold in his hand the muscles of a prepared frog; if the last person, placed at the opposite end of the room, touches the spinal cord or the crural nerves, the contractions take place; if the animal chain is interrupted, the contractions cease at that instant" [ALD 04, p. 10, author's translation].

To conclude this chapter, here is a current scientific episode whose rigor is not to be judged, but to underline its inscription both in the long temporality adopted in this work and in a scientific imagining of the neuro-electric omnipotence. In 2016, a controversial article appeared on the Frankenstein effect. Canavero, known for his research in the field of human head transplants, using Aldini and Ure's experiments, took up the following theme:

> Data from two centuries ago prove that a fresh cadaver, after hanging or decapitation, can be mobilized by electrical stimulation for up to 3 hours. By administering spinal cord stimulation by applied paddles to the cord or transcranial magnetic stimulation to M1 and recording evoked potentials, it should be possible to test fusogens in fresh cadavers. Delayed neuronal death might be the neuropathological reason. [CAN 16]

It is doubly relevant to the history of cadaver experiments: on the one hand, a fresh cadaver can serve as an experimental substitute for the animal model; on the other hand, it allows us to explore the neuropathological foundations of the process of dying. Canavero placed his research in the context of isolated neuronal cell cultures:

> Testing on brain dead organ donors in the interval between declaration of brain death and initiation of organ harvesting is an ethical option. Reviewed data from two centuries ago prove that another ethical option is available, that is, experimentation on fresh cadavers. As discussed, movements can be elicited in a fresh cadaver by electrical stimulation for up to 3 hours (and possible more). This suggests that the cerebral cortex and its projections to the spinal cord and the cord itself remain viable for up to 3 hours postmortem. This includes both the cell bodies and the synapses. [...] spinal cord stimulation above the level of attempted fusion or transcranial magnetic stimulation to M1 followed by recording motor evoked potentials (MEP) should be enough to confirm neuropshysiological conduction. [CAN 16]

While the scientific dimension seems to fall back into the spectacular and the *announcement effect*, it seemed relevant to point out the continuous resurgence of the electric imagining as a material for life.

Thus, an electrical culture, plastic and polymorphic, was born and evolved during the 19th Century: a spark of life, the guarantor of a moral order, the substrate of individual temperaments, it marked the history of physiology, medicine, philosophy and literature. The next chapters will look at how these different stages are broken down and how they give rise to knowledge, practices and techniques that are still relevant today.

2

From Physics to Electrifying Physicists

Electricity has played an anthropological, political and scientific role in the medical field, especially during the revolutionary period. Its symbolic impact has been a key element in understanding, despite repeated therapeutic failures or very uncertain results, medical electricity has been able to survive through the centuries, always newly shaped by the specific expectations of medicine. To ask whether electricity is a new goddess is not to question its connection to metaphysics, but to emphasize its materialistic significance, in the context of an environment where the very definition of human nature is undergoing a transformation. In the context of the Enlightenment and philanthropic medicine, it played a social and political role and became a treatment reserved for the most in need. As a revolutionary symbol, it could be seen as a concrete instrument of medical secularization and the notion of political and scientific progress:

> While electricity is a beneficial and regenerative therapy for children, it is also a therapy for the state. In his *Nouveau dictionnaire français* [SNE 95], Léonard Snetlage illustrated in 1794 the fortune of the term with the revolutionary formula: 'the electric fire that sets all the hearts of the freedom soldiers ablaze' [DEL 06, p. 40]. [ZAN 17, pp. 219–221, author's translation]

Electric shocks were located next to political and ideological shocks:

> Long live the Revolution, whose energetic force imprints on our nation, with an electric stroke, the horror of kings, the love of laws and the Republic. [ANT 94, p. 19, author's translation]

Figure 2.1. *"Thus the electric spark of freedom will overthrow all the thrones of the crowned bandits,"* Bibliothèque nationale de France, *Paris, De Vinck, 4209. Gallica.bnf.fr. For a color version of this figure, see www.iste.co.uk/cherici/clouds.zip*

At the beginning of the 18th Century, knowledge about this natural force accelerated and took a more systematic turn. Electrifying physicists applied it to themselves, in often painful and impressive self-experiments, and to others to study the links between the body and electricity. Many were already questioning its therapeutic potential. These scientists, while exploring the natural laws of this energy floating in the atmosphere, were considering its use in medicine, particularly in the context of paralysis, the primary target of electricity. In this chapter, we will discuss the links between physics and medicine, and the shift from one to the other, in two stages. From a natural flow, electricity quite quickly became an artificial remedy. First, we will look at how electricity can be used in the field of medical applications, without leaving the field of physics. Then we will see that the application of this medicine was linked to the design of electrical machines and the development of techniques.

2.1. Physics, knowledge of laws and nature of electricity

From the first attempts of metallic electricity to control, animate or heal the body, the idea that this energy was close to a power of the human being over itself was developed in the first third of the 18th Century, alongside the physics of electricity. Experiments oscillated between suspicions of charlatanism, notions of electrical martyrdom and treatment trials, but also between building and improving electrical machines. The history of these machines, built to heal, is deeply linked to the history of medical electricity. But what are we talking about when we approach this electrical treatment, before Galvani's experiments and the beginnings of electrophysiology? Do the years 1740–1790 represent a pre-scientific period for electricity?

Epistemological questions emerge:

– the rapid popularity of electricity, considered to be economical and accessible to all, correlated with its equally meteoric failures, leads us to wonder about the heuristic value of this treatment that had difficulty functioning:

> The imbalance between the precision of the analysis of the issues in terms of experimental physics and the speed of the questioning of the medical implications may give the impression to the reader that the application of electricity to medicine would be necessary and not problematic. [ZAN 17, pp. 8–9, author's translation]

To what extent was the medical practice of electricity shaped by physicists?

– the notion of therapy became blurred: this medical physics, marked by analogies with the mechanisms of nature, took on an experimental structure in terms of "trial and error":

> It is more a matter of trial and error and experiment rather than reason that regulates the effects of electric mosettes on living bodies. Perhaps an extraordinary shock of an indeterminable degree on a singular occasion could render part of the movement to a paralytic member, as we are told in the *Journal des Savans* from May 1748. But until now electricity has not made a great fortune; and even if it is a remedy, there will always be great difficulty, and even danger in using it. [MOR 48, p. 196, author's translation]

The study of the different stages of the history of electricity, marked by short but frequent periods of discredit, must be correlated with the understanding of the soul, in terms of the materialistic integration of the faculties within the brain. For example, the beginning of brain localizations being deeply linked to electrical therapies with the psychological sphere as a target. The cerebral organ, at the end of the 18th Century, became the base of all the nerves and the symbol, by analogy, of an electric machine. Developments in medical electricity were also marked by a polysemy of knowledge that crossed the fields of physics, mechanics, physiology, botany and anatomy, as well as by a plasticity of practice, visible through the diversity of machines, types of electricity used, or experimental protocols developed. The first electrifying people were often technicians, physics demonstrators, scientists of the mechanics of nature, who looked at the links with the body. The permeability of disciplines that marked the 18th Century, and the encyclopedic movement, facilitated these interactions. But there seems to be more... Does the increase in physical knowledge about electricity explain, in itself, its appropriation by medicine? The rise of materialism was undoubtedly key to understanding the links between the notion of man-machine and man-electricity. This is something we will come back to.

Stephen Gray (1666–1736) [GRA 90, pp. 31–32, 38, 41], a dyer known for his experiments in the fields of physics and astronomy, is considered one of the first to have carried out a systematic study of conductive substances – among which he included the human body, as early as 1729. This step was fundamental in paving the way for the inscription of humans in the study of natural forces. He also demonstrated – by suspending a child whose foot remained in contact with an electrical machine – that when in motion, the subject showed signs of static electricity. He noted several phenomena related to the still partially unknown laws of electricity, such as the fact that electrical attraction propagates in a vacuum or that the surface of water in a wooden bowl swells when an electrified body approaches. His experiments, in addition to being taken up by Jean Antoine Nollet, were frequently cited in the literature of medical physics until late in the 19th Century:

> He suspended a young boy from silk cords in a horizontal direction, whose feet he applied to a rubbed glass tube: he immediately saw that the young man's head attracted light bodies at a distance of 8 and even 10 inches: clear proof that this isolated young man had received the electric fluid from the tube, and was truly electrified. In carrying the tube to the head, and the light bodies to the feet, the electric force was more reliable; and it was insensible, unless he placed the light bodies below the head and brought the tube above it: phenomena which Mr. Gray did not dare to try to explain, but whose cause probably consisted of the known power of the tips which the hair exerts. [PAE 88, p. 150, author's translation]

Then the French chemist and physicist Charles-François de Cisternai du Fay (1698–1739) [CIS 33a, b, c, d, 34a, b] confirmed the possibility of electrifying all bodies, including the human body. He also remains known for his hypothesis of the two types of electricity: resinous and vitreous. He had also observed electrical attraction and repulsion effects.

It was in the 17th Century that Otto von Guericke [GUE 72], a German scientist and inventor, developed the first friction machine to produce electricity:

> The device had the appearance of a simple sulfur globe mounted on a wooden shaft that was rotated by a mechanism made for that purpose and on which a person pressed a rag or a dry hand. Electricity was caused by friction and its various properties were clearly visible: attraction, repulsion, conduction and luminosity. If a metal object was moved towards the globe, a small spark would occur, accompanied by a clicking sound. If light bodies such as feathers approached, it would attract and repel them. If you touched it with your finger, you'd feel a little sting. [GUE 09, p. 175, author's translation]

The invention by Petrus van Musschenbroek (1692–1761), professor of physics in Leiden, of the first capacitor model called the *Leyden jar,* marked a decisive step in the electrification of Europe. In a letter dated April 20, 1746, addressed to his friend the French physicist René-Antoine Ferchault de Réaumur (1683–1757), Musschenbroek describes the experiment that shook him like love at first sight and he highlighted the pain he felt and the danger he faced. He showed that electricity, which is transmitted throughout the body, can appear on the surface of the body in the form of sparks.

Jean Torlais relied [TOR 63] on a memoir by Dorsman and Crommelin [DOR 46] and questioned the true creator of the Leyden jar. The two scholars claimed that the real inventor was Jürgen von Kleist (1700–1748), who experimented with it on October 11, 1745 and published his findings in 1746:

> It is not known why von Kleist had placed an iron nail in a glass jar and, holding the jar in one hand, brought the nail closer to his electric machine, which was a glass globe electrified by friction. Having tried to take the nail, he felt a strong shock. He also found that the effect was increased if mercury was added to the jar. [...] During this time, the same discovery was made in Holland, by Peter Van Musschenbroek (1692–1761), successively, professor of mathematics and physics in Duisburg, Utrecht and Leiden. [TOR 63, p. 212, author's translation]

Yet it was Musschenbroek's letter that brought Europe's science to boiling point and gave rise to a huge field of physical and medical experiments, using electricity as the raw material. Torlais, in his article, highlights the fact that the discovery was made at similar times by physicists, yet the experimenters were not working together. It was Musschenbroek who learned about the instrument, described it and disseminated its effects in academic circles. These two experiments correspond to a period when studies of forces in matter were fundamental to understanding the world and humanity's place in it. It is therefore not surprising that in different geographical locations, experiments on electricity were carried out with similar results.

Thus, Musschenbroek recounted in his letter the terrible experiment he carried out with his colleague, Cunéus. Thus, two silk threads were suspended from an iron barrel which received by communication the electricity from the glass globe, rotating rapidly on its axis, while it was rubbed. At the other end a brass wire hung freely, the end of which was immersed in a glass vase, partly filled with water, which he held with his right hand while with the other hand he tried to draw sparks from the electrified barrel. Suddenly his right hand was struck so violently that his body was shaken as if by lightning, twisting in a violent spasm. Unbearable pain was felt. The participation of his whole body in this accidental electrification marked the beginning of a long history between medical electricity and movement. For Nollet, his co-worker must have been the first to feel the shock. However, it was Musschenbroek's responsibility to have shown the arrangement of the capacitor and described the device. His letter was read and translated on April 20, 1746 in Paris, during a session of the *Académie des Sciences* (French Academy of Sciences), and was commented on by Abbé Nollet. He approached this seriously and conducted experiments on electricity to confirm or deny the facts. Thus, in 1746, he described the experimental preconditions for the Leyden experiment. Given the number of accidents, the methodological issue was important for experimenters trying to reproduce it:

> 1. Care must be taken to ensure that the glass vase containing the water is clean and dry, both outside and inside, at the part that remains empty. 2. He must hold the vase, touch it by the place that contains the water. 3. Instead of water we can use mercury, and other liquids that are neither sulfurous nor greasy. You can even use iron filings, sand, etc. 4. Any vase other than glass, or porcelain, won't be sufficient. 5. Instead of holding the vase in your hand, you can place it on a metal stand, and so if you only hold a finger applied to the glass or the stand, you feel the blow. 6. If the chain is interrupted, or if two of the people forming it each hold a stick of sulfur, Spanish wax, resin, etc., at one end, the ordinary effect does not take place. [NOL 46, p. 133, author's translation][1]

1 "It is an indispensable thing that the hand which touches, before the spark is excited, should not cause the iron bar to lose its electricity, for if it does, it is useless to try to make the bar spark with the other hand; and it is a long known fact that it is easy and quick to de-electrify an iron bar by touching it with the hand" [NOL 46, p. 199, author's translation].

A proponent of self-experimentation, Nollet caused a shock by replacing the Leyden jar with what he called a glass vessel:

> [...] that the electric matter which emanates, flows more easily and more abundantly in the vacuum than in the air of the atmosphere: I have also noticed in the same place, that the glass vessel from which the air has been purged, and which receives the electric emanations of a rod of iron inside, promptly acquires a very great virtue; which follows quite naturally from the first effect. [NOL 48, p. 425, author's translation]

The history of electricity is marked by an experimental scientific structure, which explains why physicists, then doctors and electrophysiologists kept experimenting on themselves in order to determine the positive and negative effects of this force. This trend towards direct experimentation corresponded to a rationalization and an attempt to objectify and describe the effects felt from electricity:

> Other stories, moreover, were even more impressive, such as Winkler's[2], where there were great convulsions throughout the body, such violent agitation of the blood that one feared an attack of hot fever. In Franklin's experiment of the Conjurers, which was only a variant of the Leyden experiment, the company was very large; some complained about the strength of the shock, while others considered it very moderate. [TOR 63, p. 215, author's translation]

Nollet quickly became one of the leading electrifiers in Europe. After some popular experiments, he experimented with possible electrical applications in treating the body, including paralysis. By making his own body an object of exploration for the effects of electricity, he became an automaton whose parts reacted, without will, to this force. He extended this automation of the body to human chains and thus entered into a medicine marked by Cartesian dualism, in which bodies could be disarticulated and studied like machines. This philosophical context allows us to understand the logical structure behind the fact that medical electricity was immediately interested in paralysis. Indeed, while electricity caused involuntary movements on a reified body, then it was likely to render the movement in a therapeutic context:

> Electricity could be used to provide a stark illustration of Cartesian dualism. By graphically demonstrating how mind could be divorced from body, electricity provided a key element in inaugurating a sharp

2 Johann Heinrich Winckler (1703–1770), particularly skilled in physics, is known to have studied the transport of electricity, thus laying the theoretical foundations of telegraphy.

> disjunction during the eighteenth century in the ways in which people thought of their bodies and themselves. The Abbé Nollet's famous (or infamous) experimental demonstration of electrical conduction by shocking a line of Carthusian monks into leaping into the air simultaneously might have been designed with the French Royal Court's amusement and edification in mind, but it was also a telling example of Cartesianism in action. [RHY 02, p. 93]

May we speak of Nollet's experimental medicine? If experimentation consists of asking a question about physiological phenomena, designing the technical conditions of the experiment and applying it, then the investigations of electricity from 1740 onwards respond to this structure. This point can be correlated with the experimental dimension that marked the research of physicists. Thus, Nollet had impressive flywheel and gear-driven machines built, and he was involved in the design. In his treatises, he shares his experiments and describes how, gradually, the electrical phenomenon was tamed, channeled into metallic circuits and bottled. Electricity then went to the streets, or rather to the gardens and parks, where it made a spectacle of itself to the people, becoming a major attraction at the Saint-Germain and Saint-Laurent fairs.

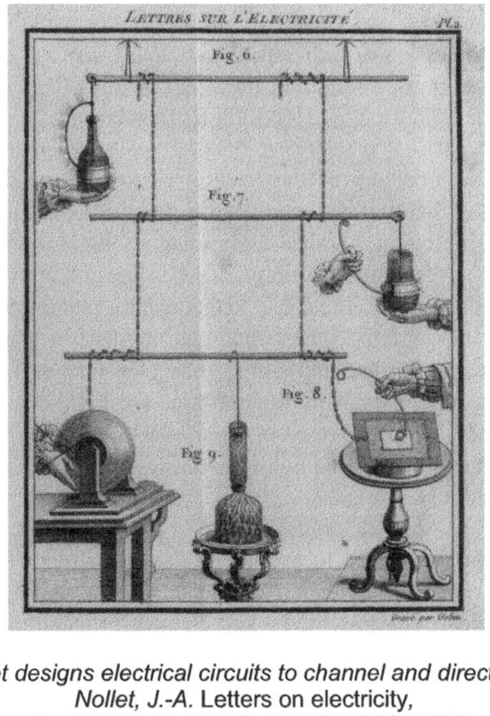

Figure 2.2. *Nollet designs electrical circuits to channel and direct static electricity. Nollet, J.-A.* Letters on electricity, *op. cit., vol. 1, pl. 4, fig. 14-17, 1753–1757*

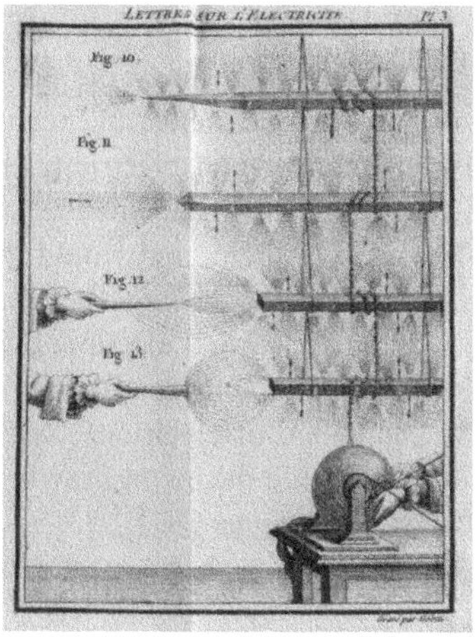

Figure 2.3. *The circuits thus designed lead to the application of static electricity to the human body. Nollet, J.-A.* Letters on electricity, *op. cit., vol. 1, pl. 3, fig. 10-13, 1753–1757*

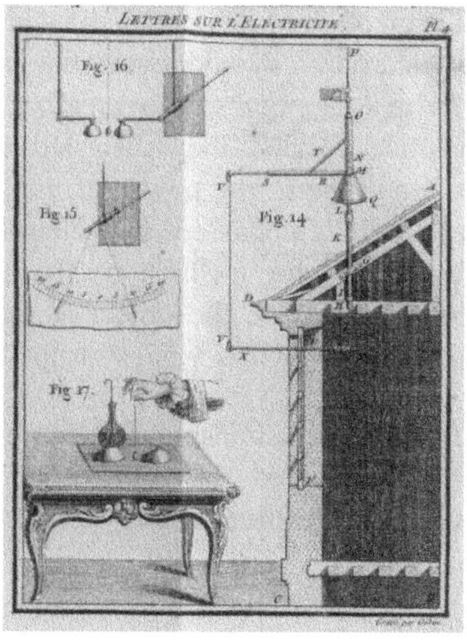

Figure 2.4. *The physicist imagines and improves his famous static electricity machines. Nollet, J.-A.* Letters on electricity, *op. cit., vol. 1, plate 2, fig. 6-9, 1753–1757*

In his *Lettres sur l'électricité,* published in 1753 and mostly addressed to Franklin [NOL 53–57], Nollet discussed the advantage of his contemporary's experiments. He warned about the dangers of the lightning rod experiment, carried out by Franklin and widely reproduced:

> After noticing that matter which comes out of an electrified body, spins more easily, and from further away the point of a needle, than such a body which is rounded at the end, and recognizing moreover a certain analogy between thunder and electricity, he dared to believe that pointed iron rods, erected in the air under a stormy cloud, would draw the matter of lightning to them, and would make it pass without brightness, and without danger even into the immense body of the earth, where it would remain absorbed; occupied with this thought, he planned to test if isolated spikes, on resin or silk supports, might, in stormy weather, give some sign of electricity, not doubting that what he had imagined would be a safe means of drawing fire from thunder, and preventing its disastrous effects. [NOL 53–57, v. 1, p. 8, author's translation][3]

The kite experiment was probably not done the way the story tells it, featuring Benjamin Franklin holding a kite to collect electricity. Nevertheless, the latter developed the role of spikes in the conductivity and concentration of atmospheric electricity.

Nollet contributed to the creation of a materialistic and interventionist imaginary that marked the history of electricity, by addressing the experiments reproduced by Messrs Dalibard and Delor in Marly[4] in stormy weather. While their experiments helped to spread the positive results, they also led to the fantasy that "the thunderbolts of heaven will now be in the power of men [...]" [NOL 53–57, v. 1, p. 10, author's translation]. Nollet emphasized the scientific facts that flowed from Franklin's experiments but called for mistrust as to their consequences which, bordering on the marvelous, should be the object of caution. Disputes over the creator of the experiments emerged [NOL 53–57, v. 1, p. 15].

In addition to stating that to study atmospheric electricity, there was no need for thunderstorms or hail, Nollet detailed the effluent and influent materials that represented different levels of electrical energy, whereas Franklin considered electricity in its uniqueness:

---

3 "[...]; I learn from letters from Bologna and Florence, that those who made them there, almost repented of it, because their curiosity was satisfied beyond their desires, by the violent shaking they experienced in trying to draw sparks from the iron electrified by thunder" [NOL 53–57, p. 16, author's translation].

4 "The King, who was kept informed, wanted to see these experiments and on February 3, 1752, at the home of the Duke of Ayen in Saint-Germain-en-Laye, in his presence, Delor, Buffon and Dalibard gave some demonstrations that earned royal applause. It remained to be seen whether, in stormy weather, sparks could be fired from an iron rod stuck in an insulating stool. This was the well-known Marly-la-Ville experiment on 10 May 1752" [TOR 56, p. 342, author's translation].

> This idea is that the operations of electricity do not depend, according to the generally accepted opinion, on a single positive power, but on two distinct powers, both positive and active, and by the opposition with which they act, so to speak, one against the other, that they produce the varieties which distinguish electrical phenomena, so that the body, which is called positively electrified, is not simply impregnated with a greater dose of electrical matter than in the natural state, and that he who is said to be negatively electrified has no less; but that the first is coated with a larger portion of one of these active powers, and that the second is coated with a larger portion of the other natural power, is electrified only because these two powers are in equilibrium. [NOL 53–57, v. 3, pp. 84–85]

Experimentalism and empiricism were organizing themselves in place of a system of electricity that continued to change in form and applicability. Moreover, questions about the nature of electricity or electrics were not unrelated to the medical applications that would be made of them. Should we talk about electricity in the plural? Or can we consider its polymorphic character in nature, in its visible and invisible scales? To outline answers to these problems, scientists were looking at its presence in the different scales of nature, compound bodies, organic and inorganic matter. The scientific and philosophical question that arose in this issue of the different forms of electricity was that of the existence of different physical laws for organic and inorganic bodies. The epistemological stakes were considerable and concerned the physico-chemical unity of living organisms. When Nollet said that it was not necessary to have a thunderstorm to capture electricity, he addressed a central point of this work: experimenting with electricity is like making visible an invisible phenomenon that is nevertheless present everywhere. This force did not need special weather conditions to be captured. Nollet used the Leyden jar to show, not two essentially different electricities, but two electrical powers:

> [...] in general electricity consists not of a single power of this kind, but in two distinct positive powers, which act in opposite directions [...] which cross each other. [NOL 53–57, v. 3, p. 87, author's translation]

This is in contrast to Franklin, who recognized only one form of electricity, as Jean Torlais points out:

> Franklin only admitted to a kind of electricity contained in all bodies. During the electrification, a new distribution of the fluid was produced as it passed from one body to the other: the first was electrified positively, the second negatively. [...] This theory was opposed to that which Nollet had set out in his *Essai sur l'électricité des corps* published in 1746. Nollet maintained that there were two kinds of

> electricity that he called effluent and influent: the first kind of electricity rose from the electrified body to a certain distance, while such influent material came to the electrified body apparently replacing the material that came out of it. [TOR 56, p. 340, author's translation]

By developing the design of these two electrical powers, circulating proportionally to each other, Nollet also developed the possible circuits, as well as the modes of transmission of metals to the body and from the latter to the outside:

> They electrified an iron pipe about four and a half feet long and two and a quarter inches in diameter, and as they continued to electrify it, a person standing on the floor took in his hand an iron punch, cut into a round pyramid, which was as big as a finger, towards the head which was rounded, and its point was very sharp. The person holding the instrument, turned the point towards the electrified tin pipe, and moved forward little by little until the point began to glow. When the distance at which this happened was noticed, a glass pane was held in front of the punch, nine to ten inches wide, and pierced with a small hole in the middle to receive the point. Whenever this was done, it was necessary to get much closer to the conductor, to make the tip glow. [NOL 53–57, v. 1, pp. 258–259, author's translation]

In these letters, Nollet referred to his texts published in 1746 and 1748, and in particular to his treatise *Essai sur l'électricité des corps* [NOL 46], from which he quoted a few passages to show the priority of his research over Franklin's. Thus we can see that in 1746 he already asserted the continuity of electric circulation; otherwise, the body from which the electricity is drawn would be exhausted. He seems to say that if these experiments allow static electricity to be channeled, they also use an electricity that is specific to the body, which explains why the body is permanently electrifiable:

> These experiments prove quite clearly: 1) That electric matter rises from the electrified body, and it gradually carries itself to the surroundings to a certain distance, since it carries away the light bodies which are on the surface of the electrified body, and that it supports at the height of eighteen inches at the most, above the electric tube, the small sheet of metal which it carries away. 2) That such matter comes to the electric body, apparently to replace that which comes out of it; for a body does not exhaust itself to be continually electrified, and how can it not exhaust itself in the end, if nothing repairs the emanations which it supplies? The corpuscles or body parts that remain applied to the electrical surface, while the others are

> removed, are sensitive marks of the existence of this matter, and of the direction of its effort. 3) That these two streams of matter, which go in opposite directions, exercise their movements at the same time; since the same electrified body attracts and repels everything at the same time. [NOL 46, pp. 79–80, author's translation]

The term *corpuscle* that he used was not trivial and was reminiscent of the debates on the nature of light. By analogy, or because investigations into the forces of nature faced recurrent questioning, this discussion entered into the quarrels over the theory of waves and corpuscles[5]. For example, Nollet explained that: "Electrical matter comes out of the electrified body in the form of bunches or plumes, whose rays diverge greatly from each other" [NOL 46, p. 144, author's translation]. Questions about the form(s) of electricity also concerned the notions of fluid or energy. Based on the analogy with nerve fluid, the notion of fluid was reinforced with Galvani's work on animal electricity. In 1748, Nollet evoked rays when he discussed descriptions of electricity and its circulatory movements between bodies and metals:

> [...] that the eruptions which are made of electrical matter outside the electrified body, (eruptions on which all phenomena depend), take their strength and their value, as much from the speed acquired in an environment favorable to their movement, as from the number of rays which come in all directions to the point of competition; because a very thin wire, or a very thin and very narrow blade, can well by its length give rise to the accelerated movement of the electric material, but then there are too few rays that run at the same time from the same place. [NOL 48, p. 304, author's translation]

Furthermore, in 1763, electricity was described in terms of "...electric fluid [is] composed of sulfurous, flammable, light-like particles [...]" [FOR 63, p. 289 t. 2, author's translation].

Thus, while he studied and questioned the natural form of the electric fluid, he quickly followed the 18th Century trend of studying the mechanisms of this energy with empirical research on its effects on the body and its physiology. In 1748 in *Recherches sur les causes particulières des phénomènes électriques* [NOL 48], in an excerpt from the registers of the French Academy of Sciences (1746), the experimenter emphasized paralysis as a preferred therapeutic target, which could be

5 In the 17th Century, the Dutch physicist Christian Huygens proposed a complete theory of light in wave terms and demonstrated that light waves form a wave front propagating in a straight line. However, his theory contradicted the corpuscular theory of light established by Isaac Newton.

correlated with the links between accidental electrocutions and the movements caused by them. Indeed, those who experienced electric shock on themselves evoked these electric convulsions of the body that escaped the individual will and reduced them to the state of automatons. The adverb *naturally* used in the quotation below takes up this idea:

> We have already electrified paralytics and people with a few limbs missing; it is an idea that comes quite naturally to mind, that a jolt such as one feels in the Leyden experiment, could well resurrect the more or less forbidden movement in an affected part; I suppose here the detail of an essay that has only just begun, and whose success is still too dubious to deserve to be announced. [NOL 48, p. 49, author's translation]

The notion of resurrection, also mentioned, reinforced the idea that electricity had the "power" to restore what had been lost, in this case the mobility of limbs. Moreover, this quotation highlights the caution from the French Academy of Sciences regarding the first failures of electricity applied to human pathologies. Thus, Nollet proceeded to electrify soldiers paralyzed in April 1748 in the *hôtel royal des Invalides*[6]. These trials, which proved to be painful even for practitioners, were abandoned because the results were so uncertain and provisional. Daleur, a subject of the trial below did not regain his flexibility:

> Daleur was electrified consecutively from April 9th until the 16th of the same month, every day for 4 hours. [...] To electrify these patients, they were made to sit on a board suspended with silk cords, and their feet were supported with resin cakes, or with a sort of stirrup attached to the board which served as their seat: their body was surrounded by an iron chain, one end of which responded to the glass globe by means of which the electric virtue was excited. [...] when the sparks were fired for a certain time, the patient was applied to the Leyden experiment [...]. [NOL 48, pp. 406–407, author's translation]

In these therapeutic trials, electricity was conceived as "...a useful remedy for paralysis, [...] perhaps for many other diseases, which are located in the nerves or muscles" [NOL 48, p. 413, author's translation]. However, it sometimes proved unreliable, which Nollet attributed to the severity of the injuries, which could make the electrical force useless.

6 The *hôtel royal des Invalides* was a place intended for the accommodation and care of officers and soldiers.

In 1756 Franklin published the accounts of his experiments carried out in 1749 in Philadelphia, which were distributed throughout the world [FRA 56]. Through imaginative experimental protocols using natural electricity, he developed model experiments to explain and understand the flow of electrical particles:

> On these accounts we suppose electrified bodies discharge their atmospheres upon unelectrified bodies more easily and at a greater distance from their angles and points than from their smooth sides. Those points will also discharge into the air, when the body has too great an electrical atmosphere, without bringing any non-electric near, to receive what is thrown off: For the air, though an electric *per se,* yet has always more or less water and other none-electric matters mixed with it; and these attract and receive what is so discharged. [FRA 56, pp. 57]

The use of spikes, as a suitable conductive form, could be found in the early days of electrical medicine with plumes, which were used for the localized application of static electricity.

Figure 2.5. *Static electricity and its use in kocalisated therapy. Vigouroux, P.* De l'électricité statique et de son emploi en thérapeutique, *Paris, Baillière, 1882, Plate VI, fig. 3, p. 40*

Franklin investigated the classification and division between electrical bodies *per se* and non-electrical bodies because he suspected, by virtue of the existence of atmospheric electricity that could be bottled or channeled into circuits:

This differentiation, initiated at the beginning of the 18th Century, formed the basis of the idea of electrical circuits. Indeed, the identification of conductive bodies, both organic and inorganic, determined the possibilities of experimenting with the concept of conductivity. Franklin was thus helping to classify and rationalize the various substances in relation to their ability to conduct electricity. The establishment of knowledge of electricity involved not only work on the word but also on calculations. How can you calculate a dose of electricity? How can you conceive the quantity that circulates in a circuit whether it is incoming or outgoing? To illustrate these questions, Franklin gave a calculation as an example: if we start from 20 as the common quantity of electricity for each surface of the jar, that we have 1 for each turn of the globe, then we will have 21 inside the bottle and 19 outside; then in the second turn 22 for the inside and 18 etc., then in the third turn 22 for the outside and 18 for the inside. And this up to 0 on the outside and 40 on the inside. A conductive body would be able to restore the interior-exterior circulation. Calculation may seem trivial, but it bode well for the role that mathematics played in shaping this knowledge. Thus mathematical language embodies a form of objectivity in the same way that technology does. Sensation and common sense tend to be sidelined by the mathematical tool that structures the limits and representations of knowledge about electricity. Here is another illustration explained in terms of experimentation:

> At the same time that the wire and the top of the bottle, &c. is electrised *positively* or *plus*, the bottom of the bottle is electrised *negatively* or *minus*, in exact proportion: *i.e.* whatever quantity of electrical fire is thrown in at the top, an equal quantity goes out at the bottom. [FRA 56, pp. 2]

For example, Franklin's innovative conceptualizations included the design of an electric battery – by connecting the inner armature of each jar to the outer armature of the next [TOR 56, p. 341]. Thanks to his argumentative and analogical imagination, it represented the beginnings of a phenomenotechnical physics that artificially reconstructed natural phenomena. The Bachelardian concept of phenomenotechnics, in which the concept of phenomenon brings us back to nature while the concept of technology brings us back to culture, highlighted the dependence of science on technology. When this term first appeared in 1931, it illustrated the implications of mathematical thinking. Method, rather than language, it provoked the instrumental conditions for the systematization of reality. Thus the knowledge of physics depends on the instrumental conditions of the sciences and leads to a constructed realism. In *Noumènes et microphysique*, Bachelard (1884–1962) explained the scientific construction of reality as follows:

> Mathematical physics becomes the noumenology by which new phenomena are not simply found, but invented, but constructed from scratch. [BAC 31–32, p. 55, author's translation]

The integration of phenomenotechnology into the history of electricity was crucial to understanding how the constitution of a natural phenomenon into knowledge models the development of science and enables correlating theories with experiments. Beyond the physics of electricity, the reification of this force as a scientific tool, as an object of knowledge, was a key element in understanding its appropriation by medical thought. Deeply related to the questioning of the integration of humans into the mechanisms of nature, the fact that this force was "domesticated" and put into circuit impacted not only the knowledge of the phenomenon but also its implications on all natural bodies. For example, when Franklin was trying to understand storm phenomena and the role of electricity in them, he technically reconstructed the movement of clouds in relation to the locations of high buildings:

> In Letter IV from September 1, 1749, he was curious to follow the path of Franklin's thought: he electrified one of the trays of a copper balance with silk cords, suspended from the ceiling by a string. The trays rotated due to twisting. A punch was placed in the floor. The electrified basin moved towards the floor and while the distance was suitable, it discharged its light on the punch. While a needle was placed on the punch, the basin discharged its fire through the point, rising higher than the punch. The analogy was immediately obvious: the movement of the basin reproduced the movement of the clouds, the punch the mountains and the tallest buildings. We understand how electrified clouds can be drawn down to the distance they need to be electrified. [TOR 56, p. 341, author's translation]

The controversy between Nollet and Franklin lasted no less than 13 years. Two points can be underlined:

– Nollet got the idea of the similarity between lightning and electricity. An idea taken up and developed by Franklin;

– Franklin supported a unit of electricity while Nollet supported a design of two electricities, or two forms of the same electricity. So, for example:

> The argument between Franklin and Nollet had led to a division of the world of physicists into two camps, the supporters of Franklin opposing those of the abbot; and until 1766, that is to say for 13 years, the discussion continued at the French Academy of Sciences between Jean-Baptiste Le Roy and Nollet and outside it in the form of Letters, always on the question of the two electricities. [TOR 56, p. 347, author's translation]

2.2. Medical physics: philosophical issues

In the context of a search for the unity of living things, where forces are shaped by the language of science, Franklin evoked the effects of electricity on the nervous system:

> The shock to the nerves (or convulsion rather) is occasion'd by the sudden passing of the fire through the body in its way from the top to the bottom of the bottle [...]. [FRA 56, p. 4]

Figure 2.6. *Nollet did many public demonstrations to show the links between the body and electricity. Nollet, J.-A.* Essai sur l'électricité des corps, *op. cit. 1746, Fig. 1*

The stakes of medical physics, which deals in parallel with the problems of the nature and uniqueness of electricity[7] in relation to physiological mechanisms, are as much technical as they are philosophical:

> Common fire is in all bodies, more or less, as well as electrical fire. Perhaps they may be different modifications of the same element; or they may be different elements. [...]. If they are different things, yet they may and do subsist together in the same body. [FRA 56, p. 47]

7 "This observation which seemed so surprising in 1691 ceases to be so as soon as we know that the matter of thunder and that of electricity are the same, and that magnetism is only an effect of electrical matter. No one will find it hard to believe that the bell towers of Chartres, because of their great elevation in the middle of a vast plain, were and are often struck by thunder" [FRA 56, v.2, p. 141, author's translation].

Franklin thus integrated into his research on the physics of electricity, the implications of the latter on the living. He provided answers to questions about the conditions under which electric shock is, or is not, lethal, especially on animals:

> A pigeon that we struck dead to appearance by the electrical shock, recovering life, drooped about the yard several days, eat nothing, though crumbs were thrown to it, but declined and died. We did not think of its being deprived of sight [...]. [FRA 56, p. 64]

The animal model played an extremely important role, especially following the studies on electric fish, in the theme of electricity as a property of organized bodies. In addition, Franklin took up and discussed Nollet's experiments on human chains, less for their spectacular dimension than for their impact on physics, the challenge of which being to understand the organic and material conditions of conductivity:

> The Abbé says in page 69 that he can Electrise 100 men standing on wax if they hold hands and one of them touch one of these surfaces with the end of his finger, this I know he can do while the Phiol is charging, but after the Phiol is charged I am as certain he cannot. That is, hang the Phiol to the Conductor and let a man standing on the floor touch the coating with his finger, while the globe is turned, till the electrical matter spews out of the hook of the Phiol or some part of the conductor, which I take to be the certainest sign that the Phiol has received all the Electric matter it can, after this appears let the man who before stood on the floor step on a Cake of wax, where he may stand for hours and the Globe all that time turned, and yet have no appearance of being electrised. [FRA 56, v. 2, p. 255]

In correspondence with Peter Collinson, member of the Royal Society of London, Franklin related the effects of shock on humans. "*He*" in this quote refers to Collinson:

> In making these experiments, he found, that a man could, without great detriment, bear a much greater shock than he had imagined: for he inadvertently received the stroke of two of these jars through his arms and body, when they were very near fully charged. It seemed to him a universal blow throughout the body from head to foot, and was followed by a violent quick trembling in the trunk, which went off gradually in a few seconds [...]. [FRA 56, pp. 303–304]

Franklin empirically developed electrical stimulation treatment, a treatment known as Franklinization. Furthermore on paralysis, he concluded that it was not very effective. On the other hand, in 1752, he successfully treated a patient with convulsions. Stimulation was then applied to the limbs affected by these movements.

In 1744, the German physician and physicist Kratzenstein (1723–1795) [KRA 46] already recommended the application of electricity to medicine, arguing that it had an effect on the pulse and on the regulation of internal fluids, but it was the Leyden experiment that acted as a catalyst for therapeutic expectations. The context of the appropriation by the medical field of new techniques, as well as the development of a materialistic philosophy, required the integration of new machines and new techniques. This appropriation was the result of a theoretical construction, a model of the human being in the process of changing, of envisaging modeling himself, to rethink his place in nature and to be able to intervene on the maintenance of the harmony of his animal economy.

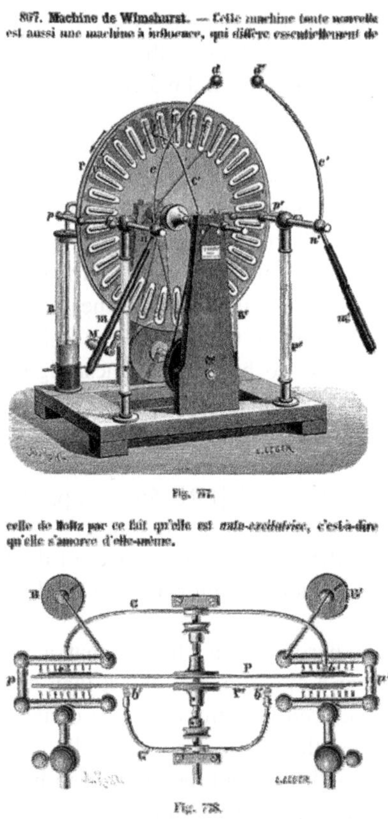

Figure 2.7. *The Wimshurt machine, created in 1882 [GAN 94]*

COMMENT ON FIGURE 2.7.– *This totally new machine was influential, differing from Holtz's machine in the fact that it was self-exciting, in other words, self-initiating.*

Thus the attracted electricity, channeled from its presence in the environment, entered the body in the form of fluid, an ebb and flow. During the 18th Century, the static electricity that could be administered, corresponded to the phenomenon produced by rubbing bodies: if the body was a good conductor, the electricity manifested itself at all points; if it was a bad conductor, it manifested itself only at the area rubbed. The machines were therefore used with static electricity. In principle, they consisted of glass disks rotating between felt cushions, as in the Wimshurst machine.

For educational purposes, it enables the observation of the effects of static electricity, such as the production of sparks, the charging and discharging of a capacitor, and the ability to carry out numerous experiments that require additional equipment, such as an electroscope or an electric tourniquet.

Thus, three types of electrical machine application can be distinguished as early as the 18th Century:

– by artificially produced sparks;

– isolation electrification;

– by means of discharging a capacitor.

Numerous experiments were conducted with the first capacitor from Leiden. It can even be correlated with the start of shock therapy, as treatment was provided, particularly at the Salpêtrière, from electrical shocks caused by adding a capacitor to the machines used. There was much debate about the appropriateness of their therapeutic use [NOL 53–57, v. 1, pp. 202–203]. Thus, continuous electrical stimulation was recommended in the last third of the 18th Century, instead of shock treatment:

> It is not the same with the regular, continuous and uniform course of electricity, which seems to restore to the motor and nervous fibers their tonic tension that they have lost without exposing them to any shaking capable of tiring them. [MAR 73, p. 12, author's translation]

It should be noted that while the use of the Leyden jar was envisaged in the medical field as soon as it was discovered and because it moved the body in spite of itself, its negative effects on the nerves were also envisaged. Two currents of thought were opposed here: on the one hand, electrifying physicists were inclined to use this new instrument insofar as it seemed to give back the lost mobility to the paralyzed limbs; on the other hand, their opponents were inclined to a continuous stimulation of the type provided by the electric bath, which was harmless but also judged useless by the former:

> It follows that if you use a capacitor that can become enormously charged, you can produce disastrous effects on the human body, which

always has about the same amount of electricity. [VIG 82, p. 34, author's translation]

On the other hand, more symbolic or more artistic experiments were also carried out. Thus Boze [BOZ 54], professor of theology at Wittenberg, electrifies subjects in the darkness, adorning them with a halo of light so as to make them resemble saints. There was talk of *beatification by electricity*. Boze was also the creator of the "electric kiss", which featured a person charged with static electricity, standing on an insulating stand, and reaching out his lips to another person.

Figure 2.8. *"Only once, how foolhardy, to Venus on the pitch I gave a kiss. The pain came close. My lips trembled, my mouth turned, my teeth almost shattered" [BOZ 54, p. 54, author's translation]*

The electric character of the spark became a symbol to express the affinity between two individuals. If one can notice the context – in retrospect not very scientific – of this experiment, it nevertheless contributes to the study of the effects of electricity between individuals, as well as to the affirmation of the inscription of human beings within the laws of a nature whose mechanisms are in the process of materialization and secularization. Indeed, during the 18th Century, the theories of knowledge, based on materialism, highlighted the need to return to knowledge that was beyond dogma and founded an empiricism where the emphasis was on experimentation. This position was potentially critical of metaphysical positions, such as the crucial role accorded to the soul. Between 1745 and 1770, a materialistic and monistic medical application of electricity developed. This point is not unimportant in terms of the links between electricity and the history of brain exploration. Materialistic scientific models were not only mechanistic, they were also chemical, vitalist, biological, physiological, medical and electrical. They tended to reflect the inherent dynamism of matter. The science of electricity and its application to the medical field was born from the convergence of two programs of knowledge: one on the forces that animated matter from the deepest part of its structures; the other on the links between the soul and the body, the interdependence of the first to the organization of the second. From these considerations, it is easy to understand why physicists, who studied the origin, form and principles of natural electricity, were also those who concretely envisaged its transformation into an artificial medicine and its integration into the pharmacopoeia.

While between 1730 and 1750, many experiments were carried out in spectacular fashion, their social and scientific impact could not be denied. Thus, electrical displays, such as tables designed to host electrifying dinners or human chains in different environments, helped to popularize electricity, made it known and made it accessible to all. This entertaining atmosphere accompanied the therapeutic infatuation of this jar, which was seen from the outset as a therapeutic instrument:

> You know, Sir, that in the Leyden experiment, the most essential condition is that there should be what is called an electric circle, that is to say, a body, or an uninterrupted sequence of several electrifiable bodies, which touches the jar or its substitute, on the one hand, and on the other, the conductor who carries the electricity to this glass vessel: the test I performed six years ago on more than two hundred people at a time, arranged in two parallel lines [...].[8] [NOL 53–57, v. 1, pp. 206–207, author's translation]

The properties of electricity were barely known and many electrifying physicists proposed to study its effects on animals and humans. Baltic amber, known as succinite, was already a traditional element of the pharmacopoeia, completing our reflection on the craze for electric fluid as a treatment tool. Electricity crosses the ages, its knowledge never ceasing to overcome epistemological obstacles to acquire the most diverse applications in the fields of human activity. Not only does it represent a stage in the growth of knowledge in the field of physics, but it is shaped by medical activity, bottled and contained in machines that brought medicine into

8 Nollet imagined monks forming a conductive chain around their abbey. He thus showed, while everyone was jumping at the same time under the effect of the electric discharge, that it spread at an extraordinary speed. His experiments with chains of soldiers are also extremely well-known.

the engineering age. In the words of Bachelard, we can say that this force is an issue of the unity of nature, of its understanding, as much as a broadening of the frameworks of knowledge:

> The mind has a variable structure from the moment knowledge has a history. Indeed, human history may well, in its passions, in its prejudices, in all that is a matter of immediate impulses, be an eternal renewal; but there are thoughts that do not begin again; these are thoughts that have been rectified, enlarged, completed. They do not return to their restricted or shaky areas. However, the scientific spirit is essentially a rectification of knowledge, a broadening of the frameworks of knowledge […]. [BAC 69, p. 173, author's translation]

From the secularization of medicine in the Enlightenment sciences, the stakes of mastery of electrical force were linked, especially in medical physics [GUI 77][9]. Knowledge marked the hope of an integrated interventionist medicine within a nature understood in technical and dynamic terms.

Figure 2.9. *Nollet, J. A.* Recherches sur les causes particulières des phénomènes électriques, *op. cit. 1749, pl. 2*

9 The term "medical physics" was used recurrently.

Jean Morin, priest and professor of philosophy at the *collège royal de Chartres*, in 1748 in his treatise *Nouvelle dissertation sur l'électricité des corps,* proposed to learn how to build "[...] a very simple and very portable electrical machine" [MOR 48, p. 27, author's translation]. This chapter is all the more representative of the importance of machines in the conquest of electricity, since Morin, beyond mechanical considerations, was interested in and described, in this same volume, the effects of electric shock on the body:

> A disbeliever wanted to draw the spark, and did so without my knowledge (for I did not want the experiment to be continued, for it frightened me so much myself) he was struck in the moment at the coccix, where he felt such great pain, that for several minutes he made loud cries, and the general trembling lasted more than a quarter of an hour. [MOR 48, pp. 115–116, author's translation]

The period from 1745 to 1770 represented a process of medicalization of the protocols and methods of physics. But these developments, in their sustainability, were not self-evident. Indeed, medical physics very quickly failed, as the paralyzed patients being treated did not regain their mobility. Moreover, it was considered extremely dangerous and was viewed with suspicion by practitioners themselves about its potentially lethal effects. It was also known to cause overactivation of the nerves for a more or less long period of time.

Jallabert in his treatise entitled *Expériences sur l'électricité, avec quelques conjectures sur la cause de ses effets* [JAL 48], starting from the definition of electricity and then developing the possibilities of its medical application, symbolized this movement of a physics that was becoming medicalized. The links between the properties of electricity[10], its passage through the living and the effect on the properties of bodies were abundantly questioned. The possibility of developing electrical therapies was one of the reasons why he emphasized their simplicity and economic aspects:

> Living beings easily receive electricity; and if it can be usefully given to them, it will be very easy to transmit it with a single globe to

10 For example, an interesting excerpt from Jallabert's physical experiment to determine the speed of electricity: "the speed with which electrical matter moves is such that all my experiments to try to determine it have taught me nothing, except that it is still infinitely faster than sound. I stopped at the bar the end of a metal chain about 1050 feet long; after various detours the other end, to which a metal plate was attached, was led over a pedestal table covered with patches of gold leaf. In order to intercept the electrical matter, a person touched the end of the chain adjacent to the bar which was being electrified; and when at an agreed signal he released it, it was impossible to observe any interval from that moment to the moment when the fragments of gold leaf were agitated" [JAL 48, p. 78, author's translation].

> several patients at once, even in their beds. It will suffice that the feet of the bunks rest on resin cakes; and that various brass threads, attached by one of their ends to the bar, reach the various beds. [JAL 48, pp. 74–75, author's translation]

While shocking was considered dangerous, the conditions under which it was fatal needed to be explored from the animal model. Fundamental research on living organisms and their mechanisms was mixed with therapeutic issues, the central question being: if it is likely to kill, at what point can it make mobility lost? From the 18th Century onwards, electricity thus made it possible to explore the biological limits of bodies and the living and to experimentally materialize the limits of life and death:

> Some of these animals were killed at the same instant of the blow which struck them; some of them survived for several minutes; others seemed very uncomfortable: and I have no doubt that by paying attention to the various means which I have indicated, either to increase the electricity of the bar or to make the shock stronger, it was not possible to give death to the strongest animals. [JAL 48, p. 116, author's translation]

It is interesting to mirror this quotation with the one in which Jallabert explored the effects of shocks on movement and thus moved from an experimental context in which he determined the lethal conditions, to an experiment designed to return movement to a subject:

> I began by giving him a shock: I tied his paralytic hand to the vase, and with the other hand I made him draw the spark. Instead of the ordinary jerks that one experiences in different parts of the body, he felt only a sharp blow to the right shoulder followed by tingling in the whole arm […]. Noguès believed that Mr. Guiot, who was present, struck him on the shoulder at the moment when the spark burst; and I could only disabuse him by making him repeat Mr. Guiot's experiment placed opposite him. [JAL 48, p. 128, author's translation]

In 1747, Antoine Louis (1723–1792) in *Observations sur l'Électricité, Où l'On Tâche d'Expliquer Son Méchanisme Et les Effets sur l'Oeconomie Animale; Avec des Remarques sur Son Usage* [LOU 47] made comparisons of individuals killed by lightning and animals killed by electric shock. He noted the similarity of the lesions observed during his dissections and deduced from this the identical nature of lightning and static electricity. He also underlined the logical structure in the application of electricity to movement pathologies:

> It was thought that the shaking and sudden revolution produced by the Leyden experiment might well be such as to revive the movement in those parts where it is extinguished by the disease known as paralysis; it is to Mr. Morand that we are indebted for this application. [LOU 47, p. 82, author's translation][11]

The application to paralysis therefore seemed to flow naturally, which disappointed practitioners of the end of the 18th Century all the more when therapeutic success was not achieved. Initially, electricity was considered either as a tool for maintaining health, in terms of balance; or as an artificial remedy, which by stimulating the nerves could relieve the inability to move:

> In the course of the ordinary treatment of paralysis I have almost always noticed more or less pain, twitching, wavy or vermicular, involuntary and convulsive movements; produced spontaneously or by some agitation of the different parts of the affected limbs and determined immediately by the effort of the animal spirits, which, not being able to penetrate, act freely and uniformly in these same parts, advance there by jumps and jerks, or perhaps in some other more irregular manner, to give rise to these various agitations, which are so many signs of a notable diminution of the symptoms of paralysis, and warning signs of some better disposition on the part of voluntary movement and feeling. [MAR 73, p. 9, author's translation]

Thus this force was considered by analogy with the nervous fluid or the *vis nervosa*[12] and enabled acting on the individual on a physical scale. In a philosophical perspective of the human being conceived as a microcosm, part of a macrocosm, electricity was thought of as a tool for regulating its internal and biological harmony:

> To avoid these awful inconveniences which abhor humanity, it was enough to help the movements of nature, by preventing the exhaustion

---

11 "The first person I shocked was a thirty-two year old girl [...] I had her apply her right hand under the jar and pull the spark with the indicator finger of the left hand. She was hit in the right shoulder, and for a few minutes she felt a heat throughout her right arm. On the left side the shock did not pass through the wrist, and the person suffered for an hour from rather acute tingling at the tip of the index finger" [LOU 47, pp. 101–102, author's translation].

12 Electricity as a nerve impulse dates back to the work of Francis Glisson (1597–1677), professor of physics at Cambridge, who showed, between 1654 and 1677, that the excitability of nerves and muscles responds to external stimuli [GLI 72]. His demonstration showed: on the one hand, the theoretical foundations of the materialization of animal spirits in the form of a fluid; on the other hand, the correlation of physiological research with the animal machine theory.

> of the efforts of the animal spirits, by supporting and exciting them even to a perfect recovery. This is precisely what electricity seemed to us to do in the most analogous way to the operations of nature. [MAR 73, p. 11, author's translation]

Moreover, the idea that electricity could regenerate contractility and irritability of nerves and muscles did not await Galvani's experiments. The links between electricity and contractions marked a history linked to the notion of a vital principle but also to that of the human machine. Bikker (1732–1801) and van den Bos (1731–1788) [BOS 57], both scientists in Leiden, showed in 1757 that the large artery of the heart and other smaller vessels contracted in living animals as a result of irritation from electric sparks:

> The paralyzed parts in which the motor power is absolutely extinguished, and where the faculty of feeling is notably diminished, or even completely abolished, are susceptible to pain, tingling and quivering, which seem to be a very good sign. The increase in heat that electricity produces in these same areas seemed to be a very favorable sign for the success of these attempts. [LOU 47, p. 110, author's translation]

However, Antoine Louis rejected the application of electricity to animate bodies and warned about the lethal effects of electricity on people, both by accident and in animal experiments [LOU 47]. The *Académie des Sciences de Paris* was extremely cautious in its conclusions regarding the application of therapeutic electricity. On May 3, 1754 [TOR 56, p. 345], during a session in which Nollet exposed the work of his correspondent from Le Tour on electrification by communication, the members of the Academy asked Grandjean de Fouchy (1707–1788) to proceed with a new critical reading of these memoirs. Although they contained an account of Franklin's experiments, their printing was postponed. This caution was also exhibited by practitioners who were victims of electrical accidents:

> The scientific press and especially the *Journal des Savants* urged its readers to be careful. We had not forgotten the death of Professor Ritchmann of the St. Petersburg Academy, electrocuted on July 25, 1753 by his lightning rod, which was, to tell the truth, insufficiently insulated. Physicists like Poncelet, like Brisson, were violently critical of the invention, which was unbelievable in their eyes. There was also reticence and hesitation, even among the defenders: the King himself, who contributed greatly to the adoption in France of what he called 'preservative bars' or 'thunder conductors', after having recommended reducing the length of the bars and not ending them with points,

> reconsidered this assertion three years later. [TOR 56, pp. 348–349, author's translation]

Jean Torlais here mentioned a Russian physicist who remained famous because of his death by electrocution. He was struck by lightning while trying to develop an experimental protocol using an electrometer to reproduce the experiment known as "the kite".

The treatment of paralysis by electricity was soon problematic, and fell into disrepute after 1765. André Marrigues (1728–1786) wrote in 1773:

> The effectiveness of Electricity, for the cure of Paralysis, is still very problematic today: the renown of its opponents always seems to prevail over the authenticity of the facts, most of which have been recognized as false, while the others have, barely retained in the opinion of a very small number of people, a weak appearance of possibility. [MAR 73, p. 3, author's translation]

How can we explain the enthusiasm for the application of electricity in the fields of physiology and medicine? It can be considered to be linked to the advent and dissemination in Europe of the anatomical–clinical method, initiated by Morgagni (1682–1771). The integration of the disease in the organs, then in the tissues with the work of Bichat, contributed to the concept of localized electrification. The Morgagnian method underwent a geographical transfer from Italy to France where it was redeveloped by Laennec (1781–1826) and perfected by Bichat (1771–1802). While in Italy, Fontana (1730–1805) and Caldani (1725–1813) turned to the techniques of electrophysiology in continuity with the research of Haller (1708–1777) and the innovations brought about by the work on metallic and animal electricity by Volta (1745–1827) and Galvani (1737–1798). At the end of the 18th Century, the principles enunciated by Morgagni [MOR 61, 37–39] were therefore more active in France[13] and Austria than in Italy:

> The successive developments and stages of the Morgagnian conception have been better known through the more directly morphological direction of the work of Bichat and Virchow and, as

13 J.B. Bouillaud, one of the founding fathers of the Parisian clinical school, contributed to the development and systematization of its principles. The scientific climate was increasingly favorable to the application of these principles on a larger scale of the population, within hospital structures. In the same period, Corvisart spread the method of auscultation, first described in 1761 by Auenbrugger in his treatise *Inventum novum*. A French translation, *La nouvelle invention*, was published in 1808. Then, his pupil, Laennec, became one of the most representative elements of the anatomoclinical school of Paris and systematized the use of the stethoscope, invented in 1819 [AUE 08].

> regards the anatomoclinical direction, through that of Laennec and the Parisian school first, then the Vienna school. [PRE 71, p. 116, author's translation]

In the history of electricity and its transition from physics to medical physics, a line of convergence can be found that correlated the expansion of the anatomoclinical method to tests on the properties of tissues, leading to a dynamic modeling of the functioning of organs in relation to each other:

> As soon as abbé Sans had told me about the processes of his electrification, I understood first of all that the direction of Electricity should be subordinated to anatomical and pathological knowledge, which it is important to have in order to draw the greatest possible fruit from this remedy. [MAR 73, p. 62, author's translation]

During the 18th Century, there was talk of *electrifying physicists* [SIG 76] who, as can be seen in the tables of contents of their treatises, went from the mechanisms of electricity to its effects on the animal economy, to determine its possible benefits:

> A cause which accelerates the movement of fluids in the capillary system of animals, which increases the intensity of their heat and which makes them sweat more abundantly, must necessarily produce more or less sensitive effects and often beneficial to the animal economy. [SIG 76, p. 399, author's translation]

Thus in 1808, Caullet de Veaumorel (1743–....) [CAU 08] explained that the sick body can be as much the element of an electrical circuit as its therapeutic target:

> Negative electricity is that same natural electricity, from which bodies are partially deprived by positive electricity; for positive electricity has the property of subtracting it from them by assimilating the subtracted part. A well-insulated, positively electrified patient is merely a continuation of a positive conductor, saturated with electric fluid, combined with a flammable gas. This state is called a positive electric bath. If this patient communicates with a negative conductor, far from being saturated with combined electricity, he finds himself on the contrary deprived of a part of his own natural electricity; it becomes rarefied by subtraction, and then the whole animal economy of the patient tends to recover the one which it is successively deprived of by the electric plate, put into activity. [...] Fluids and solids come together, condense, and the increased emanations of the

> subject subjected to this electricity are only the effects of a general attraction, produced by the tendency towards equilibrium which the electric fluid eminently enjoys. The acceleration of the pulse [...], the condensation of the whole animal economy are the result, this is what I call negative electric bath. [CAU 08, p. 81, author's translation]

If the whole body is a natural electrical circuit and if electricity is of the same nature as the nervous fluid, then electricity participates in its balance. This model of physiological harmony was imbued with a strong Hippocratic dimension, specific to the 18th Century, combining hydraulic design with the mechanistic modeling inherited from Descartes:

> Yet, if one assumes that there is some blockage in this hand or some morbific miasma that you want to dissipate by the movement of the electric fluid, you make them flow back or carry them into the body, instead of out of it: from which it follows that the point where it was necessary to make this patient receive electricity, was at the head or at the arm joint, so that from there it was carried to the sick part and carried these morbific particles outside by the shortest way; and that on the contrary, if you had used negative electricity, the way of making it receive it by hand was the best adapted to this operation; and in general, when using either of these two electricities, if you intend to produce movement in the morbific humors, you must apply positive electricity precisely at the opposite points where you want to apply negative electricity, and always, in either case, in such a way that the course of the electrical matter carries the morbific humors outwards by the shorter line. [CAU 08, p. 85, author's translation]

Medical electricity opened up two research directions:

– a first direction was given to explore normal physiological states by varying them. This medicine, which was not yet completely physiological, was conceived as a chronic and gentle practice that maintained a balance between the different systems of the human body.

Thus, in addition to the pulse, it varied body temperature and was assumed to have effects on the entire animal economy.

– a second direction concerned its effects on disease states. The point here was to give it a more punctual dimension. It was within this framework that the use of shock treatment was recommended, whose visible, obvious influence on muscle contraction was quickly noticed. Since electric shock allowed this contraction[14] and promoted involuntary movements, provided that the way it was administered was stabilized, practitioners assumed it had virtues in terms of healing paralysis.

Moreover, its action on muscles and nerves heralded the electrophysiology of the future and the exploration of the reactions of these parts to electrical force.

The history of electrical medicine was scientifically linked to physics and technically marked by the entry of engineering into the field of therapeutics. Indeed it was necessary to learn how to manipulate electric currents in order to improve the body's ailments, channel them and control their circulation. This is why medical physics was constantly improving or creating new machines to electrify the body. The possibilities of *these healing machines,* with or without a capacitor, were constantly being explored and extensively described for their therapeutic effects but also for their side effects. These instruments gave a technical, progressive and rational image of 18th and 19th Century medicine, but did they really cure? It seems that one of the elements of the success of medical electricity was the image it reflected of itself as a component of progress, in all its dimensions: social, political and scientific.

2.3. Healing machines?

Between 1745 and the end of the 19th Century, static electricity machines were constantly being improved for medical applications. Nollet was not only a designer of static electricity circuits and machines but also a supporter of their diffusion. Indeed, his catalogs of scientific instruments included exhaustive lists of the physics instruments required for these interventions[15]. He published a first catalog as early as 1738 [NOL 38, p. 113].

In 1882, Paul Vigouroux delivered an interesting review of the main machines used in his treatise on static electricity. He took the opportunity to highlight the chronic electrical treatments that enveloped the subject in an electric atmosphere without shocking them. The electric bath was, in turn, considered to be the gentlest treatment to approach the temperament of the sick and to rebalance it, but also a deception for the supporters of shocks.

---

14 "But one effect of electricity that is not indifferent to notice is that you can see various convulsive movements in the muscles from which you draw sparks" [JAL 48, p. 79, author's translation].

15 Rozenn Picaud proposed, in 2010, in his thesis *Les machines à électricité de l'abbé Nollet*, to retrace the path of his devices [PIC 10].

Figure 2.10. *Medal-winning machine at the 1878 Paris World Fair [ART 81, p. 49]*

He put forward Doctor Arthuis' machine, considered as a model and claimed against the use of electric shocks:

> It gives positive fluid, is very soft, and provides sparks that give the patient a slight stinging sensation without the feeling of shock or jerking. This machine is essentially suitable for the treatment of neuroses; nevertheless, says Mr. Arthuis, one can obtain with it sufficiently energetic effects for the treatment of paralysis and muscular atrophy. [VIG 82, p. 30, author's translation]

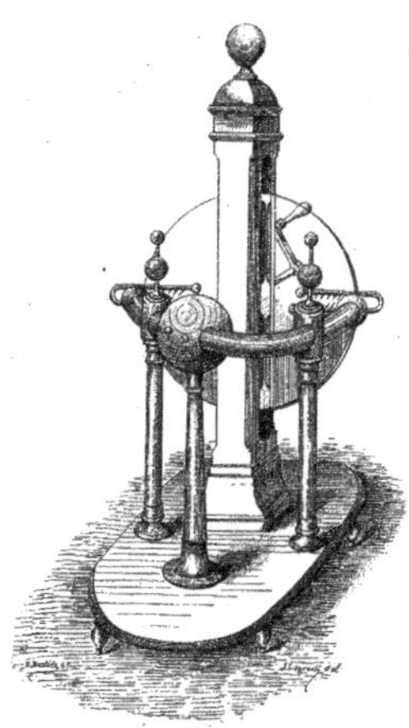

Figure 2.11. *Dr. Arthuis' machine [VIG 82, p. 28, pl. V]*

His treatise formed part of the controversy over the use of shock treatment and the Leyden jar in the medical field. The quarrel between Arthuis and Romain Vigouroux (1831–1911), in charge of the electrotherapy room at the Salpêtrière, was one of the episodes. But beyond the therapeutic arguments, these texts, illustrated by the machines used in treatment, provide an insight into the industrialization of electrotherapeutic treatment known by medical electricity in the 19th Century. Thus, Romain Vigouroux added a capacitor and treated the patients of the big hospital by means of electric shocks. These were two opposing visions of electrical medicine: on the one hand, Arthuis and Paul Vigouroux participated in a medical vision of the harmony of the organism within which the humoral theory of the Hippocratic school was still very present. The maintenance of the sympathetic functioning of the organs in relation to each other was based on the application of continuous electricity. On the other hand, Romain Vigouroux was more involved in the practice of shocking based on the idea that it was necessary to calm patients in order to slow down their episodes of crisis:

> First of all, they had given preference to violent means, electric shocks in particular [...] [VIG 82, p. 26, author's translation][16]

Hospital accessories were thus available according to the sensitivity and intensity of the treatment given by the practitioner. For example, there were isolated electrical stimulators that allowed the practitioner to avoid sharing the patient's sensations and non-isolated stimulators. This fact indicates that the choice of these accessories depended on the use of shock treatment and also revealed a philanthropic concern for the patient:

> The tip gives the current, the ball gives the spark. This exciter is hand-held and therefore makes the operator feel all the sensations experienced by the patient. In order to meet all the needs of the practice, the physician must have exciters of all shapes and sizes. [...] The figure shows that the K-chain passes through the end ring of a glass rod held by the operator's left hand. This small instrument, which I have added to my isolated exciter, is used to move the chain away from the edges of the booth. With this exciter, the doctor does not receive any of the sensations felt by the patient. [VIG 82, p. 40, author's translation]

16 "They thus determine in patients with shocks, concussions similar to those produced by the Leyden jar that has caused so many accidents" [VIG 82, p. 66, author's translation].

Figure 2.12. *Non-insulated exciter [VIG 82, p. 41, pl. VI]*

Figure 2.13. *Insulated exciter [VIG 82, p. 41, pl. VI]*

Machines like Dr. Carré's, with its glass wheel for mixing static and induced electricity, integrated Faraday's discoveries into electrical treatments. The meeting point of magnetism and electricity was due to the research by Ørsted, Ampere, and Faraday but especially Maxwell, who in 1860 unified the previous theories in order to explain, according to a principle of economy, the behavior of magnets and electromagnetic waves. This unified law was not without consequences for the medical applications of electricity. While Mesmer's animal magnetism did not lead to a lasting trajectory in the medical field, the idea of a natural force influencing metals and energies was constantly taking on new forms. Thus, in 1820, Hans-Christian Ørsted (1777–1851) [ØRS 77], by experimenting on the directing action of a current on a magnetic needle, initiated a series of research which led to electromagnetism and to Faraday's discovery, in 1831, of induction currents. He was therefore the first to test and highlight possible interactions between electricity and magnetism. Then Johann Schweigger (1779–1857) [SCH 48] invented the first sensitive galvanometer, named after Galvani, which was perfected by du Bois-Reymond (1818–1896). This instrument can detect small amounts of electric current by winding a coil of wire around a graduated compass. As early as 1821, Faraday (1791–1867) [FAR 59], a British physicist and chemist, played a key role in describing the links between electricity and magnetism. He began experiments that led him to the discovery of electromagnetic induction in August 1831:

> Faraday's discovery of the evolution of electricity from magnetism caused quite a splash in scientific circles, and he clearly felt it constituted a major coup. [RHY 04, p. 172]

This discovery was quickly incorporated into a series of instruments, particularly French ones. As Rhys Morus pointed out in 2004, his experiments on induction came as proof of what was called the electrical state. The latter can be defined as "*...a peculiar condition of matter which appeared to manifest itself only during induction, and which Faraday suspected of being 'a very important influence if not most of the phenomena produced by currents of electricity*'" [RHY 04, p. 176]. Duchenne de Boulogne, in his work *A Treatise on Localized Electrization* indicates that he preferred induction currents to galvanic current and reaffirmed this within the successive re-publications of his treatise [DUC 72]. Electrical treatment was constantly being renewed by the discoveries of physics and the development of the instruments that corresponded to them.

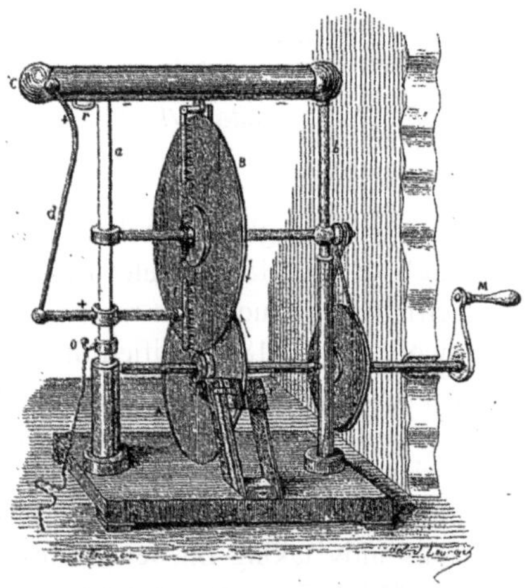

Figure 2.14. *Created in 1868 by the French engineer Ferdinand Philippe Carré (1824–1900), the Carré machine was mainly used in medicine in the second half of the 19th Century to relieve headaches [VIG 82, p. 26, pl. IV]*

The practice of medical electricity was becoming widespread in Europe with different modalities, practices and protocols. These trends were reinforced in the 19th Century. Although therapeutic targets diverged, the development of machines and the strengthening of the industrialization of medicine bridged the gap between the 18th and 19th Centuries. The challenge of improving stimulation machines led to a strong rivalry between device manufacturers. In 1803, in England, William Haseldine Pepys developed the largest galvanic apparatus that could be produced.

The electrotherapy department of the Salpêtrière, for which Romain Vigouroux was responsible, founded in 1875, was described as the site of a quasi-industrial installation of machines. Two hundred patients were treated there daily either by static electricity, by Franklinization, or by induced currents, by Faradization. Treatment thus oscillated between a global and a localized approach.

As Christine Blondel [BLO 10] explains, medical traditions were also becoming regionalized, motivating the development of devices for different uses. The practice of electricity, which was still very global in the 18th Century, became more specialized during the 19th Century, giving rise to currents specific to the countries in which they were practised. A very strong influence of experimental medicine weighed on electrical practices in France, while they took more the form of physiological exploration in Germany:

> Like their German counterparts, French electricians made little attempt to rely on electrophysiology, a laboratory science, to justify their therapeutics. [BLO 10, p. 38, author's translation]

Numerous French hospitals in Clermont Ferrand, Toulouse, Bordeaux, Nantes and Lyon, alongside the progress of industry, were opening their electrotherapy departments. Machines, considered as healing machines, were everywhere and contributed to rationalizing and mechanizing the medical practice of electricity.

Therapies and fundamental research, based on electricity, continued to reinforce the tendency to work on human analogies with the machine. It followed from this that the history of medical electricity was as deeply linked to the history of physics as it was to developments in technology. While before the discovery of the Leyden jar, the medical field was already ready to take hold of this instrument and apply it to the human body, the renewal of techniques contributing to the multiplication of electrotherapeutic applications. The symbolism of progress was so strong that it did not necessarily require successful treatment to become established:

> In the 1870s the new electrostatic machines, with influence rather than friction, revived the use of the bath and electric friction, without any more physiological justification than in the 18th Century. [BLO 10, p. 39, author's translation]

The historicity of medical electricity is complex: very active between 1745 and 1765, it fell into disrepute, notably due to too many accidents, then its practices were redeployed fairly quickly between 1770 and 1800, turning then to the understanding of mental and nervous illnesses. If we consider that experiments on life, resuscitation and the dying process fall within the historical spectrum of electrical medicine, then it can be considered to have remained very active in the early 19th Century. That is the perspective we have chosen to take. If we can consider that there was a pause in the development of electrical therapies in 1800 and 1830, as Christine Blondel [BLO 10] points out, we would like to draw the reader's attention to a complementary point. Indeed, while therapeutic applications were slowing down, explorations of bodies, following Galvani's research, were taking place in Europe. This means that it cannot be detached from a medical electricity always in search of its limits and potentials. This does not mean that medical electricity had a continuous history. On the contrary, its breakthroughs were polymorphic, technical, conceptual and geographical. To summarize the demonstration from the beginning of the 19th Century, we can highlight two branches of medical electricity: one oriented towards electrophysiology, measurement and laboratory experimentation; the other being more focused on empirical, direct and hospital medicine, both mutually enriching. The Leyden jar, static electricity machines or induction machines were all breakthroughs and new beginnings that helped to renew hope for therapeutic success:

> Since the appearance of Legendre and Morin's induction apparatus (which was approved by the *académie de médecine*), we have never ceased to apply it successfully in the cases indicated, without omitting, as far as possible, to end each of our sessions with the sovereignly restorative action of the electric fluid of Ramsden's static machine, combined with our vital magnetic fluid, or by using, if necessary, the Leyden jar which, by a happy modification, we can circumscribe and graduate the discharges to our will, at the highest as well as the lowest doses, according to the age, strength and temperament of the patients. [COU 81, p. 11, author's translation]

But whatever the technical successes, many applications remained questionable, as Louis Alfred Becquerel (1814–1862), brother of Edmond Becquerel,[17] points out:

> Electricity was applied to everything, and these applications were as often useless as they were harmful to the sick. [BEC 60, Foreword, author's translation]

17 Edmond Becquerel (1820-1891) was the originator of the principle that two dissolutions of different natures constitute an electrochemical circuit. The presence of these currents was demonstrated in biological tissues, such as veins or tendons. From these experiments arose questions about the electrical or chemical origin of currents in the body. Determining the choice of a form of electricity or current influenced the choice of instruments as much as the representation of the electrical effect on the organism.

Arsène d'Arsonval (1851–1940) played an important role in this encounter between machines and medicine, as well as in the discussions on the harmonization of methods between French and German scientists. He participated in the constitution of a biological physics rooted in the previous century. His collaborations with Étienne-Jules Marey (1830–1904) and du Bois-Reymond reinforced "[...] the links of this new experimental medicine with physics and the technical world" [BLO 10, p. 48, author's translation]. In this context, the International Electrical Congress held in 1881 marked a turning point in the official recognition of medical electricity, in the development of a universal language and in the harmonization of methods. It was part of a process of standardization of procedures and measurements, correlated with the need to standardize the use of machines. One of its objectives was to define an international system of electrical units. Throughout the 18th Century, various devices were built to estimate electrical load. The discovery of the battery in 1800 and that of electromagnetism in 1820 made it possible to approach electricity through measurement. The 1881 congress participants, 250 in number, came from 28 different countries. Famous scientists and engineers such as Tyndall (1820–1893), Helmholtz (1821–1894), Siemens (1823–1883), Mach (1838–1916) and Edmond Becquerel were brought together. One subject was obvious to everyone: that of electrical units.

That same year, Marey (1830–1904), Helmholtz[18], d'Arsonval [ARS 81] and du Bois-Reymond had fascinating discussions about the issues involved in introducing consensual electrical units into medicine. Dosing issues and measuring devices were discussed:

> In order to be able to easily control and reproduce the results announced by electro-therapists [...]; because the application of direct currents, without simultaneous measurements, can often determine and has determined accidents [...] [ARS 81–82, p. 725, author's translation]

In a dialogue, taken from the session of September 22, 1881, the technique employed by d'Arsonval is explained:

> Mr. d'Arsonval, in order to have an inductive current that is always identical, replaced the battery current by the discharge of a capacitor. This discharge, launched in the inductor wire, causes two instantaneous currents of equal quantity, but in opposite directions, in the induced circuit. For the same capacitor charge and distance from the induced circuit, absolutely similar excitations are obtained.

18 Helmholtz, the inventor of the ophthalmoscope, was able to experimentally evaluate the speed of propagation of nerve currents in the frog's nerves at twenty-six meters per second.

> Moreover, this method seems to bring about nerve fatigue much more slowly than direct discharge of the capacitor. [ARS 81–82, p. 725, author's translation]

Since the empirical and experimental beginnings of electricity in the medical field, electrifying physicists have generally evaluated the "quantity" of electricity delivered to the patient by the number of cells in a battery, by the length of a coil or by the number of crank turns given to an electrostatic machine. Technical issues greatly influence therapeutic effects:

> Mr. Marey wonders if it would not be good to propose three types of devices that can be easily reproduced: 1) The type of sled induction device by Mr. du Bois-Reymond [...]; 2) The electromagnetic device; 3) The electrostatic device. He adds that this would give unity in all research. [ARS 81–82, p. 727, author's translation]

D'Arsonval introduced high-frequency waves and opened his department within the Cochin hospital in 1884.

Medical electricity followed a trajectory, from the 18th to the 19th Centuries, from electrifying physicists to medical electricians. The alliance of experimental physics with experimental medicine can be seen in the multiplication of medical physics teaching in medical schools from 1870 onwards. The causal links of physics to electrotherapy can be summarized as follows:

> Physics, by perfecting the instruments of research, makes it possible to record with mathematical precision the operations by which life manifests itself in its most intimate characters, and it makes possible the complete study of all phenomena of mechanics, electricity, animal heat, optics and acoustics. [GUI 77, p. 5, author's translation]

The renewal of techniques stimulated research on the possible applications of electrotherapy:

> Work that would tend to be considered in the context of physics or industrial electricity, such as d'Arsonval's work on sinusoidal currents or weak currents [...] was carried out with a view to their use in physiology or medicine, before these instruments were taken over by industrial electricity. [BLO 10, p. 50, author's translation]

In 1875, John Tyndall gave a series of lectures [TYN 76][19] designed to popularize machine building and the use of electricity.

Thus, the craze for medical physics in the 18th Century corresponded to a desire to see the development of modern, technical and materialistic medicine. The bridge between the fields of physics and medicine had empirical and experimental, but also philosophical foundations. The beginnings of medical physics were part of a medicine of human automatons in which the body and mind were disconnected. This opened up an important field of experimentation and research. As more came to be known about electricity, it became clearer to physicists that it played a role in organic mechanisms. Moreover, it was not enough to know the laws of nature, but it was important to integrate Man as a subject of knowledge and intervention:

> Not only did the methods of isolating bodies become more numerous and more perfect, but electrical machines were also perfected and enabled to produce greater force [...]. The sensation that accompanies every spark that is drawn from the human body was too singular to escape the attention of Physicists: and it is probably this impression that the communication of the spark makes on each electrified person, which has given rise at the same time to several Physicists the idea that a tingling of this nature must have some influence on the vital principle. [TRO 88, p. 150, author's translation]

The legitimization of medical electricity at the end of the 19th Century was based on the practice and development of the clinical method, but also on a network of physicists, electricians and instrument builders sharing not only technical objects, but also a certain vision of experimental practice, whether it applied to matter or to the living. The alliance between medicine and experimental physics was embodied in major research projects such as those of d'Arsonval and Marey.

We have seen how a parallel movement continued to intersect research into the physical laws of electricity and the implications for organized bodies. This path was not linear and was not only hampered by accidents, due to the scientists' taste for self-experimentation, but also by the difficulties of obtaining homogeneous therapeutic results. This did not stop the development of medical electricity, which, although it stood out during the 19th Century, was still linked to physics and remained profoundly dependent on technology. How was medical electricity being renewed following the failure of its first applications to paralysis at the end of the 18th Century? Having fallen into disrepute in terms of its medical uses, it experienced a rapid resurgence linked to the importance of the brain as an organ of the faculties, the beginnings of location theory and the entry of the mental sphere

19 Tyndall, J. *Lessons in Electricity*. 2nd edition, Longmans, Green, and Co., London, 1876.

into medicine. Moreover, it could not be detached from the popular curiosity and hope placed in these new techniques, since electricity could be qualified as a popular science:

> The successful outcome of the sciences and the public demonstrations would have made the population an interested and active observer. Thus the public continued to legitimize hypothetical theories and substances such as Mesmer's magnetic fluid. In 1797, Humboldt denounced the 'impatient part of the public' waiting for the salvation of galvanism: this is proof of the importance of social factors in understanding the success of animal fluids at the end of the 18th Century. [SEG 01, pp. 83–84, author's translation]

3

Controversial Electricity Applications

In 1746, Professor Doppelmaier (1671–1750) [NOU 46, p. 439] died in Nuremberg from electrocution. He is considered the first martyr of electricity. By reconstructing a device that made a lightning conductor without insulation, Professor Bertier (1702–1783), from the Oratoire, also nearly electrocuted himself. In 1756, Benjamin Franklin wrote about the death of Professor Georg Wilhelm Richmann (1711–1753) on August 6, 1753:

> [...] is enough alone to warn that it is sometimes dangerous to approach the iron rod electrified by thunder without much care. [...] That he was not killed by the thunder that fell directly from the sky, but by the explosion of electrical matter, whose over-insulated iron bar became overloaded as his head approached it while he was making his observations. [FRA 56, v. 2, pp. 127–128, author's translation]

Figure 3.1. *Anonymous representation, Georg Wilhelm Richmann's accident (1753)*

This series of often fatal accidents marked the starting point for controversy over the benefits, dangers and side effects of using electricity, and, in particular, shocks in a therapeutic setting. These quarrels were found in the field of 19th Century electrotherapy:

> For, although Leyden jars, even those of a remarkable size, can be sufficiently loaded by small machines, and although these were used successfully in the past, and can still be used, it is nevertheless beyond doubt that such shocks, though stronger than those used today, are sometimes harmful, and must give way to small shocks which follow one another in quick succession. [TRO 88, p. 282, author's translation]

These discussions contributed to the revision of techniques, as well as to research on the type of electricity (static, dynamic, positive, negative) to be used. For example, the miniaturization of Leyden jars was recommended in order to reduce their effects, which could then be better controlled and dosed, in particular, by using several small capacitors. As early as 1747, Antoine Louis expressed his suspicions about the applications of shock:

> I am allowed to doubt that this jolting, however effective it may be believed to be, will ever do great harm to the reputation of the cinchona, & may dispense with learning, by use in the exercise of the art, the cases in which the administration of this specific procedure is advantageous or harmful. [LOU 47, p. 136, author's translation]

Until the end of the 19th Century, galvanic doctors were against the addition of a capacitor to machines used to transmit static electricity. Electricity has been the subject of debate since its first applications on human or animal subjects in an experimental and physical context. As we saw in the previous chapter, experimental physicists were interested in its effects on paralysis. This physical condition alone brought together many pathologies, physical, nervous or neurological, and was the catalyst for the first failures and an even greater mistrust of the medical applications of electricity. However, instead of falling into lasting discredit, in the manner of mesmerism, studies were turned on to the nervous system and disorders of the psychological sphere. How did this shift from paralysis to major electrical diseases, such as epilepsy, or magnetic pathologies, such as catalepsy or hysteria, take place? Did these diseases play a role in the future of therapeutic electricity?

3.1. Paralysis

In 1755, many practitioners were expecting electricity to be a remedy for paralysis:

> From the history of all these known facts, it seems to follow that Medicine should not flatter itself by taking great advantage of the new electricity experiments. However, one is not entitled to conclude that it is absolutely useless; perhaps there is only a rather rare type of paralysis which can expect some help, or perhaps there are in these diseases some favorable circumstances which have not yet been seen, & without which there can be no success. [AUM 55, p. 478, author's translation]

In addition to describing the painful effects[1] of electric shocks, Morin began a discussion about the side effects of electricity, which he considered the probable cause of nervous disorders. By exploring the physiological mechanisms of the nerves, he attempted to describe the disadvantages of electric shocks to the body. Not only did he evoke the notion of the punching bag to his own subject, but after having experienced its effects himself, he contradicted Nollet's tests on paralysis by stating that:

> [...] not only is the electric machine not suitable for curing or relieving paralytics, as Abbé Nollet imagined, but it is in itself very unhealthy; it can throw nerves into a kind of atony, cause weakness and trembling which may be difficult to cure. [MOR 48, p. 118, author's translation]

This text dates from 1748. We can see that, controversy over the positive medical effects of electricity on its first therapeutic target, paralysis, arose early on. The effect that Morin highlighted corresponded to a neurosthenic state characterized by extreme irritation combined with a state of weakness [GIA 08, p. 284] and which developed following the use of shocks. This condition, which sometimes lasted for several hours after the treatment was discontinued, raised doubts about its potential benefits. Nollet, criticized for his experiments on paralysis, criticized the *intonacatura*[2], electric cures practiced by Pivati (1689–1764) in Italy around 1746 [PIV 47]. Pivati provided care using electrified tubes and penetrating substances such as Peruvian balm. This research was notably taken up by Bianchini [BIA 50] and Winkler in Leipzig:

> [...] it can be said that he is the first one who thought of internally coating glass tubes or cylinders, with certain resinous materials and

1 "The light given by living beings must have its place here. Let one person suspended by silk cords, or placed on the pitch, touch the globe or the electrified bar; let another approach the finger; a spark goes off noisily, the action of which is equally painful to both people" [JAL 48, p. 45, author's translation].

2 The term *intonacatura* refers to coating, plastering, wrapping or covering.

> suitable for different cures. It was he who first thought he saw that the most subtle parts of these drugs, although exactly contained in the tubes, still pass through the pores of the glass, & are transmitted with the current of electrical matter into the body of a man placed on a resin cake. [FOR 63, p. 24, t. 2, author's translation]

The idea being that these tubes, if electrified, would promote the absorption of medication. Sigaud de la Fond, physicist and pupil of Nollet, took up these criticisms again in 1776[3], not to discredit medical electricity, but to rationalize its applications, which meant criticizing treatment that could not have any real benefit for the patient. The challenge of these studies was to differentiate between what was a scientific and medical application of electricity and what was close to deception:

> I will not be afraid to say that they have been grossly mistaken, both as regards *intonacatura*, and as regards electrical purges, & that no advantage can be derived, particularly from these two methods of electricity, but I cannot help but observe, too, that both methods are defective in themselves; that if those who proclaimed them have abused the public's trust, by making them hope for success that should not be expected, this is no reason to conclude that electricity cannot be of any use to the animal economy, & cannot be favorably used for the cure of several diseases. [SIG 76, p. 391, author's translation]

This highlights what was already a paradox in the 18th Century concerning medical electricity: even therapies with few results were able to develop. The revision of the techniques and machines that surrounded the electrical practice was a decisive element of its future in medicine:

> While the majority of European Physicists are convinced of the falseness of the advantages attributed in Italy to *intonacatura*, as the cylinders coated with medical materials were called, it was learned from various Letters written from Turin that the famous Professor M. Bianchini had imagined another method of using electrical virtue to transmit different medicines, but above all purgatives, to the human body [...]. [SIG 76, p. 386, author's translation]

3 "Many other physicists welcomed Dr. Pivati's discoveries in the same way. They found in Germany itself a famous partisan in Mr. Winkler who publicly defended all the rights of Electrical Medicine, against some German Doctors who dared to speak out against this practice" [SIG 76, p. 384].

Bianchini recommended the elimination of tubes coated with medical substances and held in the hands of electrified people. The underlying idea was that the body has as many entry points for electric fluid as there are pores on the skin, which would make it possible to stop swallowing drugs and imagine them passing through the tissue system. The number of pores was estimated by Leeuwenhoek (1632–1723) in 1674 at around 3 million[4]:

> Dr. Bianchi, a very distinguished member of the same Academy, & currently the first doctor of the Prince of Augusta, explains himself in his letter on Electricity, dated November 13, 1747. Many people wanted to experience at once all that Mr. Pivati had given them for certain, and I have been told by some acquaintances that they have made the experiments according to him; we cannot dismiss their skill in doubt, but so far they have succeeded very doubtfully. [FOR 63, pp. 36–37, t. 2, author's translation]

The *Société Royale de médecine* (French Royal Society of Medicine) commissioned Mauduyt de la Varenne (1732–1792) [MAU 81] to investigate the benefits of electricity applied to paralysis and subsidized the research for a period of three years. He took his work as a starting point for his investigation:

> At the beginning of 1780, I read in the Society's Sessions a Memorandum which contained the history of the electrical treatment administered for two years to eighty-two patients [...]. [MAU 81, p. 4, author's translation]

Etienne Louis Geoffroy (1725–1810), a French physician and entomologist who, in 1780, advised treatment with electricity for blindness, and Lorry and Andry (1658–1742), both Regent Doctors of the Faculty of Paris, stressed the extreme caution that must surround work on medical electricity and its applications on the sick. Their conclusions complemented the work of Mauduyt de la Varenne:

> That according to the application that I had made of electricity to diseases other than paralysis, it was probable that electricity could be used with advantage in their treatment; that it could relieve and cure those who are attacked by it, but that the facts are still too few, that they only present hopes. [MAU 81, p. 4, author's translation]

4 "Leeuwenhoek says he found 14,400 of them on the same area. A square foot of skin would, according to this, contain 207,360,000; and since, according to him, the entire surface area of an average man is 14 square feet, it would contain 2,904,040,000" [JCD 27, p. 318, author's translation].

At the end of this inventory of cases, in which electricity as a therapy held out hope for improved health, is an interesting point:

> 8. I've only used negative electricity once before: it hasn't had any effect: that's no reason to believe that it can't produce any & to deny the benefits, which Physicians who use it, say they've derived from it. Cases in which it is advertised as useful are trembling, convulsions, & in general known ailments known as nervous illnesses. [MAU 81, p. 6, author's translation]

Indeed, the application of electricity to paralysis was related to the study of nerves and their mechanisms. There was a shift towards the nervous system and the pathologies that entered into the functioning of the mental sphere. Convulsions, mentioned above, are symptoms of psychological, psychiatric and neurological illnesses that are still undifferentiated. They did not stop growing in importance during the 19th Century. Despite the controversy over electric shocks, three paradigmatic treatments of electrotherapy were often applied to nervous illnesses during the last third of the 18th Century.

– The electric bath, whose name refers to an analogy with thermalism as well as to the imaginary animal magnetism of Mesmer and his small tubs. Despite the liquid concept, the treatment was dry:

> This denomination is based on the fact that the person electrified in this way is surrounded by an atmosphere of electric fluid which has been compared to the volume of water in which a man bathing is immersed. [MAU 84, p. 6, author's translation]

– Spark electrification, the application of which refers to the theme of localization. It was developed by virtue of the refinement of anatomical and physiological knowledge of therapeutic targets related to different diseases:

> For this operation, things must remain arranged in relation to the patient and the machine, as they were for the previous operation, but for this operation, one more instrument is needed: it is called an exciter [...]. It consists of a polished copper rod, from two feet to two and a half feet long [...] during this operation, in the moments when no sparks are drawn, the electrified person is, as in the bath, surrounded by an atmosphere of electric fluid; but at the moment when a spark is drawn, all the fluid is carried and converges towards the point from which the exciter ball is approached. The manipulator can thus guide the electricity to the part it wants to act on. [MAU 84, pp. 13–14, author's translation]

– Finally, electric shocks, which, as a treatment, marked the beginning of shock therapy:

> The jar is carried close to the patient, who does not need to be isolated, but who is seated on a seat in the ordinary way: the end of the chain is attached in contact with any part of the person's body who is to be electrified; then the button of the jar hook is touched with any other point of the body of the same person who is receiving the shock at the time, it passes through the parts between the part touched by the end of the chain and the point from which the button of the jar hook has been approached. [MAU 84, p. 37, author's translation]

Two distinct medical approaches emerged from the quarrels for or against its use: an interventionist, "attack" medical approach and an approach to chronic disease relieved by constant, milder treatments. In 1872, Duchenne de Boulogne stressed that whatever the conditions under which muscular electrification was practised by the Leyden jar:

> [...] it would always be imprudent to expose the subject to a large number of discharges. [...]. Finally, this operation is always painful, because skin excitation, which is inevitable in any static electrification, increases due to the increase in electrical voltage. [DUC 72, p. 7]

As we saw in Chapter 2, medical electricity marked the entry of machines into the therapeutic and social space by targeting, initially and without much therapeutic success, cases of paralysis. The work of Abbé Sans in 1769 is emblematic of the recurrence of this type of pathology. He instructed Nollet to present the observation of Madame d'Espre's case to the French Academy of Sciences. The victim was a canoness struck with a left hemiplegia in 1768, whose stages of recovery he indicated. The minutes were signed by clergymen, notaries, physicists and medical doctors. In 1772, Sans thus published the cases of eight cures from paralytics. How was it that he could speak of a cure when many applications for paralysis had failed? What he wrote on page 117 of his treatise *Guérison de la paralysie par l'électricité* provided one element of an answer to this question. He extended the equation of nervous illness and paralysis to the many diseases whose cause was found either in the nerves or in the psychological sphere:

> Electricity tells us enough that it is supreme when speaking about nerves, since it so admirably annihilates paralysis. What do we know if it is not yet a supreme remedy that Providence offers us for the cure of many of our infirmities, especially those dependent on the nervous system? It is to be hoped that a hand skilled in the art of healing will give us an exact & faithful catalog of all the illnesses that afflict Humanity, & whose cause lies in the nerves! This will provide

> Physicists with a harvest that can crown their work with new discoveries. [SAN 72, p. 117, author's translation]

The theme of nervous illnesses enabled electrifying doctors to identify the effects of electricity that would enable them to move from the treatment of organic paralysis to that of psychological paralysis. Electricity, considered as a universal fluid of nature, gave substance to a secularized medicine and equipped it with new tools to intervene on a human scale in their physical and moral dimensions. It formed part of a medicine that inherited Hippocratic modeling, and responded to the intuition that there were material forces in physiology that caused mobility and circulated through the nerves. The idea of the circulation of animal spirits materialized and took the form of an analogy with the passage of the electric fluid through the body, at that point still considered to be what set them in motion:

> We even tried to go further, & to deduce the theory of movement & sensation, from the principles of electricity. [...] In the human body there is a very subtle & very mobile elastic fluid, which flows rapidly through all parts of the body, it is called electric fluid. [FOR 63, p. 286, t. 2, author's translation]

Thus, in the texts of Abbé Sans, we find the theme of paralysis, but it no longer appeared to be an organic disease. He also helped to reduce concerns about the dangerous and painful effects of electricity applications, moving away from the Leyden jar paradigm:

> Finally, so convinced by my own observations of the uselessness of these painful experiences that I have been able, by gentler means, to cure the illness in question, can I not say, without being susceptible to reproach, that I direct electricity in a different manner from other Physicists? [SAN 72, p. 129, author's translation]

While questioning the wayward behavior of the patients during these treatments, Sans underlined the effects of electrification on the psyche and on the faculties of the intellect. It seems that, as an instrument for rebalancing the systems of the animal economy, electricity made it possible to restore cases of reversible obliteration of mental capacities; by writing that electricity was capable of relieving brain congestion, without participating in materializing the faculties by locating them within the brain. The application of electrical treatments to all forms of mental and behavioral disorders had its origin here:

> Couldn't we rather say that these accidents are the effects of the impatience of the patients, & of the boredom caused by an electrification, which one is sometimes obliged to carry for more than

several hours? Far from being harmful to the head, electricity produces the most beneficial effects. We have seen above how much it changed favorably the intellectual faculties of the Religious of Perpignan, that the paralysis had rendered as dazed, and what fruit was drawn from it by Mr. de Monclar de Milhau whose reason was very disturbed. These annoying symptoms, which alleviate the consequences of brain blockage, were entirely dissipated by the electricity, which, by operating the unblocking of the vessels of this viscus, made the course of the solutions freer, & thus restoring functions. [SAN 72, p. 148, author's translation]

While electrical therapies experienced their first failures when applied to physical ailments, they developed in the field of nervous and then psychiatric disorders. By rediscovering the context of the emergence of medical electricity, by retracing its often complex history, it became intelligible. In the 18th Century, medical electricity was not just a fashion or the sole consequence of philosophical questions about human nature, it was marked by research that would change its nature and integrate it into the pharmacopoeia of mental illness. In the context of Mesmer's work, which, in 1775, marked the beginning of the first wave of dynamic psychiatry, and that of Chastenet de Puységur (1751–1825), electricity was applied as early as 1770 to epilepsy, hysteria and, more broadly, convulsions. Ellenberger [ELL 94], in her study of the history of the unconscious, highlights the socio-economic, political, cultural and medical stages that marked the development of dynamic psychiatry. Moreover, the fact that electric fluid was considered as an instrument to relieve congestion in the body was reminiscent of the Mesmerian method:

Since the disease was what was blocking the flow of this fluid through the body, the mesmeristic cure was to restore it by provoking a crisis in the patient that would free him from any blockage. [GUE 09, p. 178, author's translation]

The electric shock generates a convulsive type seizure. It therefore induces a clinical artifact that allows these seizures to be *replayed*. Thus, from the last third of the 18th Century, electricity was considered as a research method to treat these pathologies, either by using shocks or by suppressing it. By giving materiality, an objective and visual reality to the treatment, it allowed psychological and magnetic diseases, still very unknown in their causes, as well as in their effects, to pass into the field of technique, reason and progress. Although electrotherapy has elements of the imaginary that permeated 18th Century fluid medicine, it marked an important step in the management of nervous illnesses.

3.2. Nervous disorders

It appears that from 1770 onwards, the medical applications of electricity experienced a philosophical paradigm shift. In fact, it can be considered that from experiments on human chains and the involuntary movement of bodies, a dualistic medicine was born, in which the body was reified and reduced to an automaton without consciousness. From the moment that electricity was aimed at restoring faculties, at rebalancing nerves, behavior and reason together, it was part of the materialistic philosophy paradigm in which the physical and moral dimensions were treated simultaneously. This point was part of a physics that aimed to show the unity of the laws and forces of nature. Moreover, this paradigm shift accompanied the growing importance of the brain in research on materiality and the localization of faculties. In the human brain, movement, sensation, sensitivity and understanding are articulated, and faculties tend to be localized and inscribed. Vincenzo Malacarne [MAL 76, 80] was one of the neuroanatomists who placed their research on localization within an anatomopathological and clinical paradigm. He showed the correlation between disturbances in faculties and the state of brain structures. Electricity doctors also correlated the positive effects of electricity with "big" typical diseases for the development of electrotherapy. Thus, the dual action of epilepsy, on the involuntary and convulsive mobility of the body and on cognition during seizures, made it possible to conceive the effectiveness of electricity according to a different model from that of paralysis, concerning both the physical and the moral aspects. The therapeutic administration of electricity in nervous illnesses developed mainly from the last third of the 18th Century. As soon as the problems of the types of electrical treatment chosen, the equipment and doses were theorized, it was used to fight against nervous illnesses. Electricity does not seem to owe its success solely to its spectacular aspects. Moreover, this does not allow us to understand the Ariadne's thread between the developments, from as early as the 18th Century, of this therapeutic agent and its multiple contemporary uses. Although it was a source of wonder as well as the cause of fear on the part of practitioners, it was deeply rooted in the rationalist and materialist thinking of an interventionist medicine. The fact that natural electricity was channeled and made sensitive through mechanics made it the tool of a medicine that wanted to be objective, even if many practitioners gave in to the songs of electric sirens:

> Is it not astonishing, in fact, that lost memory is restored, that lost reason regains its former rights, that eyes almost extinguished are brought back to light, that mute tongues are set free, that limbs without feeling are restored to their natural sensitivity, that strength & movement is once again introduced into the paralyzed muscles, which we see, in a word, a sensitive image of a new resurrection of man, almost annihilated, by the sole action of an element, which is brought into play by the Electric Machine? [SAN 72, pp. X–XI]

The history of medical electricity provides an opportunity to explore the links between philanthropic medicine, particularly as applied to the emerging field of mental illness, psychiatry and the construction of new experimental protocols. The applications of electricity in the 18th Century were experimental and contributed to the foundation of two research programs:

– laboratory medicine;

– clinical medicine.

Its developments in the field of neurological and nervous system sciences are rooted in the representation of a soul that has cerebral convolutions. Convulsive or magnetic diseases represented a broad field of therapeutic trials. From the end of the 18th Century, electricity was applied to treat epilepsy and then for "madness" understood as a generalized physical, intellectual and behavioral disorder. Moreover electrotherapy, applied to nervous illnesses, in the broadest sense, correlated to the first theories of the localization of faculties, becoming itself localized on different points of the body, then of the head, and quickly taking, as a principle, the concept of localized evil on which the practitioner could intervene. For example, the German physiologist Reil (1759–1813) [SEG 01, p. 72], a professor in Halle, applied electrical vocabulary to describe the connections between the cerebrospinal and ganglionic systems: he spoke of the insulating, conducting nerves and referred to the part responsible for communication between the two regions of the nervous system in terms of a semiconductor device. Convulsions, caused by the probable disruption of these structures, were known as a main symptom of electrical diseases. But ailments such as melancholy, based on the stimulating aspect of electricity, also became the target of this treatment. Thus, electricity was applied as a stimulant and as a calming agent, by virtue of the singularity of the symptoms of the disease. This renewed the representation of a universal remedy. Jan Ingenhousz (1730–1799), a German physician, suggested the application of electricity to melancholy. Having had a few accidents during self-experimentation with electrical stimulation, he pointed out that he felt an improvement in his mood after application. Thus, in 1783 [ING 85], he proposed the use of electric shock as a cure for melancholy. But it was Abbé Sans who occupied, with Ledru, the field of nervous illnesses at the end of the 18th Century. The latter was given a hospice where a service dedicated to epileptics could be established. As early as 1777, he characterized electricity as the specific feature of epilepsy, the principle of which was a disturbance of the nervous fluid. In 1783, he specified the diseases of the nerves for which he defended the use of electricity:

> The first is caused by a tumultuous movement of the nervous fluid, which carries all its action to the origin of the nerves and suspends the internal sense: then the organs of action are strongly agitated which causes the subject to fall into convulsive movements; such as during a

> strong outburst of anger or some other strong passion, which makes him insensitive to all impressions foreign to the object that occupies him, is what happens in epilepsy. The second is caused by a kind of numbness and inertia in the nervous fluid that suspends the internal sense, and sometimes the action, as in catalepsy. [LED 83, p. 4, TOR 55, author's translation]

Physical shock, supposedly the cause of these diseases, could therefore be opposed to the administration of an electric shock. The latter, by its intensity, was able to bring about an end to symptoms and restore harmony in the animal economy. Ledru evoked "madness" as the source of his therapeutic experiments:

> I began by treating minor nervous illnesses, which were easily overcome by the application of electricity. Encouraged by the success, I continued on subjects where evil was more ingrained, and by gradation I came to epilepsy, catalepsy and madness. [LED 83, p. 19, author's translation]

Aware of the links between physical and moral ailments and the nervous system, he located the therapy on a brain and cerebellum scale:

> [...] the physical ones come from a violent shock, such as a shotgun strike, a stick, a fall from a horse, quite often from a staircase, when the spine has worn, the strong shock that one receives shakes the fluid of the nerves, & produces illnesses almost always incurable. The moral cause comes from sorrow or shock from grief caused by fear or unpleasant news [...]. I can assure you, based on the story of eight hundred randomly selected patients, that at least three-quarters of them have suffered accidents through violent contractions or shocks of the brain. [LED 83, pp. 14–15, author's translation]

While the theory remained that all body tissues were likely to be more or less affected by electric fluid and if the muscular system was the one that showed the most influence, the nervous system became a preferred therapeutic target. This was how Ledru's work was still discussed in 1825 by Most (1685–1759):

> The great effects of electricity in epilepsy are well known; Dr. Ledru's experiments are especially noteworthy. […] It results from the experiments of the above-mentioned doctors: […]. 3. That at the beginning of the application, the accesses became more frequent, but then they became rarer, and then disappeared entirely. 4. That the electricity administered during access diminished its strength and duration. 5. That it irritates and increases the free movement of nerves.

> [...]. As a preparatory means, however, it is very useful, because it changes the type of accesses, and in any disease that manifests itself periodically, much has already been gained when regular accesses have been transformed into irregular ones. [MOS 25, pp. 9–11, author's translation]

All the ambiguity of the links between electricity and its effects on the nerves is evoked in this quotation. Indeed, while it was recognized that it fatigued the nerves, it seems to be deduced that it also fatigued the disease, thus calming its effects. Yet healing is not mentioned and electricity appeared to be a calming agent to be administered in the form of chronic treatment while avoiding its harmful effects. Although electrosurgical pathologists found brain tears in animals killed by strong electrical discharges [PAL 47, p. 45], it appears that electrical treatments "calmed" convulsive disorders on the model of the transmission of natural electricity to the entire nervous system. From the lethal to the therapeutic effect, an experimental space opened up, using the central nervous system as a field of study. *A fortiori,* in the field of vaporous diseases, considered to be caused by electrical disturbances, electricity was quickly becoming a promising heuristic tool. Clinical anatomopathology formed some of the methods of these electricians. The search for material localization of lesions or electrical characteristics of the nervous system was similar to the search for lesions corresponding to mental and/or psychiatric illnesses. A significant development on the theme of the organic substrate arose in the 19th Century. The example of the works of Bayle (1799–1858) is emblematic. In 1822, he described what he believed to be a new pathological entity whose cause would be imprinted in brain matter: *chronic arachnitis*. The latter provided a scientific basis for an entire branch of organic psychiatry and suggested that mental illness was now visible. It was the foundation of the organicist paradigm, but did not survive the discoveries on infectious encephalitis. Nevertheless, from a materialistic perspective, psychiatry linked its fate to that of neurology and did not stop looking for anatomical substrates, conceived in terms of nervous system lesions, to mental disorders. Parchappe de Vinay (1800–1866), in *folie et necroscopie*, in 1841, in turn raised the problem of the need to rediscover the brain and to question its structures in order to determine the relationships between clinical and structural disorders [BAY 25, PAR 41]. Nothing should escape the observation and experimental dimension of medicine, which was found along this path of electricity:

> The reaction perhaps begins with the work of the Italians who never fully accepted the metaphysical trend and continued to remain faithful to the spirit of research of which Morgagni was the main representative. L. Spallanzani, F. Fontana and A. Corti were the first experimenters who could be truly considered innovative... [CAS 48, p. 586, author's translation]

Bertholon (1741–1800), in his chapter on *extravagant diseases*, inscribed his research in the field of the secularization of diseases of the soul, which became organic, localized within the human brain and then brought back to a dynamic dimension by means of electricity. The diseases concerned by his research were "madness", dementia, delirium, insomnia, forgetfulness, stupidity, night terrors, *canine hunger*, satyriasis and nymphomania. More than the treatments, it was the nature of mental illness as an electrical disease that was being questioned:

> [...] these first two diseases must be cured by negative electricity & the application of conductors. [BER 80, p. 327, author's translation]

While the heuristic value of a medical scheme responding to *trial-and-error* concepts was visible as early as the beginning of the 18th Century, this scheme was constantly being renewed by giving ever greater importance to experimental protocols. Epilepsy[5] became a paradigm for the application and development of electrotherapy. The term *epilepsy*, still very generic, covered many mental illnesses, such as hysteria. This fact explained many of the successes described by Ledru for the treatment of this pathology, which was significantly over-diagnosed. Most gave an example in 1825 where the confusion between epilepsy and hysteria is visible:

> Philippine Engelking from Nordholz, eighteen years old, blue eyes, blond hair, beautiful complexion and a robust and muscular constitution, has been suffering from epilepsy for three years. She had her first seizure in a church at the time of the Fall Equinox; in the second year she had three, and in the third, four. The last time, she had seizures every month, the last of which occurred on May 28th of this year; fifteen days before she entrusted herself to my treatment. From May 28 to July 8; she used my machine twenty-four times. At that time, she had only a very weak seizure on May 24th; since then, Miss Engelking has left me; she is cured. [MOS 25, p. 122, author's translation]

5 It was during the 19th Century that epilepsy became understood as a neuropathological process. While its terminology was a way of clarifying its understanding, it was often defined by its generalized nature, which, in turn, was confused with eclampsia or hysteria. Its name has undergone a great number of changes throughout the various religious and medical contexts. Two breakthroughs occurred in the course of its history:

– The moment it began to be described in terms of a science of the brain which, despite its speculative aspects, broke with divine and demonic explanations, for example in the work of Charles Le Pois (1563–1633) in the 16th Century or that of Thomas Willis (1621–1675).

– Following the work of John Hughlings Jackson (1835–1911) and the analyses of Jean-Martin Charcot (1825–1893), it was dissociated from the category of mental illness, to become a disease of the central nervous system.

Thus in the context of this electricity movement in the sphere of psychological diseases, Jacques-Henri-Désiré Petetin (1744–1808), physician and president of the *Société de Médecine de Lyon* (Lyon Medical Society), initiated a series of studies on magnetic diseases. A proponent of electrical and galvanic therapies, interested in rare cases, he devoted a large part of his work to "experimental catalepsy" (or "artificial" catalepsy). Catalepsy, as Alexandra Bacopoulos-Viau points out in her article *La danse des corps figés. Catalepsie et imaginaire médical au XIXe siècle* [BAC 12], holds a special place in 19th Century medical imagining: what was this frozen state? What did it correspond to in the context of a medicine that was experimental and materialistic?:

> Is there some form of consciousness in these decerebrate states? Is consciousness an eternal soul or simply an epiphenomenon of neurophysiological transformations? Descartes's animal-machine is evoked, the human being is compared to an automaton or a puppet whose master would be the laboratory man or the hypnotist. [BAC 12, p. 166, author's translation]

If we refer the meaning of this quotation to the history of electrical therapies, we note that one of the fundamental issues of these techniques was the control of Man over himself. Thus, in the case of catalepsy, it was a question of defining, on the one hand, whether consciousness was absent or not; on the other hand, under what therapeutic conditions electricity could return the affected subject back to the conscious state. In his twelfth experiment on the relationship between the brain and electricity, Petetin hypothesized a causal relationship between brain substances, their location and the action of electricity on convulsive disorders:

> At first glance, it seems that negative electricity should be proposed as the specificity of the essential hysterical condition & of all convulsive diseases which recognize as a principle a dominant fire in the brain & nerves; it should be able to correct the vicious disposition of the medullary extensions, which increases the activity of the electric fluid [...]. [PET 87, p. 101, author's translation]

At this point, the paradigm of electricity produced by the brain was not yet reached, but there was discussion of the paradigm of mechanical interactions between artificial electricity, applied externally, and natural electricity already present in the body:

> Of all parts of the human body, the cortical & medullary substance of the brain retains electric fluid longer, it is not uncommon for a thumb-sized portion of the spinal cord to still show signs of electricity after two hours. [PET 87, p. 18, author's translation]

What about this observation? Petetin practised nerve stimulation in a *post-mortem* context and derived a model for the parts of the brain most likely to interact with electricity. Furthermore, these interactions were directly related to the physics of electricity, as the senses, correlated with the brain, transmitted external information, including electrical stimuli present in the atmosphere. Thus, electricity circulated in nature, as through the body, impacting all its mechanisms:

> The sense of hearing is still a phenomenon of electricity, which depends less on the percussion of the soft portion of the nerves of the seventh pair than on the kind of movement that the electric fluid escaping from the sound bodies communicates to the person who animates the branches of that nerve. [PET 87, p. 25, author's translation]

The imaginary of electricity participating in the transmission of the most terrible evils permeated a great deal of research:

> [...] this is the source of contagious diseases, & most often fatal, that the electrical currents deposit in the breast. The plague that renewed itself in Milan, because an unfortunate gravedigger shook a dust-laden rope, is still a phenomenon of electricity more terrible than thunder, whose deadly force dissipates without leaving behind destructive miasmas. [PET 87, pp. 27–28, author's translation]

This pathogenic imagining of electricity, combined with hysterical imagining, led Petetin to draw an "electric" portrait of pre-hysterical symptoms. The latter, conceived as electrical disorders, were becoming prime targets for electrical therapy.

Thus natural electricity, by its impact on organisms, which could be more or less sensitive to its action, was potentially pathological, whereas, once mastered, it became an element of the therapeutic arsenal. The notions of dose and therapeutic protocol made it possible to differentiate the natural action of electricity from the effects of its medicalized application. Is it safe to say there was electrodiagnosis in the 18th Century? Potentially symptomatologically, *a posteriori* and empirically. In regards to a 12-year-old girl:

> After every effort had been made to find out whether this disease was real or faked, the patient was electrified & she was cured after seven weeks. [TRO 88, p. 264, author's translation]

Interesting physiological links were made between sensations, impressions, the internal sense and disorders generated by an increase in electrical connections between the sensory organs and external objects. Thus a dynamic operation was described between the internal electrical machine, marked by an electrical sensitivity that varied according to the gender, age and temperament of the subject and the external environment. For example, a woman's sensitivity could make her susceptible to *hysteria* [PET 87, p. 28]. The transportation of senses, which today could be described as synesthesia [PET 87, p. viii], was also described as a constant effect of epilepsy. The example, given in 1808, of a patient who could hear with her stomach made this phenomenon difficult to grasp. Synesthetic phenomena were conceived as the result of an imbalance in the electric fluid which, by migrating unevenly into other organs such as the epigastrium, would account for these strange facts. *Hysteria* was characterized by a disorder of the ideas and the body, made manifest by convulsions similar to those caused by an electric shock. However, while shocks could mimic naturally occurring pathological disorders, if the strength of the shock was tamed, it could reduce the effects of these disorders.

In addition to convulsive nervous illnesses, electricity was also applied to disorders in which one of the singularities was the absence of visible activity. It should be emphasized that the same framework still existed, of excess physical mobility and/or, by default but extended, to the materialization of mental space. Thus, Paets van Troostwyk, speaking of magnetic diseases, attributed the following cause to them and extended the analogy between the vital principle and electricity:

> Let us now look at other diseases, which have their origin in a decrease of activity in the vital principle. We classify under this class, secondarily, the soporous diseases, such as coma, lethargy, carus, paraplegia & catalepsy. [TRO 88, p. 213, author's translation]

The idea that electricity could be a cleanser capable of dissolving cerebral pressures, with consequences on the actions of subjects, was recurrent. Thus, it could restore the good circulation of the nervous fluid and thus the communication between the nerves of movement and those of feeling. Paets van Troostwyk spoke about energy and not fluid[6]. This fact heralded the drying of electricity and the advent of the dynamic model whose effects were only visible but which could be recorded and made visible by new instruments. In this paradigm of disruption of nerve flow, St. Vitus' dance was considered relevant to be treated in a similar fashion. Morbific matter obstructing the brain pathways was expelled by electricity. As the anatomy of the brain was becoming better and better known, the applications

6 "There is no doubt that the proximate cause of soporific diseases is that pressure on the brain prevents the action of nervous energy […]" [TRO 88, p. 213, author's translation].

of electricity could be located there. Targets were narrowed down. Bertholon also discussed the use of electric shocks for mental illness:

> The electric shock given to the head is certainly very likely to calm the disorder, & to chain up the violence & fury, which are peculiar to these strong diseases. [BER 80, p. 328, author's translation]

At the end of the 19th Century, many doctors considered that neuroses could be treated by static or dynamic electricity. Paul Vigouroux described epilepsy as a neurosis whose seizures were likely to be calmed by electrical therapy:

> Epilepsy is a nervous disorder for which a wide variety of treatments have been used. Static electricity seemed to us to give excellent results and we believe that it is the first remedy to apply to this dreaded disease. [VIG 82, p. 86, author's translation]

Among chronic treatments, the electric bath was recurrent [VIG 82, pp. 86–87][7]. Thus, Arthuis insisted on the importance of the techniques used to treat hysteria and the need to eliminate shocks. The electric blast was considered to be the most effective and gentle way to relieve hysteria:

> It is an exciter that ends in a metal plate, on which a greater or lesser number of tips are implanted. This instrument offers the multiple advantages of producing a soft, soothing electric blast; of covering a large area, and of being able to embrace an entire region at once. [ART 81, p. 35, author's translation]

Romain Vigouroux added a capacitor to cause electric shocks. Pinel (1745–1825), considered to be the initiator of discussion around mental illnesses, also became interested in the first applications of electricity on nerve disturbances when he met Aldini (1762–1834) in 1801:

---

7 In the 19th Century, a medical syncretism developed which linked electric shock therapy to galvanism, but also to magnetism and hypnosis: "I have used galvanic electricity in several illnesses: paralysis of the tongue, limbs, hysteria, chronic exanthemas, goiter, etc., etc., and with the greatest success; but, on the other hand, I am also convinced of the difficulty of fixing the dose for each individual case. It is only from the experience of treating these diseases over a series of years that I have deduced a few rules on this subject, which I will one day make known. In the beginning, I administered galvanism in epilepsy, and managed to relieve the patient. Afterwards, I imagined joining electricity and metallic magnetism. To my great astonishment, I observed that this compound developed a force absolutely different in its effects and accidents, from those of each of these three substances taken separately" [MOS 25, pp. 72–73].

> I found myself in this case at the Salpêtrière hospital where I began some tests in the presence of the famous Professor Pinel. A lunatic whose arms were tied, preventing me from communicating the galvanic influence of one hand at the base of the battery on the one hand, and on the other through the arc established between the top of the battery and the metal frame placed on the subject's head. In this case, I used the galvanic application of one of the ears to the lips. [ALD 04, p. 149, author's translation][8]

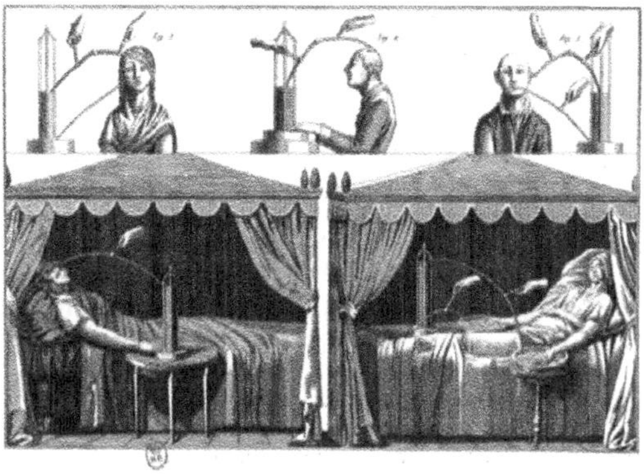

Figure 3.2. *This experience describes a treatment applied to a patient in May 1801, at the hospital of St. Ursula. Stricken with melancholy, the galvanization tears him away from the object of his reverie [ALD 04, Plate 5, figure 5]*

Pinel was no doubt challenged by the novelty of the method proposed by Galvani's nephew in conditions marked by little healing, such as melancholy. Electrical care renewed and completed the notion of reversibility of psychic disorders:

> Professor Aldini, during his stay in London, communicated to several doctors his observations and the applications he had made of this new stimulant in Paris, in the Salpêtrière hospice, in the presence of Dr. Pinel. The application of galvanism in vesania, or melancholy madness, is absolutely new and of great interest; he cured two patients in Bologna affected by this disease perfectly: the detail of the curative means he employed is all the more interesting as medicine offers very few resources against such affections. [ALD 04, p. 195, author's translation]

8 "Professor Pinel has been very zealous in my experiments; he saw for himself the excited muscle contractions in an old woman who died of a putrid fever. The interest he took in my research, committed me to communicate different views to him to relieve the unfortunate people entrusted to his skill and his beneficial care in the Salpêtrière hospital" [ALD 04, p. 199, author's translation].

This is what Pinel wrote about his encounter with Aldini in 1801:

> At the Salpêtrière hospice, I tried this kind of experiment with Mr. Aldini, directing the flow of the galvanic fluid through the ears, the end of the nose, and other points of the head: in most of these cases, a stimulating but temporary effect could not be ignored. [PIN 07, p. 126, author's translation]

Despite the French mixed opinion on mental health, Aldini described the interactions between mental illness, cognitive abilities and the tool of electricity in an extremely heuristic way. Thus, an Ariadne's thread from the end of the 18th Century to the end of the 20th Century can be drawn [PAR 04]. Aldini reportedly experienced massive facial muscle contractions after stimulating the human corpus callosum, an effect he had associated with an epileptic seizure. These facial muscle contractions were noted to occur on the side opposite to stimulation. Aldini used the voltaic pile, which provided continuous electrical stimulation that was less brutal than that of the Leyden jar. Whatever the role attributed to Aldini's experiments, the links between the application of electricity and disorders of the mind cannot be detached from the modeling of the human brain and its functions:

> The functions of the brain, as we know, are related to the operations of understanding. On the good condition of some, depends the energy of others. A fall, a violent blow to the head, often produced very noticeable alterations in the intellectual faculties; some lost their memory, others became almost stupid. There are even well-documented facts which prove that such accidents have brought about the happiest and most unhoped-for opposite changes in some individuals: they have been followed by the development in some of them of an aptitude for study which had not been noticed before in others, the development of great talents whose spark had never been seen before. We have seen these same accidents, in maniacs and people with dementia, followed by the return of reason. These observations, these reflections, and my subsequent experiences, gave me hope for the success of the administration of galvanism in insanity. [ALD 04, p. 123, author's translation][9]

9 Still within the framework of humor disturbances, which were then perceived as disorders of the vital fluid, Aldini treated melancholic subjects: "The experiment was repeated in this way several times in a row, and always with the same success. It had no harmful effect; the patient, who was questioned the next day, did not complain. His condition did not worsen; [...]. The following day and the days that followed, he was galvanized again, but more strongly, and always with a success that became more and more marked each time: his physiognomy became more animated at the sight of the device, and during its action" [ALD 04, p. 125, author's translation].

In the case of those who advocated electric shocks, as in the case of those who advocated more chronic care and inherited the fears associated with physicists' early self-experiments, the aim was to restore physiological harmony between the natural electricity present in the environment and that suspected to be in the body, by analogy with nerve fluid, using artificial electricity.

During the first third of the 19th Century an ancestral technique originating in Asia, acupuncture, was redeveloped and became associated with electrical treatments. Invented by Magendie (1783–1855), electroacupuncture is best known thanks to the work of Sarlandière (1787–1838). Magendie [MAG 16, p. 146] gave a central role to electricity in the functioning of sensations. Thus, he attributed happy effects to this technique, thanks to which it was no longer necessary to electrify the whole body or to pass the current through all organs to cure a local disease:

> The fluid could be condensed on a specific point on or below the surface of the skin. [...], it was possible to measure doses and to graduate the discharges. [REM 58, p. 32, author's translation]

The idea was to introduce electricity into the center of the suffering tissue[10]. The renewal of techniques validated new applications and supported the electrical understanding of the organism:

> But it is, indeed, the electric fluid which I cause to detonate on the needle which serves as a conductor that constitutes my curative means. The tip of the needle that I insert into the affected tissue is brought into immediate contact on the one hand with the muscle or fibrous fibers that I want to modify, while on the other hand the handle and the button that end the instrument communicate with the exciter or the insulated conductor of the machine. At the moment when I operate the electric discharge on the button that surmounts my needle, the jolt is transmitted instantaneously to all the branches or nerve threads that are distributed in the muscle or fibrous tissue that the tip of my needle has penetrated; [...]. [SAR 25, p. 18, author's translation]

---

10 On the subject of electroacupuncture, La Beaume, in 1828, raised questions about how this technique works within the body: "Do they draw from the outside a sum of electricity that they add to the one that is constantly developing: in the body, or do they draw from these same bodies an excess of electric fluid that was harmful to it? Do they thus restore, in the way the nervous fluid is, the balance necessary to maintain health? Do they also act, by themselves developing new electricity in the body, through their contact with the parts into which they are made to penetrate?" [LAB 28, p. 51, author's translation].

Electroacupuncture[11], in addition to representing a new stage in electrical therapies, can also be correlated to Xavier Bichat's work on tissues. Indeed, it should be pointed out that, in France, a set of favorable conditions [KEE 79] around 1800 allowed a group of young doctors, influenced by the conceptions and theories of Morgagni and Bichat, to correlate the clinical register with the pathological repertoire and tissue alterations. Thanks to clinicians such as Trousseau (1801–1867) [TRO 55], who gave his name to various clinical signs, Corvisart (1755–1821) [COR 06] and Laennec, the main features of a dynamic medicine based on the knowledge of tissue lesions were set out. Electrical therapies linked their developments to this experimental and clinical medicine in a sustainable way. Moreover, the fact that electrical therapy was considered applicable to nervous disorders confirmed the organic representation that electrifying doctors had of these disorders:

> [...] that the electric fluid conducted inside of our tissues, and put in immediate contact with the nerve threads or radicles of the one of these tissues that is painfully affected, causes, at the moment of the electric discharge, a tremor that spreads to the whole suffering organ; this shaking which distorts the pain, and which is felt in a direct way, is much more advantageously applied to the cure of gout, rheumatism and nervous affections, than the shocks impressed through the skin, and one conceives that if one were to take some advantage of the introduction of the needles alone, one would take infinitely more by joining to acupuncture, the good effects which one obtains from electricity. [SAR 25, p. 25, author's translation]

Gradually, from the analogy with vital fluid, the texts dealing with electrical treatment moved towards the paradigms of force and energy[12], referring to the draining of electricity. Moreover, the identity of phenomena between normal and pathological nervous mechanisms and electricity was being reinforced. Antoine Despine (1777–1852) [DES 40], referring to Petetin's work, underlined this similarity which opens up the medical perspective of the electrical control of nerves.

11 In 1840, protocols for the application of these new techniques were published: "The same rule is to be observed when you employ electricity, the best mode of using which is to place the patient on an insulated stool, and draw sparks from, or shocks through, the affected limbs. Electricity frequently does much good in such cases; but in order to obtain decided benefit from it, you must persevere for some time in its employment. It has been lately proposed to employ the stimulus of electricity and galvanism in a different way, by transmitting it directly to the muscles of the affected limbs by means of needles, and which are intended to act as conductors for transmitting the galvanic influence" [STO 40, p. 329].

12 "Electricity has been used often, but almost always with little success; because it is a dynamic force, not a cure, which then requires another body to act upon it" [SIE 30, p. 4, author's translation].

At the end of the 19th Century, the application of electrotherapy, or rather, should we say, the applications of electrical therapies, plural, polymorphic and dependent on different techniques, were made around medical syntheses of the discoveries of: Aldini, Galvani and Petetin, on animal electricity; Ampère, Prévost and Dumas on the irritability of the muscular fiber; and Malacarne, Gall and Rolando on the brain considered in its physical and moral dimensions, its dependencies and its functions.

Indeed, from attempts at paralysis treatment to the management of nervous, convulsive or mental illnesses, a greater knowledge of cerebral anatomophysiology, as well as electrical physics, was emerging. The confluence of these two fields of knowledge made it possible to understand what was at stake between the second half of the 18th and the 19th Centuries: the intervention of a medicine that answered medical–philosophical questions about human nature and could regulate its actions while explaining them technically:

> Thus, philosophically speaking, there is nothing repugnant to the belief that the living body cannot become a galvano-electric device in certain sickly states of man. [DES 40, p. 166, author's translation]

In any case, this was the representation that was given by this medical and physical movement. This is why, on a recurring basis, health conditions, such as pathological states, were understood and described according to electrical characteristics:

> These curious observations have often led me to make a connection between this irregular pathological state and what happens in electricity. The latter only manifests itself, by noticeable and apparent effects, when positive and negative electricity are isolated from each other; and when these two changes, or these two parts of the same whole, are set in motion, either by means of an electromotive machine, or in opposition to each other. Wouldn't the normal or natural state of health be the result of the fusion of two analogous principles (positive and negative) constituting animal electricity, of which the brain would be the focus, the nerves the conductors and the nervous fluid the vehicle? [DES 40, p. 48, author's translation]

It was thus all knowledge on humankind that was permeated with an electric imagining, opening up the idea of an electric nature that varied between individuals. Everyday language has also retained traces of this, except for beliefs about the dependence of mood and behavior on the environment and the climate. The themes of harmony and disharmony permeate, in a flow of Hippocratic medical thought, the considerations on the inscription of humankind in a nature understood in physical and mechanical terms:

> It is, I believe, by the effect of an immediate and *sui generis* action of electricity on the nervous fluid, that electricity finds its therapeutic usefulness in the treatment of neuralgia and in that of nervous illnesses in general: it is not, however, by increasing its mass, or by subtracting it from the economy: but rather, by equalizing it in a uniform manner and distributing it according to the normal and natural order. Thus, according to my theory, electricity relieves, and often even cures neuralgia alone, by correcting the vicious distribution of nervous fluid, or by preventing its morbid and abnormal accumulation on one point of the body, to the detriment of others. [DES 40, p. 141, author's translation]

By virtue of the notion of internal balance, applications with plumes or electric baths were preferred to shocks. The theories of the four humors and temperaments marked medicine at the turn of the 18th Century and permeated the model of an organic environment that prefigured the concept of the organism. Temperaments were what the electrifying doctor needed to have knowledge of, in terms of healthy people. While in the Hippocratic doctrine they corresponded to humors, that is to say, to vital substances in greater or lesser quantity in the organs, they seemed to be transferred to the notion of balance and electrical imbalance, the latter being at the origin of many pathologies:

> In our cataleptics, it would be animal electricity distributed in an irregular manner: that is, the nervous fluid irregularly distributed: because, in these people, things happen quite differently than in the state of health. [DES 40, p. 166, author's translation]

Disease was therefore conceived as the result of the organic disturbance of the fluid or electrical energy. The machine intervened to re-establish good circulation, reinforcing the notion of a natural electrical circuit. Thus, beyond the brain, all the organs, in the perspective of a hierarchical and harmonious functioning of the different systems of the animal economy, contributed to this electrical balance:

> This is probably due to a particular nervous disposition in the intestinal tract, which I look at myself as if I am overloaded with electricity. Fatty substances, as we know, are idio-electric and non-conductive bodies. It is undoubtedly this quality that is contrary to Nature in these patients, when it comes to the internal use of animal substances, fats and oleaginous substances: hence, more difficult and often no digestion, stomach heaviness, abdominal tension, borborygma, *globus hystericus* [...]. [DES 40, p. 147, author's translation]

Despine's remarks are reminiscent of the work of Pierre Antoine Prost (1770–1832) who conceived of "madness" as originating as an intestinal disorder. Prost was influenced by the theory of biological systems whose regular functioning supposed a sympathetic link between organs, here between the intestines and the brain.

In 1832, a new physical and technical breakthrough was initiated by Faraday, an English physicist and discoverer of induction currents. He offered medicine an energetic and easy-to-use source of electricity. The induced currents were immediately rooted in therapy, thanks to the stimulation and treatment of the muscles. Faradization, whose medical practice was systematized by Duchenne de Boulogne, acted mainly on muscle contractility and nervous excitability. In addition to the excitomotor effects, the induced current had analgesic, vasomotor and repulsive effects. Thus, electrical techniques were diversifying, in order to relieve patients from such singular ailments as nervous vomiting:

> Nervous people, especially women, sometimes experience vomiting that is resistant to the best medical treatment. In all the cases that have come before us, it only took a few galvanic and magnetic applications to cure them radically. [CRI 52, pp. 49–50, author's translation]

Electromagnetic currents came to be used in the treatment of hysteria coupled with symptoms such as hyperesthesia or, on the contrary, the anesthesia of certain parts of the body. These currents were applied to neuroses and their symptoms:

> The exaltation of skin sensitivity can be a symptom of nerve center disease, but in most cases it is a simple, very common neurosis in nervous, hysterical women, and often coexists with anaesthesia, i.e. with decreased sensitivity in other places. Over time, this over-sensitivity can spread to the muscles and deep parts of the body, so it is important to combat it from the outset. Few ailments dissipate as quickly as this one through electromagnetic currents. [CRI 52, p. 60, author's translation]

Beyond the fact that the medicine of nervous and mental illnesses took over the electric tool as early as 1770, it contributed to redefining the place of humans in nature and their very nature. While natural electricity was included in vital movements, generated them and circulated within the body, it became, in the same period, a tool for physiological exploration. Galvani's experiments reinforced these ideas.

3.3. Electricity: between the normal and the pathological

As we have just seen, from physical ailments to mental disorders, electricity became an instrument for the treatment of a whole range of pathologies. But insofar as the balance of the animal economy depended on its harmonious circulation, it was also conceived as a scale for measuring physical and mental well-being. Thus, its theoretical and practical applications oscillated between normal conceptions and pathological frameworks, each one delimiting and defining itself in relation to the other:

> In biological matter, it is pathos that conditions the logos because it calls upon it. It is the abnormal that arouses theoretical interest in the normal. Standards are only recognized as such in contraventions. Functions are only revealed by their failures. Life rises to consciousness and science of itself only through maladjustment, failure and pain. [CAN 10, p. 39, author's translation]

How can we understand these words of Canguilhem in the context of the history of medical electricity? Electricity, in its interactions with animated matter, modeled the questions and problems to which it provided answers. Knowledge that is built on an experimental basis is part of both qualitative and quantitative representations. On the one hand, electricity, because of the innate nature of its suspected presence in organized bodies, qualitatively permeated its physiology; on the other hand, the healing depended on its internal quantitative variations and the external quantity that was applied. The electrical standard field was measured and defined in relation to electrical unbalance.

The causal relationship between human behavior and the pressure of atmospheric electricity influenced, firstly, the state of health; secondly, disorders due to too much of it. Emmanuel Pallas, based on this quantitative conception of the effects of atmospheric electricity, recommended the installation of insulating beds to control the electrical input received by the nerves:

> According to all of the above, the cause of many diseases can be attributed to atmospheric or terrestrial electricity. Its morbid action is first carried out on the nerves of the periphery of the organism, which transmit it directly to the nervous tree and from there to all the tissues of organic and animal life. [...] Fortunately, we have achieved this result through the use of insulating beds, with the aid of which we can break at will the communication which continually exists between man and the earth, and thus remove the organism from the action of electric currents while placing it in another environment. [PAL 47, p. 260, author's translation]

The aim was to make the link between a Hippocratic perspective in which the individual is considered an element of nature and obeys his mechanisms, and the management of diseases resulting from interactions that have become pathological under the effects of the environment. The explanation of biological phenomena in terms of natural laws has its roots in a double movement aiming at:

– escaping the mechanism's insensitivity;

– going beyond religious conceptions where God is the foundation of natural phenomena.

Thus, during the second half of the 18th Century, teleology came to the forefront of physiological explanations. This change was due to an unsatisfactory application of the principles of the mechanistic sciences to the life sciences, according to which everything in nature can be explained by the laws of motion. The concept of life acquired a new dimension. Considered as a biological power that guaranteed the development of living beings, the electric spark then took on the meaning of a vital spark [ASH 13]. The notions of organism or animal economy become the theater of the normal and pathological links between electricity present in the indoor environment and environmental electricity. The analogy between nature as a whole and the organism made it possible to establish a correspondence between the microcosm and the macrocosm.

Alongside these teleological perspectives on the laws of nature was the idea that the brain, the location of the organic foundations of human nature, was dependent on the forces that governed the environment, which may explain certain disorders. Pallas describes the concepts of health and illness in terms of balance and imbalance. The positive effects of this insulated bed technique were evaluated through the prism of epilepsy:

> On the 20th of the same month, the day before he waited for access, R... was placed in an insulated bed; the next day a seizure did not take place at the time at which it had been occurring for several months; it was not until eleven o'clock in the morning, i.e. five hours after the usual time, that the patient had his attack, at the moment when he was in communication with the ground to visit the latrines. [PAL 47, p. 281, author's translation][13]

Epilepsy was thus correlated with environmental causes, with atmospheric electricity, which reminds us of the Hippocratic principles on the links between a subject's temperament and the climate in which he or she lived:

> Epilepsy, which is the subject of the 9th observation, was really modified in the first moment of the application of isolation; later, under the influence of the humid, foggy and stormy weather, after

13 Pallas describes here the case of a soldier seen in 1846.

> having been delayed in the return of attacks for several days, it regained its control by the conductivity of the humid air, by the force of 1 habit and by the power of the cause which gave birth to it. [PAL 47, p. 302, author's translation]

Thus, Pallas concluded that neuroses were caused by the exacerbated influence of natural electricity [PAL 47, p. 351]. Paradoxically, a clinical approach that aims to eliminate care based on electricity, but built on the imaginary links between the body and natural electricity is being followed, the idea being to prevent it from penetrating the body in too large quantities. Petetin [PET 87], taking up Leeuwenhoek's calculations on the number of pores, emphasized the fact that given the available skin surface, atmospheric electricity had a thousand doors of entry.

Bertholon, in 1780, already detailed the existing unity between natural and artificial electricity. He bases his remarks on a law of conductivity:

> The electric fluid tends to spread evenly over all conductive substances [...]. It is therefore no more astonishing that the human body, immersed in the atmosphere where a very real electricity reigns unceasingly, receives a marked influence from it, than to see these same bodies placed near an electric machine brought into play being subjected to the action of the electric fluid that one has excited & collected by this means. [BER 80, p. 17, author's translation]

In the same way, he underlined the medical challenge of controlling the effects of exchanges between these scales of electricity by knowing how to identify the idioelectric parts, electric by nature, such as nerves, and anelectric parts that obeyed the laws of conductivity. Pallas quoted Bertholon on the influence of electricity on physiological mechanisms [PAL 47, p. 345]. From the perspective of a physiological microcosm dependent on a natural macrocosm, atmospheric electricity determined normal mental acts. Thus Bertholon described its action on poetry, painting, music and all the arts of the imaginary:

> It is not only the physical appearance of a person that is affected by electricity, it also has a significant effect on morale. No one is unaware that imagination, for example, is never more brilliant than in these times when the electricity of the atmosphere reigns with more control & that the soul seems to be above itself while it is barely able to find itself, that it is annihilated in these instants when the temperature is diametrically opposed. [BER 80, p. 48, author's translation]

The innate action of electricity on nervous disorders was highlighted within the emerging sciences of the brain. This innaism made it possible to make the link, in the same interpretation framework, between natural electricity, its influence on human nature and therapeutics:

> Just as the moon allows a return to the discourse on the influence of the stars through the mediation of gravitation, electricity is part of the Hippocratic revival, as an affirmation of the influence of the environment and the atmosphere on the human body on the scale of a population. [ZAN 10, p. 13, author's translation]

One of the issues at stake in these physiological and medico-philosophical questionings was the question of the materiality of the soul. If the latter was a property of brain tissue, then medicine could intervene on the ills of the mind while philosophy could understand human nature outside of metaphysics:

> [...] as yet unshared kinds of *a priori* which subsequently impose the same values on the organic and on the spiritual, then we see that there can be diseases such as madness which are from the start diseases of the body and of the soul, maladies in which the affection of the brain is of the same quality, of the same origin, of the same nature, finally, as the affection of the soul. [FOU 65, p. 99]

Provided that the soul was understood and integrated into the electrical properties of the tissues, electrical therapies became adequate treatments for disturbances of the intellect. These were then reduced to quantitative electrical disturbances with qualitative effects on nervous and mental activity. Thus, during the last third of the Enlightenment period, the search for organic links between a mental pathology, whose clinical signs could be observed in the patient's moral and intellectual behavior, and brain lesions, became systematic. In the context of the normal and the pathological, the challenge was to determine whether the expression of intellectual faculties should be described, understood and treated in relation to intracranial morphology. In an anatomophysiological context, the multiplication of dissections of human brains led to the development of the following concept: the irregularity of anatomical structures showing that these organs were both identical and different from one subject to another from the normal state. Beyond a certain threshold these irregularities became morbid or pathological. Medical questions arose from all these studies: what consequences did these organic variations have on the integrity of the faculties of the intellect? How did they influence their development? Did faculties depend only on the latter? Did they have an organic origin?

This research contributed to the shift from a medical discourse to an anthropological one where the human being was at the center of concerns. Very interested in the phenomena of mental illness and sensitivity of intellectual faculties, Gall (1758–1828) correlated dementia to the obliteration of all human faculties:

> I had to show that, until my anatomical and physiological discoveries, there was no knowledge to determine accurately the flaws, injuries and diseases of the brain, or to judge the influence of brain injuries and diseases on moral qualities and intellectual faculties. [GAL 09, p. 151, author's translation]

His four postulates, set out in the open letter to Baron M.J.F. de Retzer [GAL 69], published in the *Nueuer Deutscher Merkur* in 1798, concerned his program on the functions of the brain in humans and animals. The project to make brain physiology a description in which human moral and intellectual qualities were described as innate was stated. The dependence of these qualities on the anatomical state of the brain was defended. According to brain structure, there needed to be as many specific organs within it as there were original functions:

> The dispositions of the properties of the soul and the spirit are innate and their manifestation depends on the structure. [GAL 11, p. 3, author's translation]

This problem of the innate, in addition to developments in brain anatomy, was moving in parallel within research on the links between electricity, brain function and mental disorders. This was not insignificant, as the static nature of anatomy was matched by a dynamic conception of electrical medicine. Thus, electrotherapy brought elements of understanding to the dynamic design of the brain. It could not be a passive organ since, in addition to generating ideas, feelings and faculties, it reacted to the introduction of electric fluid:

> The material cause of these diseases, which resides either in the brain, or in the sensory organs, is an extraordinary unsound, abnormal arrangement of nerve fibers, to which ideas, judgments and appetites respond rather than to the impression of external objects. [BER 80, p. 324, author's translation]

Thus, it appeared, in the transfer of the application of electricity, from paralysis to the psychic disorders that all:

> [...] these diseases depend on too much electric fluid, [...] that in mania there is a quantity of electric fluid, greater than the ordinary & natural quantity, & consequently, that negative electricity is very

> suitable to this disease, especially if one takes care to apply it immediately to the head, & especially to the temples. [BER 80, p. 326, author's translation]

Discussions on electrical diseases were shifting towards the definition of a human nature dependent on electricity in its normal functioning. This spectrum, which extended to all kinds of behavioral and moral disorders, concentrated the problems of a medicine that constructed the articulation of the normal and the pathological from the articulation between body and mind. From the 18th to the 19th Centuries, we witnessed the polymorphic developments of a multi-faceted science of electricity, applied in parallel at different scales of the animal economy, for:

– understanding the mechanisms of the living;

– characterizing the body's ailments;

– locating and determining nervous illnesses with a double impact: physical and moral;

– inscribing and redefining the place of the human being in a nature explored and described by physics.

Gradually, the electric tool becomes a fundamental exploratory one for defining the links between body and mind, as well as a diagnostic tool. The technical, empirical and experimental dimensions, put in place as early as 1740, contradict the idea of a pre-scientific period of electricity applied to the living during the 18th Century. Here is what we read in 1773 in the *Gazette de santé*:

> The action and nature of electrical matter is still hidden, and we have only a glimpse of its properties. [...] Medicine is not a figment of the imagination, it is based on experience, and it is only through patience and time that the most valuable discoveries are made. [GAR 73, p. 33, author's translation]

The development of circuit techniques, of new machines and knowledge of the different electrical currents accompanied domestication and increased its heuristic potential. The representation of biological electricity was partly based on the analogy with the nervous fluid, on a mechanico-hydraulic model related to the materialization of the prerogatives of a soul that was emptied of its metaphysical meaning. It was based on anatomical and physiological explorations of the nervous system. Thus, from 1730 onwards, when the body was listed among the conducting bodies, electricity gradually became a tool for exploring its physiology. That is why it was used in non-pathological settings. An instrument of materialist thought, this is how we can understand this movement, which ranged from the application to paralysis to the treatment and exploration of nervous disorders. An intellectual schema particular to the end of the 18th Century, where knowledge of natural electricity influenced considerations of its effects on the human psyche, resulting in the modeling of electricity as a fluid capable of restoring reason, communication between the senses and the order of ideas:

> The dominant & sustained electrification of the sensorium usually deceives the intellectual soul, offering it images that are not communicated to it by the external sense, & observation proves that the barbaric method of castigating madmen, of overwhelming them, to the point of extinction, so to speak, of the vital principle, has had the greatest success in diminishing spontaneous electricity, by restoring the electrical relations between the sense organs. [PET 87, p. 63, author's translation]

It was the entire animal economy that was electrified, the body becoming a circuit in which an electric fluid circulated:

> After about twelve minutes of electricity in the bath, I drew a spark from the sole of the foot, oh wonder, the dyspnea ceases at once [...]. The dyspnea was to return at seven o'clock in the evening at the moment when the seizure began, one electrified the young person, six minutes being enough to make it disappear, and fifteen days passed without it reappearing but one was careful to electrify the patient morning and evening for half an hour, and only in the bath. [PET 02–03, pp. 39–40, author's translation][14]

Petetin took the example of anger, which, by forcibly expelling blood from the heart into the arteries, charged the whole body with a burning electricity. Between 1787 and 1803, he tackled galvanic rupture and experimented with various electrical treatments. He did so in order to define the extent to which these therapies made it possible to reduce the effects of disturbances that were themselves linked to personalities which were more or less subject to innate electrical characteristics. The behavior was thus materialized in the form of an electric fluid. This data on human nature was quantitative and led to a research program, which developed after 1840, on the control of behaviors by electricity.

14 The patient in question here was about 15 years old, and was characterized by a sensitive and very irritable blood temperament, according to the author.

The electro-muscular study of facial expressions from passion to reflection by Duchenne de Boulogne [DUC 62][15] reinforced the analogy with the organism's electrical mechanics, due to the potential deregulation:

> We can also see that these muscles are not only intended to represent the image of passions, feelings and affections; that certain acts of understanding can even be reflected on the face: it is, for example, with the greatest ease that the human physiognomy can be written on – and this only by the partial contraction of one of the motor muscles of the eyebrow! – reflection, the most important, the noblest state of mind, the one that seems the most abstract, and meditation, which is the mother of great conceptions, which, in some men, is, so to speak, the dominant passion. [DUC 62, pp. 47–48, author's translation]

His electrical investigations allowed him to reconstruct and delimit the muscle groups involved in each expression, but also to understand the consequences of facial paralysis on the psychology of the subject. We're talking about the *Duchenne smile.*

Thus, from the research and experiments conducted with electricity from 1746 onwards, three fundamental points stood out, in order to make the way electrotherapy developed within the brain sciences intelligible:

– the intuition that each subject carried within him a natural electricity.

This therefore already existed in its normal state, participated in the redefinition of an electrical living being and was susceptible to imbalance:

> Is the guy that Molière ridiculed in such a spiritual way to be found? Are there any imaginary patients in the usual sense of the word? We don't think so. On the contrary, we believe that hypochondria is always a symptom of a more or less serious condition, often ignored, but that a thorough examination can always reveal it. [...] For us, who treat a large number of hypochondriacs, we can affirm that nothing works better than electricity and magnetic currents, which easily dissipate all the nervous accidents that make the patients despair, even when the disease has its primitive center in the brain. For just as the brain makes its discomfort felt in all nerves, so the actions imparted to the nerves change the state of the brain, especially when these actions

15 A pioneer in the use of electricity as an instrument for physiological experiments, Duchenne, thanks to alternating current, precisely stimulated only one muscle cluster at a time, enabling him to describe and locate several conditions such as *Duchenne muscular dystrophy*.

> are produced by electric fluid, similar to nervous fluid. [DUC 62, p. 90, author's translation]

– the idea of adopting electricity as an artificial remedy designed to rebalance the electric fluid circulating within the body;

– during the 18th Century, the brain was considered the center of all nervous networks, faculties and moral behavior.

From the interlacing of these three points two conceptions developed: on the one hand, electricity was an exploratory tool of the central nervous system which allowed its organic form to be correlated with its artificial form; on the other hand, it generated human nature, in its normal state, and disorders in the event of disharmony between the functions of the different organs. This allows us to understand why electrophysiological explorations and electrotherapy developed alongside one another.

We have analyzed how, around 1750, controversial experiments were conducted and disseminated within a context of polemics, experiments and counter-experiments. Then we took this further, starting in 1770, discussing the shift towards pathologies of the central nervous system in connection with a dominant place given to the brain as the organ of materialized reason. Finally, we have highlighted, beyond electricity as an artificial remedy, the fact that, from the end of the 18th Century onwards, it became an exploratory tool for understanding physiological mechanisms. Thus, between 1746 and 1780, it was discredited because of its accidents and its failure to cure paralysis, and it made a fresh start in the emerging brain sciences. If a continuity can be traced between 1770 and 1850, in relation to research on nervous diseases, an important scientific breakthrough must be questioned: the discovery of animal electricity by Galvani. This founding episode for electrophysiology, as well as for the new deployments of electrotherapy, had to be isolated within a dedicated chapter.

What are the links between galvanism and Haller's studies on the properties of the living or with vitalism? What did the discovery of animal electricity bring to the knowledge of human nature? What did it bring to the modeling of electrical and nervous diseases?

4

Animal Electricity: Between Medicine and Physiology

Research on animal electricity, produced by the organism, refers to the dynamic metaphor of an invisible phenomenon made visible through its animating functions. In this perspective, the brain, at the center of research into the origin of this fluid, was thought of by analogy with an electricity-generating machine. These metaphors were renewed until the 20th Century.

At the end of the 18th Century, the work of Volta, Galvani and Fontana brought physiological research into the experimental era of electrophysiology. Caldani and Fontana were among the first physiologists to practice stimulation directly on animal brains. The issues at stake were many: on the one hand, it was a question of identifying the role played by the brain, the organ to which the properties of human nature were attributed; on the other hand, it was a question of confirming suspicions that the body was not only a conductor but also a producer of electrical fluid. Developments in electrophysiology thus have their roots in the late 18th Century, marked by polemics over animal electricity, considered as the material and agent of nerve conduction, and metallic electricity, but also shaken by the advent of medical electricity applied to nervous disorders. How does medical galvanism fit into this story?

4.1. Understanding life: heuristic experiments

Galvani began his work following Caldani [CAL 57] and Fontana, who demonstrated around the 1750s that muscular contractions occurred following weak electrical stimulation:

> Caldani was the first to introduce electricity as a means of producing stimulation into the study of the physiology of the nervous system, conducting extensive research on frogs to study the action of electrical stimulus on the crural and phrenic nerves. [MAN 72, p. 210, author's translation]

While Fontana [FON 57] developed the distinction between exciting and efficient causes of muscular contraction and proposed an important mechanical and dynamic analogy with the role of the spark in the ignition of gunpowder, Caldani, in an epistolary dissertation addressed to Haller, dated October 30, 1756, at the Academy of Sciences in Bologna, wrote his text *Sull'insensività, ed irritabilità di alcune parti degli animali* on November 25, 1756. By experimenting on the dura mater, he agreed with Haller's theories about his nervous insensitivity. Proponents of the vitalist doctrines [AND 13] and of the Halerian experiments, Caldani and Fontana introduced electricity as the instrument of stimulus, seen as the technical opportunity to provoke contraction, the cause being irritability. Hallerism was spread and discussed in Italy from the first translations by the Swiss physiologist, Bologna being one of the centers of these polemics. The first works on Hallerian irritability were carried out by Urbano Tosetti (1714–1768) [TOS 55] and G.V. Petrini (1725–1814), both professors in Rome. After these experiments came those of A. Cocchi (1695–1758) and G.T. Tozzetti[1] (1712–1783), the liveliest discussions taking place between 1755 and 1757 on the properties of bodies: how to define life materially? Does the material have the conditions for its setting in motion?

Many Italian, French and British naturalists hoped to unravel the mysteries of a dynamic life independent of metaphysics. When, in the 17th Century, the microscope completed static anatomy in its visible structures, it revealed an organic universe invisible to the naked eye though in motion. This was not without consequence for research on the inherent life force generated by the very organization of matter. A physiological, chemical and philosophical enthusiasm then arose:

> The nervous system and muscle revealed by the microscope prompted a reevaluation of ideas about how muscle, nerves, and the brain function. [...] Stensen also looked at brain tissue under the microscope. [CAM 16, p. 51]

1 "Giovanni Targioni Tozzetti (1712–1783), a Florentine physician, is in fact a learned encyclopaedic scholar. Director of the Botanical Garden, having travelled in Tuscany on behalf of the Florentine Botanical Society and the Regency, he is also Prefect of the Magliabechiana Library and collaborates in the Giornale dei Letterati" [MAR 18, p. 40, author's translation].

Thus anatomy gradually electrified in its physiological dimensions from the second half of the 18th Century. The treatise published in 1775 by Fabien Gautier-Dagoty bears witness to this:

> Living and animated bodies are constantly being electrified by the lungs, always in motion, and whose resting causes death. Air supplies the lungs with electrical matter, as it does in all other electrics, by stripping itself of the fire parts it contains. Arterial blood, which comes from the lungs, is the conductor of this electricity. The impulse from the left ventricle of the heart carries it throughout the brain by the flowering of the arteries, as in the Leyden jar, and from there it spreads to all the nerves. [GAU 75, p. 9, author's translation]

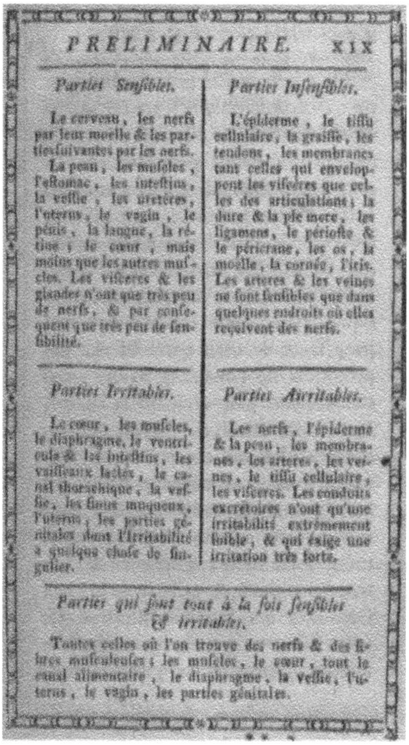

PRELIMINAIRE. XIX

Parties Senſibles.

Le cerveau, les nerfs par leur moelle & les parties ſuivantes par les nerfs.

La peau, les muſcles, l'eſtomac, les inteſtins, la veſſie, les uretères, l'uterus, le vagin, le pénis, la langue, la rétine; le cœur, mais moins que les autres muſcles. Les viſceres & les glandes n'ont que très peu de nerfs, & par conſéquent que très peu de ſenſibilité.

Parties Inſenſibles.

L'épiderme, le tiſſu cellulaire, la graiſſe, les tendons, les membranes tant celles qui enveloppent les viſceres que celles des articulations; la dure & la pie mere, les ligamens, le perioſte & le péricrane, les os, la moelle, la cornée, l'iris. Les arteres & les veines ne ſont ſenſibles que dans quelques endroits où elles reçoivent des nerfs.

Parties Irritables.

Le cœur, les muſcles, le diaphragme, le ventricule & les inteſtins, les vaiſſeaux lactés, le canal thorachique, la veſſie, les ſinus muqueux, l'uterus, les parties génitales dont l'irritabilité a quelque choſe de ſingulier.

Parties Anirritables.

Les nerfs, l'épiderme & la peau, les membranes, les arteres, les veines, le tiſſu cellulaire, les viſceres. Les conduits excrétoires n'ont qu'une irritabilité extrêmement foible, & qui exige une irritation très forte.

Parties qui ſont tout à la fois ſenſibles & irritables.

Toutes celles où l'on trouve des nerfs & des fibres muſculeuſes: les muſcles, le cœur, tout le canal alimentaire, le diaphragme, la veſſie, l'uterus, le vagin, les parties génitales.

Figure 4.1. *From Haller's work, 1755*

Moreover, the Cartesian theory of the animal machine and the heuristic analogy of anatomy with mechanics was the basis of a mechanism in confrontation, from which vitalist thoughts emerged. Baglivi (1668–1707) [BAG 00] assumed that muscles and fibers generated a force capable of contraction, independent of the action of animal spirits. The idea that tissues and muscles were capable of autonomously expressing their functions was a transformation in the way of thinking about living things. Could organic matter be reduced to a sum of internal processes? From his experiments, Haller concluded that tenderness was a property of nerves, in that only parts with nerve endings reacted to painful stimuli, while irritability appeared to be a property specific to muscle fibers, independent of nerves.

In fact, if a nerve was severed in such a way as to deprive the corresponding muscle of its sensitivity, stimulation of the muscle still produced a contraction. The same was true if the intestine or the heart was isolated from the nerves. Muscle contraction was therefore a function that did not depend on nerve action but on an autonomous property intrinsic to muscles:

> Apparently Haller's experimental procedure was simple, but it was in fact rather original and innovative. It developed some fundamental characteristics of the 'new science' that emerged in the 17th Century, such as the active manipulation of the body and the idea of the experimental apparatus, and it embodied new vistas on the scientific method. Haller carried out his experiments on many different animals in order to eliminate the accidental circumstances of phenomena and to find what was constant in a world of nature that appeared variable and difficult to investigate. [PIC 13, p. 44]

What did Haller write about electricity? In 1755, he referred to Jalabert's experiments on tendons and the possibility of drawing electrical sparks from them, while emphasizing the impossibility of inducing movement in them:

> The tendons are as little irritable as they are insensitive; no irritation with the knife, or with a mild corrosive, can cause them to convulse, nor can it move the muscle from which the irritated tendon originates. If a strong electric spark is fired from the tendons, the famous Mr. Jalabert observed that the other strongest and hardest parts of the body also gave very strong sparks. [HAL 92, p. 45, author's translation]

Haller practiced tissue irritation using various substances such as salt and alcohol. Electricity was therefore only one substance among others to test the properties of matter. Moreover, he differentiated between creating sparks and experimentally provoking movement. Nevertheless, he helped to find organic targets to stimulate movement, especially in the spinal cord, which could be reproduced with the localized application of electricity:

> The spirit of wine has barely made itself felt in Mr. Le Cat's experiments, instead of exciting a sharp pain in the skin; marks that

> the former has no sensitivity towards the latter. The convulsions are soon noticeable when the brain marrow or the spinal cord of the back is irritated. So the cause of the movement is in the last one, and brains can't do anything about it. [HAL 92, p. 72, author's translation]

Haller, in his description of the fibers that made up the animal economy, not only emphasized the heart and in particular the possibility of prolonging its beating chemically or mechanically after death, but also highlighted the involuntary properties of living things:

> This is followed by verification of its immutability in decapitated animals, animals without spinal cord and/or under conditions of artificial isolation of nerve endings. [MON 90, p. 55, author's translation]

The idea that the way an organism reacted to external stimuli was an expression of its internal strength and organization, independent of external influence, was one of the most fundamental principles introduced by Haller into 18th Century physiology and developed by Caldani and Fontana:

> Effervescence and putrefaction, fermentation and elasticity deny the supposed passivity of matter; neither should they be related to mysterious spiritual forces. [MON 90, p. 83, author's translation]

The notions of irritability, contractility and sensitivity were not without impact on the understanding of diseases, on the practice of surgery and on the development of mechanical medicine, revealing common ground for natural research and medical practice. While vitalism focused research on matter and the ways in which its properties acted in organized bodies, Haller detailed its structures and developed a fibrillar and microscopic model of muscle actions. Caldani and Fontana, used in the interpretation of their experiments, the classical notion of animal spirits, the concepts of a *vis nervosa* of an unknown nature and a *vis vitalis* produced by the brain and transported by the hollow ducts of the nerves:

> The force *par excellence* of corporeity may not be material in itself; it is intrinsic, but associated with matter in the creative act; it is in itself unknowable because experience has not grasped it, but the physiologist does not wander to achieve it, if the technical approach becomes less crude; it is obvious for the striking effects it causes at the level of macroscopic structures, but it is by definition inherent in the sub-microscopic environment. [MON 90, p. 84, author's translation]

Did Haller, Fontana or Caldani influence the approach developed by Galvani in his electrophysiological experiments and in his interpretation of the role of electrical stimulus in muscle contraction? It is less a matter of answering this question than of highlighting the existence of a context that favored the emergence of Galvani's experiments. It was an extensive scientific and philosophical investigation into the properties of matter that began in the 17th Century. Thus, vitalist theories were part of this research. Paul-Joseph Barthez (1734–1806) assumed the existence of a higher vital principle that would encompass all forms of energy. In his book *Nouveaux éléments de la science de l'homme* [BAR 58], he developed the idea that all biological phenomena in the human body are caused by a single vital principle. Matter comes to life, it is mobile, irritates, and contracts by virtue of a principle that is intrinsic to it:

> Thus I call the Vital Principle of Man, the cause that produces all phenomena of life in the human body. The name of this cause is quite indifferent, and can be taken at will. [...] There is a fairly marked gradation scale from the simplest Principles of Movement to the Principles of Life that create and maintain the organized bodies of plants and animals. [BAR 58, p. 47, author's translation]

Electricity was considered as an animating principle, even when reason ceased. Was it an instrument of life or did it generate its animating properties? Was it at all scales of matter? As long as it could be shown to exist in several forms and at different organization levels, a field of experimentation opened up on the limits between life and death, persistence or absence of movement, research on the organism. Animal electricity would then prove to be the link between natural and artificial electricity:

> In the experiments that Mr. Aldini and other observers have made on this application, not only is the vitality not quite extinguished in these parts, but also the sensory forces are violently affected in a very large number of organs by the stimulus of Galvanic electricity. At the same time as they excite the motive forces of all these organs, they multiply and exalt themselves until they form, through their cooperation, a kind of instinct, which directs and combines these movements; so as to render expressions of feelings of fear, fury, anger, etc. in a way that they can be expressed. [BAR 58, p. 254, author's translation]

Barthez's ideas were taken up by Bichat, who placed vitalism in an experimental scientific approach. If life was "*the assemblage of functions which resist death*" [BIC 00, p. 274], then the galvanic fluid could be the force that gives movement to these functions and which was extinguished in the dying process. On the basis of this reflection, the sympathizers of the doctrine of animal electricity tried to retain, to inspire, to mechanically revive this fluid beyond the visible limits of life. In the same way that vitalism was presented as a dynamic theory of matter, galvanic fluid claimed the role of the latter's animating instrument. Was the vital fluid electrical? Suspected and conceived in analogy with nerve fluid, the idea of an electric life force also emerged from previous experiments on electric fish [FIN 11]:

> But I think I must speak here of Animals that possess a spontaneous electrical virtue; such as certain fish, the best known of which are the Electric Ray, the Electric Eel (*Gymnotus electricus*), the Malapterurus (*Silurus electricus* by M. Broussonet). It is known that these fish give those who touch them shocks that are not significantly different from the electric shock that occurs in the Leyden experiment. We even managed to draw sparks from the bodies of these fish, after they had been properly prepared. The electrical force of these fish acts mainly in the efforts or movements they are seen to make to produce a shock. [...]. We can say with Mr. Humboldt that the fluid in *Gymnotus electricus* is absolutely galvanic, not electric. But it seems to me that Galvanism in Animals differs from their Electricity only by modifications. [BAR 58, pp. 325–326, author's translation]

Based on research carried out in the Mediterranean and the Atlantic Ocean on these animals, people began to think it was possible that electricity played a role in the normal functioning of nerves and muscles. This theory became experimental. Francesco Redi (1626–1697) described the electric ray's electric organ as a muscle in 1671 [RED 71] and located, inside the electric ray, the organs responsible for its discharges. In order to establish the electrical nature of the concussions caused by these animals, the physicist and chemist Henry Cavendish (1731–1810) [CAV 79] made in 1776 an artificial electric ray made of wood and then another made of leather following the analogy with the Leyden jar. His detailed experiments convinced his colleagues of the electrical nature of electric ray shocks, especially since John Walsh (1726–1795) [WAL 74] was able to produce a spark from the electric ray. This type of work contributed to Galvani's theoretical groundwork:

> The only occurrences of animal electricity readily observable in the 18th Century were the shocks produced by electric fish. [CAM 16, p. 38]

What was the impact of Galvani's experiments on the concept of animal spirits? Did the patient emerge deconstructed, "when he observed a contraction of the 'morta rana' muscles in the absence of direct stimulation" [BOS 93, p. 6, author's translation]? Or did Galvani first materialize it in the form of an animal fluid and then conceive it in terms of a neuro-electric fluid? By studying some of Galvani's experiments, we can provide some answers to these questions. If the life force emerged from the organo-chemical conditions of matter, then movement was a prerogative, with the spirit and animal economy as results:

> It made no sense to think of life as something superadded to matter – it was matter or it was nothing at all. Thelwall [THE 93, p. 41] argued that life was simply a particular state of organized matter that only needed some specific stimulus to set it in motion. The most likely candidate for being that stimulus was electricity. [RHY 09b, p. 266]

Humphry Davy, a late 18th Century chemist and physicist, took the identity of electric and nervous fluid for granted, arguing that electricity itself was in fact condensed light, providing another compelling reason to assume that the mind was gaseous, ethereal light. Chemists played a crucial role in understanding the different states of matter and the conditions of its metamorphosis. There was an assumption that life was the result of a perpetual series of singular corpuscular changes[2], causing perceptions, ideas and motion. This idea of change as the cause of the movements of matter can be found in Galvani's research on animal fluid. Two schools of thought emerged at the end of the 18th Century from the polemics between Galvani and Volta: on the one hand, Galvani and Humboldt (1769–1859), following numerous experiments, described a phenomenon as being exclusively dependent on the animal system; on the other hand, Volta, van Marumn (1750–1837) and Ackerman considered galvanic action as a general phenomenon of nature, independent of the vital force and requiring the irritation produced by the electric fluid to occur. They reached opposing conclusions, sometimes following the same protocols but interpreting the results differently. After Galvani's death, Volta unified the theories.

Thus, as early as 1780, Galvani studied the effects of electric discharge on all kinds of animals by varying experiments with different sources of electricity: positive or negative, Leyden jar and electrophoretic. It should also be pointed out that he had already adopted a sound anatomophysiological approach, practised comparative anatomy in a systematic way and studied human clinics in the Bologna hospitals, before describing his concept of animal electricity:

> In his anatomical investigation of the 1760s Galvani developed an anatomo-functional approach that already contained all the characteristic elements of his later research on animal electricity. On the one hand, he conceived the study of animals as a means to

2 In Aristotle's *Physics*, what we call motion is a very particular kind of change, the "local motion", conceived as the displacement from one point to another. This local movement is of the same kind as putrefaction, liquefaction, and corresponds to what we now call changes of state. Matter and change were inextricably linked. The animal fluid induced movements in organic matter, making it animated and allowing intervention on its different states, both normal and pathological [PEL 14].

> understand animal economy, including the functioning of the human being; on the other hand, he united the observation of anatomical structures with the investigation of functions, carried out through experimentation on the living. [PIC 13, p. 41]

Galvani completed numerous experiments on frogs between 1780 and 1791.

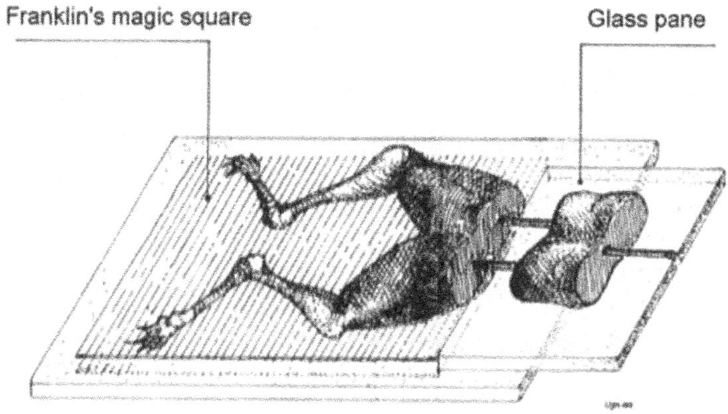

Figure 4.2. *Diagram of Galvani's experiment of December 9, 1780, designed to determine the insulating or conducting property of nerves*

But it was in 1786 that he conducted one of his seminal experiments on a frog's contraction using a metal arc [FIG 67, v. 1, p. 613]. Here is how it was described:

> The Bolognese anatomist took a frog, prepared as we have said, and hung it with a copper hook from the iron railing of the Zamboni Palace, which he lived in; from hour to hour he observed what was happening; the animal remained motionless. Impatient, he vigorously rubbed the copper hook along the balustrade to make the contact between the two metals more intimate. Galvani then saw the lower limbs contracting as soon as the copper hook touched the iron railing; several times his experience gave him the same result. [GUI 77, p. 9, author's translation]

He established three laws of muscle function based on these experiments on amphibians:

– the muscle contraction produced by nerve irritation was proportional to the minimal parts of the nerve displaced by the stimulus and the force with which they were displaced;

– irrespective of the cause of the irritation, it was almost exclusively local: it spread very poorly, if at all, beyond the place of application;

– the communication and propagation of the action of the nerve force or the induced movement to the muscle depended only on the action of the nerve.

It is possible that he suspected, from these first experiments, the existence of organic electricity. First, let's look at the experimental protocols used and then at the way in which he related his research through the study of some extracts from the English translation of his treatise *De viribus electricitatis in motu musculari commentarius* [GAL 91, 53]. Thus, Galvani used an electrostatic machine with a glass plate, placed close to the frog and from which he created sparks, then he planted a metal hook in the spinal cord and connected it to the muscle. What experimental paths, both physiological and theoretical, did he follow to conclude that there was an electricity intrinsic to organic matter? What were the analogies that facilitated his study?

Figure 4.3. *Galvani, De viribus electricitatis in motu musculari commentarius, op. cit., pl. II*

From the first reports of his experiments, he insisted on the role of chance and on the association between organic matter and metallic electricity. Which leads us to ask, what was the crucial experimental point that allowed him to bring animal electricity to light?

> When by chance, one of those who were assisting me gently touched the point of a scalpel to the medial nerves, of this frog [...], immediately all the muscles of the limbs seemed to be so contracted

> that they appeared to have fallen into violent and tonic contractions. [...]. Therefore I myself also applied the point of a scalpel to one or other crural nerve at a time when one or other of those who were presented elicited a spark. The phenomenon always occurred in the same manner: violent contraction in individual muscles of the limbs, just as if the prepared animal had been seized with tetanus, were induced at the same moment of time in which sparks were discharged. [GAL 53, p. 24]

Galvani's theory was the result of an original protocol repeated under a variety of conditions that involved connecting the limbs of a frog and a metal arc. This resulted in a variability of results, as movements were not always triggered even after a spark. This instability pushed Galvani and his collaborators, including his nephew Giovanni Aldini, to work on the concept of electrical conductivity within the living. The theme of a natural circuit in which no metallic conductor was involved emerged from analyses of the capacity of a bone, nerve or muscle to conduct electricity. Galvani initiated a series of electrophysiological explorations to determine the conditions under which this force was circulating. It was therefore necessary to differentiate between insulating parts and those which allowed easy passage and the role of metal parts:

> Now, since dry bones possess a non-conductile, but the metallic blade and the iron nails a conductile nature, we came into this suspicion, that perhaps it happened that when we held the bony handle with our fingers, then all access was cut off from the electric current in whatever way it was acting on the frog, but that it was afforded when we touched the blade or the nails communicating therewith. [...] Hence it appeared to us clearly established, what we had suspected to be true, that contact of a conducting body with the nerves is also required in order that the phenomenon should occur. But when both the body by which the nerves were touched, and the man who touched them, could be available, we applied the iron cylinder G to the same nerves, without touching it with our hands, that by this means it might be determined whether the phenomenon was to be ascribed to the man and the iron cylinder, or to the latter alone. [GAL 53, p. 25]

Galvani followed a gradual intellectual path to determine if there was electricity generated by living matter. It was a matter of first connecting the material to metallic arcs and then abandoning the use of the latter. Thus, Galvani's experimental investigations into the properties of living things followed stages where machines were used to visualize the effects of all kinds of electricity. He thus took artificial electricity as the starting point for his demonstration, with the basic question: how to isolate organic electricity?:

> the nerve conductor […] was placed as near as possible to the conductor of the electric machine […] when the electrophore was far distant from the same conductor, without any spark being elicited. [GAL 53, p. 31]

In addition, the transition to live animals was important, as was the transition from cold-blooded to warm-blooded animals[3]. These considerations can be found in the texts of galvanists such as Aldini or Humboldt. What made an animal a good test object needed to be determined. The parameters taken into account by Galvani were internal temperature and age, since it seemed that the older they were, the longer and easier they could be stimulated and prepared. In the latter, Galvani reported that "prepared animals, in whom these electric experiments were undertaken, decay and rot much more quickly than those who have suffered no electric force" [GAL 53, p. 35]. The physiologist followed an inductive approach, accumulated observations and experiments and rationalized them by recording the variation in the conditions under which they were made:

> These experiments were all performed in animals which are called cold-blooded. These things having been tested and discovered, nothing was more in my desires than to perform the same or similar experiments in warm-blooded animals, as for example in hens and in sheep. The experiment having been tried, the result was the same in the latter as in the former. [GAL 53, p. 34]

The desire to make a transition from metallic electricity to animal electricity was visible in the very structure of his book. He carried out a textual gradation that seemed to follow an experimental order of exposure: starting from environmental and artificial electrics, he confirmed the existence of organic electricity. He needed to understand the different scales of electrical activity, from the sky to matter, to begin the work on electricity and animated bodies on electrophysiology. He was therefore interested in atmospheric electricity, the effects of lightning and stormy weather:

> Having discovered the effects of artificial electricity on muscular contractions which we have thus far explained, there was nothing we would sooner do than investigate whether atmospheric electricity as it is called, would afford the same phenomena, or not: whether, for

3 "Now, indeed, these and other things having been performed and ascertained, that seemed at last to remain which promised the greatest usefulness in our experiments, that we should institute them also in living animals" [GAL 53, p. 33].

> example, by employing the same devices, the passage of lightning, as of sparks, would excite muscular contractions. [GAL 53, p. 36]

This experimental approach made it possible to deduce that the contractions produced could be produced, in turn:

– thanks to lightning;

– thanks to atmospheric electricity;

– thanks to metal conductors connected to natural electricity.

We arrive at a theoretical plan where Galvani treated the different electrics according to their origin (atmospheric, metallic) and added animal electricity as the last scale of this force: a scale that would correspond to the internalization of electricity. It was part of an integrative approach where it ranged from the electricity of the air to that of matter by experimenting with all the intermediate stages. Natural electrical phenomena were repeatedly compared with artificial and mechanical phenomena:

> Results of this sort both brought us no slight amazement and began to arouse some suspicion about inherent animal electricity itself. Moreover both were increased by the circuit of very thin nervous fluid which by chance we observed to be produced from the nerves to the muscles, when the phenomenon occurred, and which resembled the electric circuit which discharged in the Leyden jar. [GAL 53, p. 42]

Thus in 1791, the description of animal electricity represented the third stage of his experiments and corresponded to a change of scale where the different electrics communicated with each other and could all be the object of experimentation on animals. This was not without consequences for electrical therapies as medicine was able to use these different levels of electricity to interact with the body and its functions. Galvani's research ranged from experimental electrophysiology to medical electrophysiology. As a result of this differentiation of types of electricity, Galvani could not only imagine how to make them work together but also announced the existence of animal electricity:

> From what is known and explored thus far, I think it is sufficiently established that there is electricity in animals, which, with Bartholinus and others, we may be permitted to call by the general name of animal electricity. This, if not in all, yet is contained in most parts of animals; but manifests itself most conspicuously in muscles and nerves. The peculiar and non-previously recognized nature of this seems to be that

> it flows from muscles to nerves, or rather from the latter to the former, and that it traverses there either an arc or a series of men or any other conducting bodies which lead from nerves to muscles by a shorter and quicker way [...]. [GAL 53, p. 60]

The path from the nerves to the muscles thus represented a natural circuit within which electricity flowed, acting on the nervous fluid. A question arose: was it of an electrical nature? Was electricity an instrument of matter that activated a fluid generated by nerves and at the origin of actions or feelings, or was it essentially the principle of nerves? The answers to these questions had a central role in understanding animal spirits, no longer as invisible entities, but as a fluid capable of animating the machine. From these questions, Galvani concluded with an organic assemblage comparable to the compositions of the Leyden jars and made many analogies between the body and the machine:

> These things being admitted, it would perhaps be a not inept hypothesis and conjecture, nor altogether deviating from the truth, which should compare a muscle fiber to a small Leyden jar, or other similar electric body, charged with two opposite kinds of electricity; but should liken the nerve to the conductor, and therefore compare the whole muscle with an assemblage of Leyden jars. [GAL 53, p. 65]

Galvani called this "neuro-electric fluid" [GAL 53, p. 64] and attributed its genesis to the brain. He thus headed in the direction of the materialization of the faculties within this organ that became, during the 18th Century, central to thinking about the human being in behavioral, moral and intellectual aspects. From the intermediary between the soul and the body, it became the central circuit of the latter:

> Therefore we believe it equally true that electricity is prepared by action of the cerebrum, and that it is extracted from the blood, and that it enters the nerves, and that it runs through them within, whether they are hollow and free, or whether, as seems more probable, they carry a very thin lymph, or some other peculiar similar thin fluid, secreted, as many think, by the cortical cerebrum. [GAL 53, p. 67]

By placing the cerebral organ and nervous system at the center of knowledge about animal electricity, Galvani not only reinforced the idea of applying electrical therapies to nervous illnesses but also questioned the electrical nature of animal spirits, whose "obscure nature [...], long sought in vain, may perhaps appear clearly" [GAL 53, p. 68].

Electrophysiology, with an exploratory and therapeutic aim, was now inherent to research on this force present in body tissues. Galvani saw from the outset the consequences for the treatment of diseases, thus paving the way towards medical galvanism:

> [...] it seemed to me that I saw a wide field for the felicitous explanation, not only of voluntary motions, but also of unnatural and violent ones; and of various nervous maladies and their causes, as also of their relations to terrestrial and atmospheric electricity [...]. [GAL 53, p. 95]

Its argumentative intelligence needed to be underlined: starting from a typical experiment in which a paradigmatic animal model was selected, the frog, Galvani varied the conditions and technical parameters, testing the reactions of the body of the batrachian under the effects of different electrics:

> Galvani's discourse was so convincing that he tacitly overlooked a difficulty inherent in the very structure of his research, the fact of analyzing a property that belonged strictly speaking to the living, namely movement, on dead animals and with organic structures from a single animal species. [FRE 14, p. 259, author's translation]

As early as 1782, Volta[4] already spoke of the existence of animal electricity. Thus, in a letter addressed to Madame de Nanteuil, in addition to being astonished that an animal could move the electric fluid at will, condense it in one part of its body, make it rare in the other and finally launch it through conductors before bringing it back to equilibrium, he took up John Walsh's work on the eel as well as the analogy of an electric machine:

> To call it animal electricity, you have to find electricity that is essentially related to life... But does such electricity exist? Yes: it was discovered in the electric ray and in the Surinamese trembling eel that naturalists call after Linnaeus, *Gymnotus electricus*. [VOL 82, author's translation]

After reading Galvani's work on the different electrics and their actions on muscle movement, Volta repeated the experiences described there. In a memoir

4 In 1792 Alessandro Volta (1745–1827) was already known internationally. Professor of Physics, he maintained an abundant scientific correspondence with French, English and German scientists. Before the invention of the battery, he was already responsible for the electrophoretic and the construction of highly sensitive electrometers capable of detecting atmospheric electricity.

addressed to the English electrician Cavallo (1749–1809) [CAV 85], he stressed the importance of Galvani's discovery, which "contains one of the most wonderful and surprising discoveries, and the source of many others" [VOL 93, p. 10, author's translation].

But doubts based on his own experiments soon appeared to him:

– he advocated for the diversification of the animal models used in the experiments of his contemporary;

– as a physicist convinced of the importance of mathematics and formulas, he sought to evaluate and measure with his electrometers the voltage necessary to cause a contraction.

By applying electricity directly to the nerve, he found that the frog was a natural animal electrometer, since the electrical voltage required to cause the contraction was extremely low and could only be detected by ultra-sensitive devices. However, the action of the conductive arc formed by two metals and used by Galvani were highlighted, thus casting doubt on the fact that the convulsions obtained could be obtained without this metallic arc. Volta, working on his battery, came to the conclusion that galvanism referred to artificial electricity and did not activate other substances:

> By touching the two ends of this device with his fingers, Volta experienced a shock similar to that of electricity; hence he concluded that electricity and galvanism were identically one and the same thing. [LAB 28, p. 217, author's translation]

Volta thus opposed Galvani and gave a physical version of the phenomenon, insisting on the dependence of muscular contraction on a metal arc formed by two distinct metals. Thus he called metallic electricity what Galvani called animal electricity and highlighted the indispensable presence of metals in the production of electricity:

> When two different metals, he said, are in contact with each other, as a result of this contact, due to the heterogeneity of nature, there is the development of electricity, which is how he developed the notion of metallic electricity. If the metal arc for Galvani only served as a conductor for the electric current that flowed from muscle to nerve and from nerve to muscle, it was a producer for Volta. [GUI 77, p. 12, author's translation]

To this Galvani responded experimentally and concluded, as Ambroise Guichard relates, that:

> A frog's thigh with its nerve was placed on an insulating tray; a second similar thigh was placed near it; the two nerves were applied to each other; as soon as the circuit was closed, contractions occurred. Since no foreign substances were used, it was impossible to deny the frog's own electrical current. [GUI 77, p. 13, author's translation]

The first battery built in France was the Robertson battery, made of copper and zinc parts mounted in stages. The effects were declared "surprising" [CAS 03, p. 11]. Volta designed his battery model in March 1799 after first developing the so-called cup apparatus. It consisted of a set of cups, liquids and two metals. He compared his battery to the electric ray's natural electrical organ in preference to the Leyden jar and called it an artificial electrical organ. Based on the principle of the recognized effect of bringing two different metals into contact, he formed several stages of alternately silver, zinc and copper metal discs and separated them by a wet element:

> That the electric fluid, neither in the singular organs of the electric ray, [...] nor in those common to other animals, is in no way animalized as the Galvanists claimed, nor does it undergo any change or alteration: that it is pure, authentic and simple electricity, excited by new objects and devices, conventionally called electromotors, in no way different from the Leyden jar with which we have constantly compared it. [VOL 14, p. 141, author's translation]

Galvani, following the invasion of northern Italy in 1796 by the French revolutionary armies, refused, unlike Volta, to take an oath of allegiance to the new Cisalpine Republic created by Bonaparte, losing his position at the University as well as his residency and died in 1798. He therefore did not witness the durability of his discovery. He is credited with updating animal electricity, showing an infinite number of possibilities to vary the different scales of electricity between them in order to explore, understand and then intervene on normal and pathological biological mechanisms. The theories of the two scientists, Volta and Galvani, remain extremely complementary but aim to explain a series of phenomena involving life, movement and electricity differently. Where Volta[5] focused on showing the action of metallic electricity on bodies, Galvani concentrated on the electrical properties of organic matter. These controversies led Galvani, but above all Giovanni Aldini, to spread the concept of animal electricity but also to vary the

5 Concerning Volta's work, Arthuis explained in 1881 that: "In a word, the battery seemed more convenient than rotating machines and it was preferred: static electricity was partly abandoned and replaced by direct currents" [ART 81, p. 16, author's translation].

experimental protocols. Volta's objections sent Aldini back to the paradoxes of his uncle's theory and made him push back the limits of electrophysiology: while he did not hesitate to use the voltaic pile, he experimentally took the notion of the natural circuit further. Volta emphasized that the metal frame should not be regarded as merely conductive, but as a driving force and a generator of electricity. One of the metals pushed the electrical fluid through the conductors that were the animal tissues, while the other metal attracted the fluid to itself. This idea was extremely important in order to understand to what extent the development of machines that acted on the electrical fluid and the concepts of positive, negative, static or dynamic electrics were decisive for the development of medical electricity but also for the rationalization of electrophysiology. Where Volta thought as a physicist, Galvani completed a process of biologization of electricity that began after 1740. Indeed, the polemics between Volta and Galvani can be interpreted as a symptom of the links between the mechanism and vitalist thoughts:

> It follows from the fact that the two alternative paradigms, vitalism and mechanism, not yet being clearly differentiated, may have led to speculation. A very good example is the debate between Galvani and Volta on the nature of animal electricity: the two scientists had fruitful exchanges between research and animal experimentation, physics and physiology [...]. [FRE 14, p. 253, author's translation]

Galvani is credited with finding negative and positive electrics at work within the body:

> Galvani, on the contrary, thought that all animals experienced an electricity inherent to their economy, secreted by the brain, which resided in the nerves that communicate it to the whole body; he placed the main reserves of this fluid in the muscles, considering each fiber as having two surfaces and as possessing the two electrics, positive and negative, each of them representing a small Leyden jar whose nerves were the conductors; each time this bottle was discharged, a muscular contraction responded. [CAS 03, p. 8, author's translation]

While Volta's perspective was the understanding of the action of metallic electricity on matter; Galvani showed the integration of organized bodies within an environment subject to interactions between forces such as gravity, magnetism or electricity:

> Volta's pile, which originated the chemical battery and its dynamics/movement, sustains itself as a closed system; Galvani asserted the existence of animal electricity and common electricity. [SOE 13, pp. 218–219]

The theories of animal and metallic electrics both encountered contradictors, notably in the work of Fabroni[6] who in 1792 refuted them with chemical explanations. Based on the interactions between metals and alloys, he gave a new origin to electrical phenomena and stated the idea that electricity was generated by the chemical action of oxygen on metals in contact, especially when the exciter arc was formed by two different metals. It was, according to him, also produced by the chemical interactions between the fluids in the animal's body and the metal of the exciter arc when the conductor was unique. Chemical designs complemented the knowledge of metal and animal electrics. While the history of medical electricity is punctuated by periods of enthusiasm and mistrust, it is also deeply linked to the history of other sciences. In addition to its links with chemistry, Dalton's theory of atoms in 1803 represented a scientific breakthrough that enriched our knowledge of electricity. Dalton (1766–1844) symbolized simple and compound bodies with symbols to which he gave an *x*-weight of matter and presented his new nomenclature in *A New System of Chemical Philosophy* [DAL 08]. Extremely heuristic, this classification opened up a field of experimentation on ions or electrons as named in 1894 by Stoney (1826–1911). From the first cells, seen through a microscope, to atoms, a dynamic biological universe was revealed where everything was influenced by the constant collision of particles with each other, which led to research on the production of electricity by cells. At the beginning of the 19th Century, anatomy, vitalist theories and experimental research on the functions and structures of matter followed lines of convergence that led them to conceive of an electrical and then electromagnetic unit in nature.

As we have seen, Galvani considered the potential medical applications of this research on the human body. Very quickly, medical galvanism was spread by the doctors who supported its theories and was applied to nervous, mental and behavioral disorders.

4.2. Medical galvanism

Cassius, in his *Précis succinct des principaux phénomènes du galvanisme*, highlights the application of animal electricity stimulation to epileptic seizures:

> Geiger[7] again cites experiences with epileptics; he reports that he obtained changes in the paroxysm of epileptics, which then became less frequent. [CAS 03, p. 33, author's translation]

6 Giovanni Fabbroni (1752–1822), chemist, physicist and naturalist, was also a colleague of Fontana and taught in Pisa and Florence.

7 Cassius refers to the treatise *Dissertation sur le galvanisme et son application* by Geiger [GEI 02].

The German naturalist, geographer and explorer Humboldt, following the example of curious electrifying scientists, experimented galvanic excitation on himself, especially on the tongue, an extremely sensitive organ with nerves for warm-blooded animals:

> The sensation that galvanism excites in me does not seem to have the slightest resemblance to the sensation that the electric fluid causes. It is a particular kind of pain, which is less acute and less poignant than that produced by the electrical fluid. In this sensation, I distinguish a strong pulsation and a feeling of pressure, accompanied by prolonged firing. Firing is much more pronounced when the wound is covered with a silver frame and irritated with a zinc rod than when a zinc plate covers the wound and a silver clip is used to establish communication. [HUM 99, p. 328, author's translation]

Note that in addition to describing the self-effects of galvanization, Humboldt stated that, depending on the stimulant used (zinc rod, silver clamp, etc.), the excitability of the organs was increased. According to a causal effect of nerve function in relation to aroused sensation, scientists experimented with exacerbated tastes. In the context of language experiments, this gave a central role to the lingual nerves. This fact could be generalized with the nerves of the different sense organs. In 1794, two anonymous works were published on the role of the conducting arc and perceived as a response to Volta. The first was awarded to Galvani [GAL 94], the second to Giovanni Aldini [ALD 94]. Proponents of animal electricity were leading to more and more human and animal experiments to improve the means of channeling this fluid, to show its reality before gradually applying it to physical and mental ailments. Thus, in a work with an unknown publication date, Vassali Eandi (1761–1825) initiated a series of chemical tests to define the conditions and substances that promoted the strength of the conductive chain:

> Having the nitric acid-saturated water from the above-mentioned experiment, I mixed the sulfuric acid-saturated water with it, and I made the same pile with the wet cardboard in this mixture: then I had much stronger effects. [...] I had observed, with all those who are engaged in these experiments, that the action of the pile decreases in proportion to the drying of the discs, so I had no doubt that by slightly moistening the discs, I would have had small effects, and I was convinced that what happens to cardboard discs, should also happen to discs of other bodies. [VAS 02, pp. 7–8, author's translation]

From these technical trials, the scientist proceeded to galvanize the frog by trying to make the limits between lethal and therapeutic doses intelligible. Medical perspectives emerged from these considerations, taking up the idea of medical electricity, as it has been explored since 1746:

> The galvanized frog whistled from mouth to anus; a dozen shakes, however, were not enough to kill it completely. The lizard presented me with similar phenomena, it never stopped breathing, the only sign of life it still had. [VAS 02, p. 8, author's translation]

Numerous treatises proposed, in the form of a post-galvanic survey, to explore the field of the forces of nature, among which animal electricity was integrated in order to understand the amplitude of its applications on humans:

> Whether we speak of organic electrometry, or underground electrometry, we see it as the art of knowing the influence that the substances mentioned[8] have on certain individuals with special sensitivities, who not only act as vectors for electricity, but also produce it and set it in motion. [AMO 08, p. 8]

But the person who widened the influence of galvanism was Giovanni Aldini. Influenced by Haller's research, intrigued and stimulated by Galvani's, he conducted his experiments, between medicine and spectacular public demonstrations, in order to highlight the physiological and therapeutic potential of animal electricity. After Galvani's death, he set out on a European tour to defend the concept and convince the scientific community of the usefulness of using galvanism in the field of medicine. Sticking to the concept of bimetallic electricity, Volta argued that the production of muscle contractions did not occur because of the existence of inherent animal electricity but simply because the mercury used was not sufficiently purified and contained traces of at least one other metal. In response, Aldini initiated a series of experiments in which no metal was used to induce muscle contraction, thus employing totally organic conductive circuits [PAR 04, p. 578]:

> Aldini was thus a key player in a battle that led to a major neuroscience paradigm shift, that is, the transition from the 1500-year-old galenic concept of *animal spirits*, which were thought to be stored in the brain ventricles and coursed within the hollow nerves to induce muscular contraction, to that of *animal electricity*. [PAR 04, p. 583]

He also appeared to be one of the first to treat patients with personality disorders by using electrical stimulation[9]:

---

8 The substances covered by this investigation program were caloric substances, fire, light, magnetism, electricity and galvanism [AMO 08, p. 3].

9 These experiences were rediscovered by Gustav Fritsch (1838–1927) and Eduard Hitzig (1838–1907) during the 19th Century.

> Aldini also treated patients with personality disorders and reported complete rehabilitation following transcranial administration of electric current. Aldini's work laid the ground for the development of various forms of electrotherapy that were heavily used later in the 19th Century. [PAR 04, 576]

Aldini's trials, notably with Pinel, may have played a role in the design of brain stimulation applied to psychiatric disorders. His attempts to relieve melancholy highlighted the technical possibility of a treatment based on the notion of the reversibility of "madness":

> Aldini says he also saw some relief, but it was momentary and fleeting, in people deprived of their reason, and plunged into a state of foolishness. These experiments were attempted in several hospices in Paris, and it is believed that a slight improvement in the condition of the patients during Galvanization was noticed. [CAS 03, p. 35, author's translation]

Aldini's work is part of the path of medical electricity within brain structures, but it should not make us lose sight of the singularity of the applications of electricity to the nervous system and mental illness in each era. The fact that this can be seen as a common thread does not erase the technical obstacles, the developments of the different medical schools and the emergence of more or less favorable contexts:

> However, it should be noted that Aldini's stimulations, although applied to the head, probably could not stimulate the cerebral cortex through the bones of the skull with sufficient precision and intensity to cause peripheral muscle contractions which were rather caused by the diffusion of electric current in the muscles surrounding the stimulation (especially the facial muscles). [MIC 13, p. 320, author's translation]

Because of the side effects of shocks and the pain that followed, Aldini no longer applied these applications to the ears, preferring the top of the skull. In addition, he was also a self-experimenter which led him to an idea of the side effects of galvanizing the head, such as insomnia:

> First, the fluid seized a large part of the brain, which experienced a strong jolt, a kind of shaking against the walls of the bone box. The effects increased even more when I led the arcs from one ear to

> the other. I felt a strong action on my head, and a prolonged insomnia for several days: a phenomenon that was also experienced by those who took part in these trials. [ALD 04, p. 122, author's translation]

In 1804, Aldini offered a summary of the issues related to galvanism:

– galvanizing is independent of metals: "muscle contractions were excited by the development of a fluid in the animal machine, conducted from nerves to muscles without the assistance and action of metals" [ALD 04, p. 1, author's translation];

– the application of galvanism serves to excite the life forces;

– these applications can be useful in medicine, especially as a stimulant.

The last consideration explains his trials on patients diagnosed with melancholy. Effects of galvanization on behavioral changes, associated with the functions of the various target organs, were regularly reported. Here is an example described in 1803 of the effect of these treatments on the digestion organs and appetite:

> The intestines, which are connected to the mouth by means of a metal chain, undergo sharp contractions in the region of the stomach; these contractions felt by Cassius were followed by a singular energy, which gave him a great appetite for several days; several physicists who have repeated this experiment on themselves have had more or less similar results. [CAS 03, p. 24]

Contemporary to the work of Galvani and Aldini, doctors were refocusing on medical galvanism and supporting physiological research. Although it seemed to be accepted that galvanic current was not essentially different from electric current, some people claimed that it had the best effect on the animal economy. This hypothesis, based on its intrinsic origin in tissues, gave it a central role in the development of electrotherapy:

> Galvanism penetrates the nerves more easily and more deeply than electricity; it follows them as its best conductors, while the electrical fluid seems to spread more evenly over the surface and throughout the mass of organized bodies. [GRA 01, p. 22, author's translation]

The following observations reinforce this theory:

– as we have already seen in Humboldt's self-experimentation, the particular irritation of the nerves of the taste organ, manifested by an exaltation of the phantom flavors felt, supported the fact that the nerves conducted the galvanic current which, in turn, exacerbated the function of the nervous network on which it was directed. The same type of experiments on the optic or auditory nerves were conducted by Grapengiesser (1742–1799) and gave rise to intense visual sensations.

– the effects of galvanism are intense even on organs that have recently been separated from the animal's body and are still considered to possess their vitality.

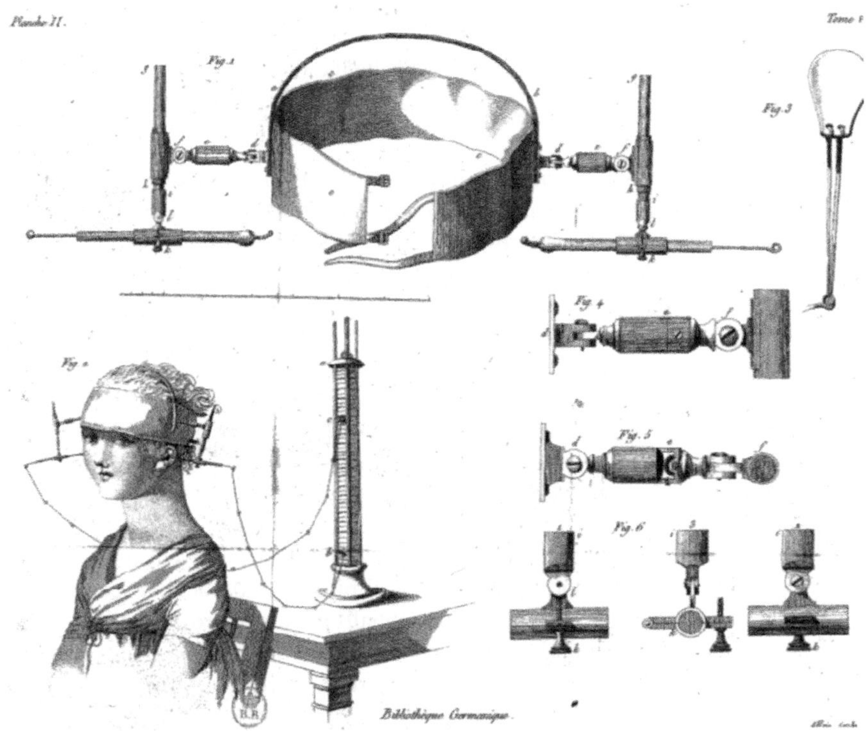

Figure 4.4. *"A machine that keeps the conductors in the ears. It was constructed in such a way that conductors could be guided in all directions. It consisted of a whale arc a. b., both ends of which passed through a ribbon c c c, which was looped behind the head. Each end of the arc supported a brass arm with two f hinges, and ended in a hollow cylinder, into which a glass tube was inserted, which could be lowered and raised at will. Another hollow cylinder with a pressure screw k was held by a hinge l to the glass tube g h, the screw k serving to increase or decrease the capacity of the cylinder more or less depending on the size of the conductor passing through it." [GRA 01, p. 2, pp. 133–134, author's translation]*

Thus, galvanic fluid appeared to spread much more easily than the electrical fluid at the heart of organic structures. It acted powerfully on the nervous system, causing violent concussions, although its action on the electrometer was weak. It accelerated blood circulation and acted on the organized bodies as a powerful stimulant. For this reason, "all these properties should make it more stimulating and more effective in the treatment of diseases" [GRA 01, p. 26, author's translation]; especially in desperate cases of asphyxia where the risk of convulsions was very high:

> for it is possible that the galvanic current conducted by the vessels that pass through the skull could make its way into the brain and along the spinal cord, which it follows as its best conductors. [GRA 01, p. 39, author's translation]

To what extent did physiological research influence medical action? Galvanic research on the link between the different organs and the nervous system enabled the design of therapeutic interventions, sometimes urgently needed. Aldini pointed out that medical galvanism could stimulate the nervous system, highlighting organ functions and generating effects on behavior and mental disorders. His work permeated research on the links between temperament, nervous disorders and the brain. The temperament related to neurasthenia, for example, was considered in terms of the imbalance of nervous energies. Thus, Beard (1839–1883) [BEA 69], an American neurologist, described the concept of neurasthenia in 1869. He considered that each individual was more or less naturally rich in the nervous reserves he/she possessed. Some would then have only a minimal energy reserve, while others would be abundantly supplied. It is interesting to note that Beard used analogies with society and the financial world to describe the functioning of neurasthenia, and through it, the functioning of the nervous system:

> Among the special exciting causes of neurasthenia may be mentioned the pressure of bereavement, business and family cares, parturition and abortion, sexual excesses, the abuse of stimulants and narcotics, and civilized starvation, such as is sometimes observed even among the wealthy order of society, and sudden retirement from business. [BEA 69, p. 218]

The model of mental and nervous health based on the concept of energy accompanied the development of new electrophysiological techniques and the hope of deriving therapeutic effects from action on brain electricity. While it was accepted that "the science of galvanism is one of the branches of physics; and the galvanic fluid or agent is that species, that modification of electricity which is developed by means of chemistry" [LAB 28, p. 213, author's translation], medical galvanism stood at the confluence of physiological research into the principles that generate life and medicine. Its application was experimental and clinical. From the moment

Galvani, Aldini or Humboldt developed experimental protocols based on organic electrical circuits, showing that nerves and muscles could be excited *post-mortem* without the help of metals, electrophysiology linked its destiny to electrotherapy. Where electricity ended, galvanism began. Although some countries were developing these techniques according to medical and empirical singularities or more experimental and physiological singularities, at the basis of the application of galvanism to nervous diseases was the notion of cerebral localization of faculties linked to the idea that mental physiology depended on the anatomy of the brain:

> According to the law of electromotive forces developed by the contact of different bodies, is it not possible to believe that the ashen brain substance and the white substance contribute powerfully, by the contact of their extremely large surfaces, by the difference in their composition and probably also by the difference in their temperature, would it not be permissible, I say, to think that this contact contributes to the production of a considerable quantity of electric fluid, modified in such a way as to give rise to such admirable phenomena as are seen with varying degrees of development in all beings who are endowed with a brain? Wouldn't the cerebrospinal fluid, recently discovered by Mr. Magendie, be to the elements of the brain cell what the fluid interposed to the elements of the metal cell? [LAB 28, p. 47, author's translation]

Petetin also fed his research with the understanding and treatment of convulsive or magnetic diseases through galvanism, which he considered as a variable of the electrical fluid. He took up one of the theoretical pillars of galvanic medicine:

> The electric fluid that is spread in space and enclosed in the pores of the body is everywhere in equilibrium with itself; if some cause breaks it, it immediately recovers. [PET 02–03, p. 10, author's translation]

Based on this idea, he described the advantages of using galvanism when the quality of the atmosphere did not allow the use of artificial electricity, especially in cases of catalepsy or hysteria:

> galvanic cells were used with the same success; the patient could not distinguish between the shock she received from that of the Leyder jar, and cataleptic or convulsive attacks that gave way just as quickly. [PET 02–03, p. 55, author's translation]

While galvanic electricity had advantages for medical use in magnetic diseases, its relationship with static electricity contributed to a global theory of the influence of natural forces on each other:

> Two electric forces are thus developed in the galvanic column, as they exist between the surfaces of the Leyden jar; that of zinc is similar to the electricity of glass, that of copper to the electricity of resin; the first is centrifugal, and the second centripetal; they are born of each other, support each other, and subsist continuously. [PET 02–03, p. 14, author's translation]

The question of shock treatment, which arose with the use of the Leyden jar in static electricity machines, arose again with the application of galvanism. The idea was that only shock could calm the crises permeating 19th Century medicine. Shock therapies did not arise from discoveries related to electricity, but preceded them. In fact, from the moment that the ills traditionally linked to the soul descend into the brain structures, medicine started looking for ways to restore lost mental health by intervening on the brain. The heuristic value of the research on galvanism was in particular in joining "madness" and reason, localized within the cerebral apparatus, to criteria of quantifiable variations of animal electricity and nervous fluid. In medical galvanism, the project of a dynamic retranscription within the brain space of the state of brain functions was nestled. To the extent that human cerebral functioning became an object of study and medical intervention, then the question arose as to the nature of the interventions that could be made there, particularly with a view to "curing" disorders of the mind:

> it is not enough to make the electric fluid pass peacefully through the various parts of the body, even to carry it directly, by means of a special conductor, into the nervous channels: it is often indispensable to give the sick system concussions, shocks proportionate to its state. [LAB 28, p. 26]

Furthermore, galvanism reinforced the body's status as a conductor of electricity, since the human and animal anatomy, composed of fibers, nerves and muscles, seemed to offer the best natural structure for conducting this fluid. Having demonstrated that the vital harmony existing between all parts of the body depended on the communication that the nerves established between the regulating organs and those that were subject to them, it was necessary to show that the fluid by means of which this communication took place behaved absolutely as the galvanic fluid did. For example, experiments were conducted in which the sectioning of certain nerves connecting the brain to organs such as the stomach prevented normal functioning:

> If, by operating on the section of an essential conductor of the vital principle, of a nerve that leads to the stomach, for example, I prevent the flow of the southern fluid to this viscera, and that of the boreal fluid to the brain; if I thus suspend the exercise or even if I only reduce the activity of the stomach's functions, I will have proved, I believe, that the life of the stomach is maintained mainly by the action of the brain; that this action is transmitted by a nerve and that this nerve is the conductor of an agent which is and can only be the nervous or vital fluid, since on the interruption or non-continuation of its current depends either the weakening or the interruption or the non-continuation of life. [LAB 28, pp. 72–73, author's translation]

The challenge of this research was to show the identity of the galvanic fluid with the vital fluid, which in a medical perspective allows us to think that stimulation by galvanic currents contributed to the rebalancing of nervous disturbances.

The anatomophysiological diagram, inscribed in the context of the history of galvanism, brings us back to the analogy between the brain connected to all the organs subject to its functioning and a machine producing electricity:

> Some philosophers believe that the brain itself is a kind of galvanic cell, in which the nervous influence is particularly generated, and with the help of an intermediate conductor (the nerves) is transmitted to all parts of the body. Others argue that nerve fluid and galvanic fluid are identically the same (though differently developed) because of the similar effects produced by their influence on vital functions; and the numerous experiments on the eighth pair of nerves of various animals would support this view. [LAB 28, pp. 223–224, author's translation]

Galvanism not only opened a heuristic path to neuro-electrophysiology but also to the knowledge of chemical mechanisms at work in organized bodies, thus linking chemistry, physics and physiology:

> Some physicists have reported the galvanic phenomenon to an electrical action; others to a chemical action. The last hypothesis was considered the most likely, because it satisfactorily explains the excitation of the galvanic cell by the action of the dilute acid decomposing the metal surfaces and thus developing the galvanic fluid. It has also been noted that the galvanic cell is both an electrical and a chemical device. [LAB 28, p. 226, author's translation]

Electrotherapy rooms were open in many European countries and Italy saw the establishment of psychiatric institutions offering the systematic application of galvanism:

> A modern traveler, who recently visited an institution developed in Aversa, Sicily, to receive the insane, reports, according to what the doctors of that institution told him, that several hundred people there have been cured of melancholic madness by the application of galvanism. [LAB 28, pp. 229–230, author's translation]

The asylum in Aversa is considered an innovative institution in terms of both care and treatment of the sick. Its establishment, between 1806 and 1815, took place during the period of *decennio francese* [ALB 15–16], under the influence of two major figures of the Napoleonic Empire: Joseph Bonaparte (1768–1844) and Joachim Murat (1767–1815), the Emperor's brother-in-law. Intrinsically linked to the French presence, Aversa's establishment was made official by decree on March 11, 1813. Influenced by philanthropic medicine, it responded to specific problems related to mental illness. In the Hospital for the Incurables in Naples, 400 people diagnosed with insanity were mixed with the wounded and infirm and "The winter of 1812 was marked by a very high mortality rate (a quarter) for the madmen admitted to the *Incurabili* hospital" [PAR 58, p. 330, author's translation]. This mortality rate pushed the authorities to find a solution and could be seen as the prelude to the creation of the *Manicomio* of Aversa mental hospital. France played an important role in the diffusion and institutionalization of galvanic treatment, through an empirical clinical practice where disorders as diverse as head tremors or melancholia became therapeutic targets. The conceptual common point between these disorders was the idea that, at the origin, there was a nervous and/or electrical imbalance which needed intervention. In the same way that surgery reduced a tumor, the galvanic doctor intervened on ailments whose origin was in the brain:

> Tremors of the head, hands and legs, resulting from a high degree of nerve irritation and an irregular flow of nerve influence from the brain to the organs of movement, have generally given way to galvanic influence. [LAB 28, p. 278, author's translation]

Among the targeted diseases, catalepsy[10], sleepwalking, epilepsy, "general debility" or St. Vitus' dance were all treated with galvanism. In turn, neurological, psychiatric and/or psychological, the fact that medical galvanism was applied to such a range of disorders implied several facts: on the one hand, it was an active principle general enough to be able to bring relief to many pathologies. On the other

10 Richer considered that the electrical treatment applied to catalepsy "has the most analogies to muscle hyperexcitability" [RIC 85, p. 299].

hand, it was designed according to an analogy between the importance of the nervous fluid in the body and that of the animal fluid whose circulation was to be restored, increased, or channeled. This point clearly made galvanization an exploratory tool, which, by alleviating certain ills, provided a body of knowledge about how the individual functions in his/her physical and mental dimensions. Thus, the different areas of biological and mental life became electrocentric.

4.3. Electrocentric life

The polemics around the nature and origin of electricity in the living were extremely heuristic for the understanding of physiology and, in line with Haller's research, for the materialization of animal spirits in properties of matter. In the 18th Century, irritability and contractility were still attributed to animal spirits, invisible entities derived from galenic medicine that, from the brain to the muscles, produced muscle contractions, while in the opposite direction – from nerve endings to the brain – they produced sensations:

> Boerhaave thought that this fluid had a material nature even though it did not produce any increase in muscle size (a fact found by Glisson). In fact, the nerve fluid was 'imponderable', i.e. that is, without a perceptible mass, in a way similar to other natural fluids such as light electricity, magnetism and heat. This fluid circulated in the body and especially in the nervous system, and was supported by the analogy with blood, whose circulation had been established, thanks to the work of William Harvey. [PIC 13, pp. 43–44]

Beccaria (1716–1781), author of the differentiation between natural electricity (meteorological, torpedo fish) and artificial electricity [BEC 53] channeled by machines or bottled, was very cautious about the electrical nature of animal spirits and the existence of organic electricity. However, his work undoubtedly contributed to opening an experimental research program to take this questioning further, nodal in nature, to understand the differences between the two electrics.

Let us digress to discuss a dynamic and chemical model of the organicity of animal spirits, developed by Thomas Willis (1621–1675) in the 17th Century, notably in his treatises *De Anima Brutorum* [WIL 72] and *Cerebri Anatome* [WIL 64]. The point is not to assert a link between the anatomist's research and the history of electricity, but to show how he laid the foundations for a new way of thinking about the action of animal spirits in the normal framework as well as in morbidity, particularly in the field of mental disorders. Indeed, the latter were no longer attributed to supernatural or religious causes by Willis but to chemical and circulatory disorders affecting the relationship between nervous fluid and animal spirits. In *De anima brutorum*, he developed the model of a soul, enraged in a brain disturbed by the alteration of these links resulting in a state of frenzy or the pure and simple evaporation of the latter:

> This juice, by its fluidity, diffuses the spirits throughout the entire nervous system, while by its viscosity it keeps them in a certain order [...] and continuous series, so as to prevent their dissipation. For it seems that, without such a juice, animal spirits could not stay within the boundaries of the nervous system, but would vanish into thin air. [WIL 64, p. 95, author's translation]

The animal spirits, conceived in physical terms of rays, impregnated with the chemical disturbances of the blood, confuse their "irradiation" from which arises the correlation of mental symptoms and chemical properties as well as a dysfunction of sensorimotricity:

> The result is an 'explosive copula', which properly enables muscle contraction, similar to the deflagration of gunpowder; if the explosion is amplified and multiplied, the discharge evolves into spasm, shaking, convulsion: thus in epilepsy, thus in frenzied madness [...]. [CON 78, p. 206, author's translation].

Delusions and ramblings like convulsions were due to explosive chemical issues. A microscopist of the soul, Willis initiated a tradition by stating that:

> animal spirits made their way into the brain and the chemical changes they were subject to controlled everything that happened in our lives, from emotions to sleep, perception and locomotion. [ZIM 14, p. 262, author's translation]

Willis, by correlating animal spirits with chemical or circulatory effects, contributed to the evolution of their conception into physical concepts. In the same way that the first medical and philosophical reflections on the cerebral localization of the faculties of the intellect engaged anatomists and galvanic physicians to make the link between the circulation of animal electricity and the individual mental and nervous state, the materialization of animal spirits opened up to analyses of physiology in terms of circulatory and physical movements.

Thus, Diderot questions in *D'Alembert's Dream* [DID 12] the nature of the structures enabling the gush of consciousness from matter. The brain would be like a spider in the center of a web. The metaphor of the canvas is reminiscent of the concepts of wires, which are driven by the brain:

> On the one hand, with La Mettrie [LAM 65] and Helvetius [HEL 59], the psychological question of the nature of the soul or spirit and their relationship with the body is privileged, and from there we constitute a science of Man. […] On the other hand, with d'Holbach [HOL 70] and Diderot, the approach consists of starting from the examination of matter, or the 'great entirety of nature' (d'Holbach), or the presupposition of the unity of all things (Diderot), with the related themes of the heterogeneity of matter, its essential movement, sensibility and life. [BOU 96, p. 24, author's translation]

However, from Galvani's experiments, we can isolate three known causes for the production of muscle contractions in living animals:

– a concentration in the muscle cell induced by the link with the brain: this first cause would make it possible to understand voluntary movements;

– an overload that can be qualified as accidental, it would be caused by an external agent or irritation forcing electricity to descend violently from the brain to the muscles: this second cause would be at the origin of the reflex movements;

– an experimental discharge produced following the application of an external agent to the nerve or brain and determining the flow of electricity from the inner surface of the muscle through the nerve and back to the outer surface of the muscle.

These three causes have in common the postulate that all movement results from the circulation, voluntary or involuntary, of an electricity already naturally present in the body. As mentioned above, Galvani was aware very early on of the impact of his discoveries on medicine and knowledge of human physiology. So he experimented on human body parts:

> This morning, in our hospital of St. Ursula, in which the Professor of Surgery is the learned and my most distinguished colleague Doctor Gaspar Gentili, excellent master of surgery, I tested, with my customary devices, an amputated leg and arm, immediately after the operation, in the presence of the aforesaid professor and other physicians and men of learning, and the flexor muscles of the thumb and of the adjacent digits [fingers] were seen to contract, both of the

> hand and of the foot, and in consequence the aforesaid digits to move. [GAL 53, pp. 96–97][11]

All of this led him to conclude, with regard to organized bodies, that:

> 1. The muscle is a Leyden jar. 2. The nerve acts as a simple conductor. 3. Positive electricity flows from the inside of the muscle to the nerve, and from the nerve to the muscle, through the exciter arc. [GUI 77, p. 11, author's translation]

Humboldt contributed to the electrical representation of organic actions and movements:

> Humbolt has proven that contractions can be excited in an animal by placing nerves and muscles in certain positions with respect to each other, without using any metallic substances. Based on this principle, a battery was formed consisting of alternating layers of muscle fibers and portions of the brain, separated by a porous body soaked in salt water. [LAB 28, pp. 223–224, author's translation]

This prefigured the open electrophysiological research following the enunciation of the concept of animal electricity:

- exploring nerve mechanisms;
- the limits of life and death[12];
- the anatomophysiology of the nervous systems of the smallest organisms:

> This irritation, exerted on the nerves that distribute themselves to muscles or organs with muscle fibers, excites contractions and

11 Galvani, L. *Commentary on the effect of electricity on muscular motion, op. cit.* pp. 96–97. "Galvani had nothing more at the heart of it than testing an arm and a foot cut off by a skilful surgeon in the public hospice of St. Ursula, and for this purpose he exposed the nerves and muscles and placed them on a frame, so that the nerves communicated with mercury and the muscles with warm water [...] but the contractions of the foot were much stronger than those noticed on the hands, either because the nerves in the feet were more considerable, or because the hand, put to the test, had been affected by a longer illness" [CAS 03, pp. 18–19, author's translation].

12 Here is an example of a resuscitation experiment on a linnet performed by Humboldt: "She had already closed her eyes, she was lying on her back, and the mechanical irritation of the tip of an excited pin near the anus produced no effect. I hastened to place a small blade of zinc in the beak and a small piece of silver in the rectum, and immediately afterwards communication was established between these metals with an iron rod. What was my astonishment, when at the moment of contact the bird opened its eyes and stood up on its legs flapping its wings! It breathed again for six or eight minutes, and then it exhaled quietly." [HUM 99, p. 333].

> movements that are very obvious: on the nerves intended for certain senses, it produces impressions similar to the sensations themselves; it also increases secretions by altering the nature of the fluids concerned; this sort of irritation has still sometimes been used successfully to bring asphyxiated animals back to life; finally, naturalists who have made use of it in the examination of the structure of insects and worms, have succeeded in discovering, with the aid of this new means, the hitherto unknown nervous system of several of these animals. [LAB 28, p. v, author's translation]

Presented as fundamental for the study, exploration and stimulation of the animal economy, organic electricity by acting directly on the nerves could act and modify the functions of the organs. Electrophysiology thus presented itself as a new means of exploring the nervous system through a long history of intertwining physiological and clinical research. Humboldt was involved in research on physiological continuity from animal spirits to nerve fluid to electricity. Galvanism was then considered as the causal agent of the action of nerves on themselves and on the environment with effects on sensations, ideas and movements. While he refuted, as Volta did, the assimilation of nerves to Leyden jars by designating this functional analogy as the result of the imagination, he considered galvanism as an experimental theory capable of making it possible to understand the principle of life. In 1799, he presented a series of experiments conducted in Paris on excitatory substances such as metals, carbonaceous bodies and liquid or moist conductive substances. Thus, in addition to frogs and lizards, he used substances such as mercury, while diversifying metals[13]. He also classified them in tables that aimed to arrange all the metals and substances that could interact with electricity.

With Galvani's experiments, the electricity entered the depths of the tissues, it became intimately linked to them, participating in their composition. An internalization of electricity was witnessed as electrophysiology and electrotherapy was able to make people interact with each other using electricity applied and dosed by machines developed in the 18th Century and electricity specific to living beings. Thus, the latter became a favored instrument to explore the principles of life:

13 "When animal organs are brought into immediate contact with mercury, there is no need for a high degree of irritability to occur; this is proven by the many experiments I have done on this subject, which many people have witnessed. [...] On a cool morning in the spring of year 3, I prepared a common lizard (*Lacerta agilis*), whose thigh I placed on a glass slide; I touched an exposed muscle with a dry finger of my right hand, and the nerve with a silver pincer which I held in my other hand: there were very strong contractions; they were especially strong when the contact began between the finger and the muscle" [HUM 99, pp. 55–57, author's translation].

Excitability is generally a property of animal and vegetable substances, a prerogative enjoyed by organized matter; galvanic irritation obviously acts only on organic parts with sensitive fibers: it presupposes a reaction of the vital force, and belongs to what Hufeland calls vital action. [HUM 99, p. 13, author's translation]

TABLE des substances conductrices et des substances isolantes du fluide galvanique.

SUBSTANCES actives dans la chaîne galvanique, excitatrices et conductrices de l'électricité animale.	*SUBSTANCES inactives et isolantes dans la chaîne.*
Tous les métaux à l'état de régule.	Les métaux oxydés.
Les sulfures métalliques, et les minéraux contenant des métaux non-oxydés.	Les sulfures métalliques, et les minéraux contenant des métaux oxydés et diversement colorés.
Le charbon végétal.	Toutes les espèces de gaz.
Le charbon minéral.	Les os des animaux dans l'état naturel.
Le graphit.	Les poils des animaux.
La blende charbonée.	Les feuilles et les tiges des plantes recouvertes de leur épiderme.
La pierre de Lydie de Naïla.	Les fibres du bois.
Le schiste alumineux.	Le verre, même chauffé.
Le schiste inflammable.	Le succin.
La manganèse grise et noire.	Le blanc d'œuf durci.
La chair musculaire, les membranes, les nerfs, les ligaments et les vaisseaux des animaux frais ou cuits, rôtis ou desséchés.	La cire.
Les morilles et les champignons exhalant, dans l'état de putréfaction, une odeur cadavéreuse.	Tous les sels secs, et les substances dépourvues de carbone.
Le blanc d'œuf.	L'huile.
L'eau, le sang, le suc des plantes.	Les raisins.
Les parties de végétaux contenant du tissu cellulaire frais, mais dépouillées de l'épiderme.	Les gommes.
L'esprit de vin.	La flamme.
Le vin, la bière.	Le vide.

Figure 4.5. *Humboldt offers a classification table of the conductive and insulating substances of the galvanic fluid [HUM 99, pp. 175–177]*

Humboldt took up the animal model of the frog, which became the paradigm of a natural electrometer:

> I detached the thighs of a frog which heralded a high degree of natural excitability; I promptly uncovered the crural nerve on the right side, and placed it and the whole limb on a dry glass blade; I touched the nerve and muscle at the same time with a piece of fresh muscular flesh attached to a handle of Spanish wax, which is an insulating substance, and there were, to my astonishment, strong contractions. [HUM 99, p. 30, author's translation]

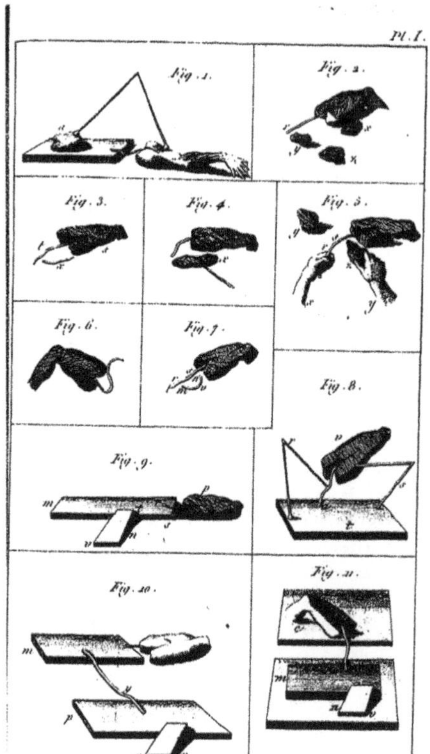

Figure 4.6. *During his experiments in 1799, Humboldt, in addition to frogs and lizards, used chemicals and diversified metals [HUM 99, p. 18, pl. 1, fig. 1 to 21]*

By exploring the interactions between human tissues and electricity, the scientist wondered whether these tissues could exhale a conductive atmosphere that would condense around them.

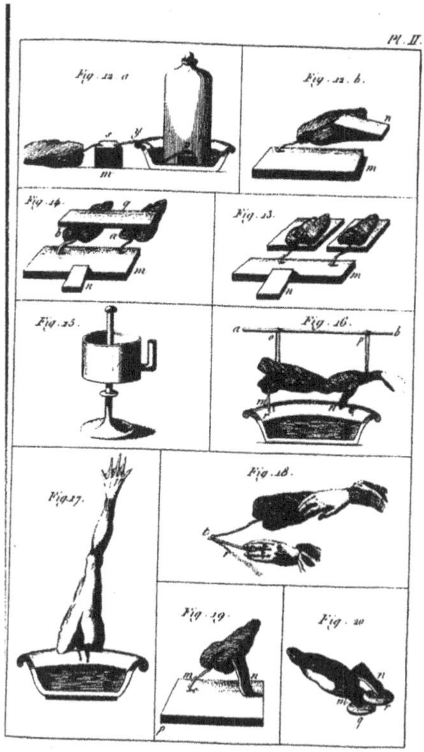

Figure 4.7. *Humboldt also switched from cold-blooded animals to human body parts such as the arm shown here [HUM 99, p. 20, pl. 2, figs 22–30]*

So, do all bodies have the same amount of electricity in their tissues? Do they all have the same ability to fit into the experimental design of a natural circuit? Is this variability intrinsic or extrinsic and due to atmospheric electricity?

> While composing a chain of several people, I noticed a phenomenon that seemed too striking for me not to report here. With seven or eight people together, I sometimes observed that the muscle movements only took place when one of them in the chain came out; and the non-conducting person was often only discovered when they were all successively taken out of the chain. I have seen cases where this person wet his hands unnecessarily and without making them conductive, although in other circumstances this was very effective, as well as watering the floor on which they are placed. I also found the same person, at certain times, conductive, and at other times insulating. [HUM 99, p. 151, author's translation]

Weren't these variations in conductivity, reception and the fall of electricity caused by the internal state of the tissue? While animal electricity was part of the properties of the material, it followed the pathological changes among others. By virtue of the tissue-chemical-animal electricity connection, different degrees of conduction needed to respond to pathological processes and changes in animal economy. Potentially, this made galvanism a diagnostic tool:

> Could one not believe that parts whose rigidity, in the healthy state, opposes any shaking, are so modified by a pathological state, or by other circumstances, that their increased elasticity makes them capable of propagating the shock they have received? [...]; for almost any irritation of muscle or nerve fibers must certainly relate to chemical laws, to changes in the mixture of the material. [HUM 99, p. 160, author's translation]

Humboldt [HUM 99, p. 327], in a process of rationalizing galvanism, drew an axiom from his experiments of excitability: if the force of irritation x is equal to the product of the stimulant y and the incitability z then:

$$x = yz$$

Thus, the stimulant or irritant force caused the increase of z while leaving y invariant. Using experiments to explore the potential of animal electricity opened up new avenues of research into the fate of this energy after the cessation of life. Beyond the impact of medical galvanism on future resuscitation medicine, while the organs remained excitable after voluntary mobility had ceased, electrophysiology led us to question the transformation steps that this electricity caused on deterioration of the body, as well as on all *post-mortem* modifications:

> That after the cessation of the life of all organs, there would always, and necessarily, be a continuation of the development of electricity, as well as of all the elements of life; but that when these elements can no longer contribute to the exercise of life, as it was before, modifications are established in the electromotive forces which are in an uninterrupted state of action, as well as in their effects? These forces would, then, take a new direction; different chemical decompositions and compositions would take place; other bodies would appropriate these new compounds and become an essential constituent part of them, and soon they would be annihilated to the last trace of such an admirable existence. [LAB 28, p. 50, author's translation]

Experimental protocols such as galvanizing applications were extremely polymorphic. We can consider that:

> There was no fixed meaning for galvanic experiments at the end of the Enlightenment. What galvanic experiments might mean tended to depend on where they were conducted and by whom. Even then, they could be differently read by the various participants. Significances might be quite local as much as universal in their scope. Galvanic experiments often meant different things to different groups and individuals. Galvanism's radical and materialist implications were certainly not self-evident. They had to be argued for – both by supporters and by detractors. [RHY 09b, p. 266]

According to the Halleian paradigm, the galvanic agent was willingly regarded as a stimulant. Based on this principle, the types of experiments were diversified with warm-blooded or cold-blooded animal models. Aldini, for example, connected the head of a recently killed ox, pushed a finger moistened with salt water into its ear, and with the other hand supported a prepared frog so that its spinal cord touched the top of the ox's tongue. He observed very strong convulsions in the frog. For the success of his experiment he used a battery of 100 pieces of silver and zinc separated by small boxes dipped in a solution of muriate of soda.

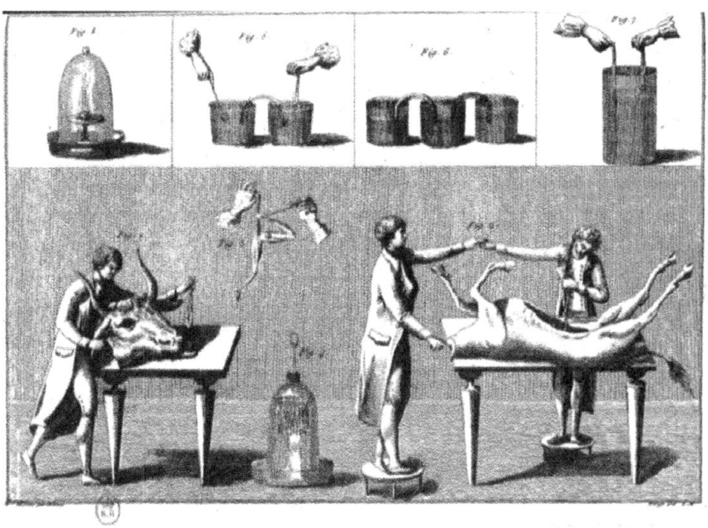

Figure 4.8. *Aldini separately stimulates the head of a connected ox and the body of the animal, while connecting them to a prepared frog, a battery and a chemical solution [ALD 04, pl. 1, figs 1-8]*

Then, after separating animal from battery, he confirmed that animal electricity was not due to communication or transfusion of general electricity, but to specific electricity. He thus contributed to modeling a nature obeying electrocentric laws:

> [...] that there is a principle in the animal machine which physicists can, in their experiments, excite and direct as they please by certain means, but which wise nature brings into play in the living being in a hidden and even more wonderful way. This is therefore a powerful fluid, formed, developed, and driven by the action of animal forces, since these parts, separated from the common reservoir of general electricity, have nevertheless by themselves the faculty of reproducing it and making it circulate in a manner likely to excite muscular convulsions. [ALD 04, p. 6, author's translation]

We're talking about Aldini's *animal pile*. The latter was made from skinned frogs:

> Professor Aldini conceived the possibility of producing galvanic phenomena, establishing an animal chain by the simple correspondence of muscles with nerves, without any interposition. [CAS 03, p. 15, author's translation]

In correlation with Humboldt, whom he readily quoted, he took up the idea of a galvanic atmosphere emanating from organic matter. In order to prove the existence of such a bubble around the subjects, he wondered whether it would not be possible to excite pain by applying only a surgical instrument next to a nerve branch, without touching it directly. In the context of the transition from exploratory galvanism to medical galvanism, Aldini recommended that this may contribute to the explanation of some phenomena of sensations such as synesthesia, sometimes described as pathological, notably by Petetin. If animal fluid emanated from bodies to form an electric atmosphere, then it was no longer liquid but rather designed in analogy with the forces described by physics. Aldini compared Leyden jars, and the voltaic pile which he did not hesitate to use in his experiments and animal substances. It provided comparative knowledge on electricity and galvanism from the interactions between human and electrical machines:

– galvanism, like electricity, was related to the atmosphere that surrounded them and to which they contributed;

– artificial electricity accelerated the putrefaction of animal substances whether with the use of batteries or Galvanic devices;

– the propagation of galvanism approached the speed of flow of the electrical fluid even over long distances;

– in the same way that the force of the electric current was increased according to the surface area of the conductors that transmitted it, the galvanic current escaped with a force proportional to the surface area of the conducting arcs that established communication between the opposite poles;

– the Leyden jar, after having been unloaded, did not give any more sparks but when left aside for a few minutes, it recharged itself;

– just as the action of the galvanic column was destroyed if the alternating order of the metal plates that constituted it was disturbed, so too, by changing the arrangement of several animals forming a system, it happened that, in some cases, muscular contractions ceased to take place.

The expressions *animal machine*, *human machine* and *human-animal machine* punctuated Aldini's treatises. In 1804, they were counted more than 25 times. The demonstration that the physiological mechanisms of animal electricity[14] went deeper, in terms of actions and behaviors, than the physical mechanisms of electricity found in nature was made. While few galvanists contradicted this comparison, can it be said that there was only one and the same force in nature but on different scales[15]? In Pallas' view, we find the opposite idea:

> The sensations that animal electricity causes us to experience are different [...] from those excited by artificial electricity; the organ, despite the force of its concussions, rarely gives rise to crackling sparks; nor do we observe the phenomena of attraction and repulsion; and it does not act on the electrometer like other electrical bodies. According to this, it would seem that organic electricity does not entirely resemble the electricity that I will call inorganic, either artificial or natural [...] [PAL 47, p. 59, author's translation]

14 In order to determine the differences and similarities of effects between galvanism and electricity, Father Vassali Eandi experimented on the effects on various animals: "However, by comparing them, we can see that insects, such as flies and butterflies, are very much affected by a faithful action of galvanism, when positive and negative electricity does not seem to have any effect on them, but only sparks. Galvanism also acts on lizards, frogs, birds, when they are not even affected by the electric bath. The galvanic torrent, although it is quite credible, acts so much on the frogs that it makes them whistle like birds" [VAS 02, p. 4, author's translation].

15 Many galvanic doctors affirmed the identity of the two currents beyond the effects and sensations they caused: "The continuous sensation of shaking caused by galvanic current undoubtedly differs from that caused by electric current; but is it enough to establish an essential difference between the two causes? Volta's latest work on this subject, and the fine experiments he made in the presence of the commissioners of the institute, no longer allow us to assume this; there can be no longer any doubt today that the difference observed in this respect is due solely to the way in which the driver takes charge in one case and in the other" [GRA 01, pp. 21–22, author's translation].

While Aldini's experiments were diversely received, as Iwan Rhys Morus explains, he places Galvani's research in a post-revolutionary political context and stakes, particularly in relation to materialism and atheism. His experiments on Forster are regularly presented by radical writers as evidence of the material and electrical basis of human life. Galvanism was both a fashionable fad and the pet peeve of conservative ideologues. For conservative satirists, Aldini's experiments, like Beddoes' pneumatic chemistry, were symptoms of revolutionary delirium. Its public performance refers to at least three issues:

– a medical issue related to bringing a subject back to life;

– an exploration of the mechanisms of life where the electrical and material nature of the vital principle was highlighted;

– a philosophical issue on the origin of consciousness and the materiality of human nature.

In a British context, Aldini was particularly disturbing, as he no longer allowed electrophysiology on human bodies to be reduced to the fields of the extraordinary and magic:

> We can see Davy's disdain for Aldini's 'disgusting' galvanic manipulations as being close to the origins of efforts to remove the sensational from experimental natural philosophy that remained a preoccupation in some circles for much of the 19th Century. [RHY 09b, p. 273]

Thus caught between convulsions and cataleptic states, medicine at the end of the 18th Century was marked by the electrical conception of animate and inanimate life. The scientific jolt provoked by the affirmation of the electrical nature of the mysterious animal spirits was close to the revolutionary jolt in terms of both its immediate and long-term impact. The electrocentrism that arose from all the galvanists' research tended to show that: "What was certain, however, was the ease and near-obviousness of the transition from the field of physics to that of physiology" [FRE 14, p. 252, author's translation]. In the field of physiology, galvanism managed to overcome a deeply rooted obstacle in the metaphysical conception of the animal spirit, in favor of an electricity that was specific to organisms. As De Frenza explains in his article on electrical experiments, until the middle of the 18th Century, the galenic derivation hypothesis describing movement and sensation assumed that natural spirits which entered through food or were refined in the brain became vital, animal entities. There were therefore two main alternatives: to attribute a somewhat immaterial nature to animal spirits or to consider them, as Willis did in the 17th Century on the basis of the Cartesian mechanism, as material entities but without knowing what they corresponded to. Galvanism enabled taking the step to make them electric and to integrate them into a dynamic body schema.

Between 1828 and 1834, Nobili (1784–1835) took up Galvani's experiment and showed with the help of a galvanometer that the contraction was produced by a current specific to the frog:

> The frog's current has a certain direction and strength, and can be destroyed by directing it against another of equal intensity. If two frogs are prepared in the usual way, and only the galvanic circuit is formed with them, by putting the nerve of one frog in contact with the muscle of the other, the frogs both contract [...]. As long as the frog contracts a little under its own current, it enjoys all its sensitivity [...]. [NOB 28, p. 235, author's translation]

Nobili is credited with the creation of the astatic galvanometer introduced in 1825 and characterized by being independent of the magnetic field. In 1844, Matteucci (1811–1868) [MAT 44] discovered the existence of a current in whole or cut muscles, directed from the animal's feet to its head. He demonstrated its presence in humans. For the first time he measured an electric current of biological origin, confirming the conductivity of nerves and adopting a position where animal electricity was both produced by organic matter and of the same nature as artificial electricity. He therefore used two modes of measurement: one, physical with the galvanometer; the other, biological with the frog's foot.

In 1877, questions were still being raised at the medical school in Angers, such as: do electric currents exist in the nervous system? Is the nerve impulse a phenomenon identical to the electrical agent [GUI 77, p. 11]? These problems, addressed in a recurrent manner, indicated the scientific, medical and historical importance of the galvanic legacy. Thus, the political and institutional legacies of discoveries related to animal electricity are important. In London, a voltaic society and a galvanic society clashed and in France, the *Prix du Galvanisme* (1802–1815) and the *Société Galvanique* (1802–1809) were created. Stéphanie Deprow in her thesis entitled *Un héritage des Bonaparte : le prix du galvanisme (1802–1815) et le prix Volta (1852–1888); L'État et l'encouragement à la recherche sur l'électricité* [DEP 08] details the political context of the 19th Century and how it encouraged research on electricity, notably through the distribution of prizes. The aim was to stimulate scientific discoveries at a time when the Nobel Prize[16] had not yet been created:

> Galvanism constitutes a very special moment in the history of electricity, as it corresponded to an appropriation by doctors and

16 Its creation dates back to 1901.

> chemists of a field that until then had been reserved essentially for physicists, medical applications being quite contested. With the confrontation of the know-how specific to these various ways of approaching experiments on mineral or organic matter, the conditions were therefore favorable for the emergence of all sorts of discoveries and scientific battles. It is within this disputed field that Bonaparte intended to intervene. [DEP 08, chap. 1, author's translation]

Many of the participants belonged to the *Société Galvanique*, which was founded shortly after the Galvanism Prize. Its aim was to help develop better medical applications of electricity. Equipped with a laboratory and the *Journal de galvanisme et de vaccine*, the Society played a considerable role in the constitution of a group of French physicians, physicists and chemists working on galvanism such as Fourcroy (1755–1809), Laplace (1749–1827) and Gay-Lussac (1778–1850). Thus, research on galvanism, combined with a broader program of electrotherapy, developed in two ways. On the one hand, in the field of applications to ailments of the mind and nerves; on the other hand, in the exploration of the organs of the nervous system to record their natural electricity, measure it or understand its mechanisms.

5

Between Electrotherapy Rooms and Laboratories: Specializing Electricity

Despite the fact that research on animal electricity opened up a wide field of experimental investigations in electrophysiology, physicians who, like Léon Danion, asserted that "*the true guide to electrotherapy should be electrophysiology*" [DAN 89, p. 137, author's translation] were rare. As we have seen, the medical field took over electricity as early as the middle of the 18th Century. This field, then developing in different directions, adopted a trial and error scheme and this was applied to many forms of illness. It was also used to explore the physiological mechanisms of the living and declined in terms of animal and galvanic electricity. From the outset, the two aspects of research on this force were intertwined, between biological explorations and clinical applications. However, the trend towards specialization that spread through the various fields of the life sciences during the 19th Century provoked a disciplinary and even geographical separation between two of its branches. On the one hand, electrotherapy was developing, in the form of gentle stimulation or shock treatments; on the other hand, laboratory research was developing, the aim of which was to explore the central nervous system and to record the various electrical activities. Electrotherapy, well developed in France, was marked by empiricism and its dependence on technology; while laboratory research on organic electricity constructed questions to which it brought elements of answers. Research and treatment separated in order to come together again. Electrical therapies used the calming aspect of electric shocks, already reported during the 18th Century, while laboratory electricity was used to stimulate and recreate the mechanisms of muscles and nerves. This history is complex: while electrical therapies were gradually becoming treatment tools and attempted to establish differential diagnoses accompanying the progressive determination of nervous, psychiatric and neurological diseases, electrophysiology measured, recorded and explained, by virtue of these measurements, the actions of currents at work in organic matter. While the fundamental work of electrophysiology was sometimes little considered by French electricians, this did not mean that these two fields did not influence each other:

> In fact, studying the effect of electricity on living things can be done in many ways, depending on the intention. The problem of elucidating the nature of excitability and the underlying processes is different from the problem of investigating its possible use in medicine. For a doctor, knowledge of the laws of excitability, or even that of simple correlations or modification of excitability according to the pathology making electrodiagnosis possible, as well as the revelation of the therapeutic effects of electricity, could prove to be quite sufficient. In this sense, the studies of electrophysiologists carried out in the laboratory in animals seem to have little to do with the history of medicine. [DUP 01b, p. 90, author's translation]

Several factors come into play to understand this division, appearing to be partially methodological. Indeed, French technicism, accompanied by an extremely empirical medicine of nervous diseases and quite close to the way Ledru practiced it, differed from the research approach carried out in laboratories on the measurement of the nervous impulses. Clinical applications of electricity were therefore conducted alongside the increase in knowledge of the electrical and chemical physiology of the human brain and were respectively guided by different intentions:

> In fact, electrophysiologists and electrotherapists use the same machines: batteries, induction coils, galvanometers and rheostats, and they work on essentially the same material: muscles and nerves. But doctors claim responsibility for the specificity of the living being, of the patient with a disease and a history, when faced with biological elements dissected on a laboratory bench. [BLO 10, p. 42, author's translation]

5.1. Electrical therapies: emergencies and interventionism

The deployment of electrotherapy accompanied that of hospital structures where important services were dedicated to it. Thus, a first phase of the electrotherapeutic craze took place from 1770 to 1800, then a second phase from 1840 to the end of the First World War. Indeed, electrotherapy developed during the First World War, marked by the need for emergency medicine. Finally, a third period of resurgence ran from 1930 to 1950 for the understanding of psychiatric illnesses in France. Once again, the impact of the Second World War must be underlined. The development of electrotherapy slowed down in modern psychopharmacology:

> Psychiatric disorders moved from a purely neurological interpretation to a mixed, neurological and psychological interpretation. In this period psychotherapies and modern psychopharmacology appeared. The neurological interpretation of psychiatric disorders took on a neurochemical coloring concomitant with the discovery of monoamines and their agonists and antagonists. [MIC 13, p. 324, author's translation]

The increase in the number of rooms dedicated to this treatment, particularly in asylums, and the hope of alleviating diseases as old as hysteria and epilepsy, aid understanding in the scope and success of this treatment at the end of the 19th Century and during the first half of the 20th. The many mental illnesses, still undifferentiated from one another, made excellent targets for an electricity that seemed to have an effect on the whole organism and made it possible to mark some differences:

> In 1845, the chief physician of the Deux-Sèvres Asylum tried galvanic action on the incomplete body of the deprived, in need of salvation [fools and idiots]. Working at Maréville, he continued to electrify both the prostrate to pull the person out of their inertia and excite them to calm them down. He was also not against a little electric jolt when it came to discipline. [QUE 12, p. 263, author's translation]

5.1.1. *Deceptive diseases*

In spite of short periods of postponement, electricity followed its path to the center of the brain until it redefined human nature in relation to the action of this force on its capacities and its moral and psychological dimensions. Electrotherapy was applied to calm seizures in the medical sense but also in a more coercive sense. The fact that electric current could be weakened according to softer or more aggressive approaches explains the diversity of the applications used. While in the 19th Century, electrical therapies were applied to all kinds of mental illnesses, a diagnostic caution arose, relative to the fact that certain pathological denominations seemed, in reality, to cover various illnesses. Thus, electricity was developed as a "calming" instrument within the asylum establishments, but also as a tool to control behaviors, especially female behaviors and their tendency to hysteria asserted by electrifying and galvanizing doctors. We saw the contribution of this research to fantasies about the powers of electricity in the first chapter. Within the framework of the history of electrotherapy applications, we wish to develop the turning point taken in 1840 in the electrification of behaviors, mental disorders and more general behaviors. Around 1840, the trend towards female hysterization, still traditionally associated with uterine mechanisms, led physicians such as Milligen (1782–1862) [MIL 48] to correlate the expression of passions with hysterical manifestations, literally inscribed in female nature:

> In this revolution, the energies of the brain or the sensorium of woman are less called upon than the sympathetic system of the nerves; and hysteria, in its multiplied and anomalous forms, warns us of the rapidity and exaltation of the progressive development of all the functional organs of her sex. [MIL 48, p. 44]

This fact, linked to the development of electrical treatment, allows us to understand the application of galvanic currents to the different forms of hysteria. Either they had a psychological origin, the electric treatment allowing then to shake the spirit incarnated in the cerebral matter; or the cause was organic, taken in the influence of the uterus on the brain and understood on the basis of the model of analogy between the body and the machine:

> Another doctor made the electrical mechanism underpinning such views quite explicit: The ovaria, uterus and mammae (breasts) form, as it were, a reproductive pile, the circuit being completed by the nervous system; women's bodies were just like galvanic batteries. [RHY 11, p. 89][1]

While hysteria and epilepsy were still undifferentiated, electrical therapies were seen as a general approach to treatment. Thus, Alfred Becquerel, in 1857, questioned the diagnosis of epilepsy which had existed since the middle of the 18th Century:

> How and at what times was electricity used in such circumstances? Was it during the attack, was it in the interim? Were they real epilepsies? I believe that any reasonable doctor must wait for new facts and especially more serious observations before admitting the possibility of curing such a terrible disease with electricity. [BEC 57, p. 248, author's translation]

Epilepsy is a heuristic disease both in terms of its contribution to the knowledge of normal brain electricity and for the knowledge of its pathological phenomena. It enabled the establishment of a sustainable paradigmatic framework for

1 He cites the work of Dr. Tilt [TIL 53, 9].

understanding the action of electricity in the brain, but also contributed to the concept of shock therapies, since shaking the brain seemed to calm its disorders:

> Pathologies are not passively simple fields of application of theories, but have an active heuristic function in construction and dynamic oppositions. [DUP 18, p. 138, author's translation]

Later on it was confused with a group of convulsive diseases of organic origin as in the case of certain types of encephalitis or which were caused by psychological disorders as in hysterical manifestations. This explains why the results of electrical therapies were extremely inconsistent and difficult to grasp in the treatment of subjects considered epileptic. By considering hysterical conditions as epileptiform, practitioners were faced with the confusion of nervous states that had yet to be distinguished from each other. Successful treatment of one of these conditions could perhaps be considered one of the first forms of electrodiagnosis, particularly in the context of differentiating between organic hemiplegia and hysterical or traumatic hemiplegia [TIB 77]:

> Not only do I assert, without fear of contradiction, that cases of organic hemiplegia have frequently been reported in hysteria, but I submit that such errors were once inevitable because there was no sure way to distinguish between these two types of paralysis. It was accepted that hysterical hemiplegia could mimic the facial features of organic hemiplegia and that extrinsic features, such as the age of the subject, the presence or absence of a concomitant heart condition, a history of syphilis, the alleged stigma of hysteria, the circumstances in which the paralysis arose, were the only data to establish the diagnosis. [BAB 09, p. 5, author's translation]

The empirical weakness of electrotherapy, while it called for great caution, could also be considered a strength. The fact that it could be designed as a treatment instrument which, by virtue of the variability of the methods, devices and currents used, could be adapted to the different ailments makes it a tool for differential diagnosis:

> Convulsions and hysterical contractures are certainly an accident against which the usefulness of electricity can be conceived *a priori*. It is perfectly rational to use it, either as a hyposthenizing agent, or to operate on the antagonistic muscles, or finally to use it as a substitute method, substituting a convulsion of a new kind for hysterical accidents. [BEC 57, p. 249, author's translation]

In 1859, Dr. Auzouy (1819–1879) [AUZ 67] made a mixed assessment of the applications of electricity in the Pau asylum, mentioning waves of empirical experiments on all types of mental illness. He concluded that there was a generalized action of electrical treatment, which by the physical pain it produced, generated a let-up in the deliriums, temporarily bringing the subject back to a feeling of reality. Epilepsy, or at least the symptom cluster that comes under this name, has been treated electrically since 1770. However, the more we knew about this pathology, the more doubts would about the advantages of electrotherapy:

> The action of electricity is very dubious. Some improvements have been reported by the use of weak continuous currents applied to the sympathetic and surrounding areas of the bulb; but the few observations are hardly conclusive and are relative to the low level of illness. However, Mr. Voisin recommended focusing his action on the bulb and not on the sympathetic nerve as Benedikt did; for this he placed one electrode on the chest and the other in a point where it could, according to him, have an effect on the bulbar region, on the face, chin or tongue behind the V lingual. [AXE 83, p. 868, author's translation]

Thus, the work of John Huglings Jackson (1835–1911) or Jean-Martin Charcot (1825–1893) on epilepsy allows us to clarify its links with electricity from two points of view: physiological and internal, clinical and external:

> But what seems certain is that the real center of epilepsy is in the brain, and that the essence of the condition is to be found, first of all, in a particular disorder of the brain tissue, which from time to time manifests itself in a kind of explosion, i.e. an epileptic seizure. […] Electrotherapeutic experiments have been carried out at various points to combat epilepsy. […] Later Althaus published a series of success stories and Benedikt spoke out in favor of electrical treatment of epilepsy. […] All your efforts should be directed, in the first instance, at suppressing the epileptic change in the brain. [ERB 84, p. 541, author's translation]

As knowledge about epilepsy as a neurological disease became more precise, an electrical approach to psychogenic diseases was also developing, which was likely, but with great variability, to respond to these therapies. The idea was that electric shocks, by imitating epileptic seizures, calmed psychogenic disorders:

> [...] the successes of electricity against hysteria are not exactly brilliant, but they cannot be predicted in advance and they vary

> widely. It is above all in hysteria that wonderful cures, the suppression, somehow magical, of seemingly serious disorders, paralysis, etc., appear the earliest. [ERB 84, p. 545, author's translation]

The history of electricity as a remedy was far from continuous between 1740 and the end of the 19th Century, not only because electrical treatment was the subject of controversy but also because the same names did not cover the same pathological realities. The understanding of diseases evolved both in their relationship with the body and with the subject. Electricity as a remedy was a concept built over time, through technical progress but also through societal upheavals such as wars. We already find in the classical uterine conceptions of hysteria, questions on the modalities of electricity to be applied as a remedy:

> When therefore the sources of organic life, and especially the uterine source, are the center of an abnormal excitation, they react on the cerebrospinal axis, they thus determine general phenomena but they only have an effect then on the starting point of the nervous impulse[2], an impulse which they precipitate so violently on the organs of movement. It is this impulse that lead, copper, mercury, etc., preparations can alter and even destroy, it is also this impulse that electrical currents can powerfully modify. [DUB 33, p. 450, author's translation]

In the 19th Century, simulation phenomena, contextual traumas, psychogenic illnesses were being deciphered and modeled at the rate of changes in representations of the human brain. The advent of brain electricity completed a process in which the brain dried up, mechanizing itself and becoming the center of mental disturbances. Moreover, between 1874 and 1886, several theories related to consciousness were strengthened. The memory sciences, marked by new studies on dual or multiple personality disorders, signaled the end of a consciousness considered unified. Polypsychism [HAC 98] reinforced the idea of mental dysfunction. At the same time, the concepts of trauma or psychological fractures participated in the medicalization of disorders, formerly attributed to the soul:

> A century ago, trauma was shifted from the body to the mind, just as the multiple personality was emerging in France, and memory sciences were making their entrance. [HAC 98, p. 289, author's translation]

2 Nerve impulse refers to the electrical current that travels along nerves and neuron extensions.

This medicalization, accompanied by interventionist medicine, was becoming electrified. Thus, much of the research on mental disorders was directed towards treatments that used the body as a foundation. Neurology, the localization of brain functions, electrical theories of mental illness can be considered as the science of a mind intertwined in and generated by matter. This preponderance for anatomy and physiology, when it comes to understanding the mechanisms of psychological disorders, allows us to understand the importance taken by electrotherapies.

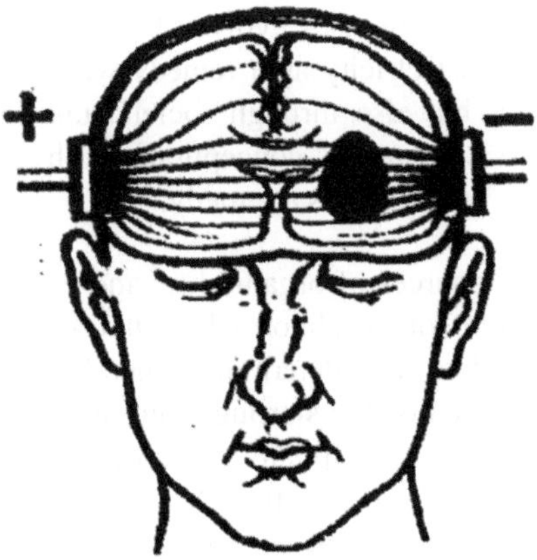

Figure 5.1. *Diagram of the best application of electrodes to lead to a pathological site situated deep in the left cerebral hemisphere. Taking place in the area with the densest and most active wires [ERB 84, p. 39]*

Moritz Benedikt (1835–1920) researched the differentiation between epilepsy and hysteria or hystero-epilepsy in order to develop two different treatments. In an attempt to determine the starting point of the various seizures, he correlated the application of galvanism, preferred over other methods, with the idea of being able to rationalize the field of convulsive diseases according to the model of a one disease therapy:

> If access is from the periphery (sensitive scars, vasomotor phenomena in the territory of the great sympathetic), galvanization will be applied successfully. [WIE 88, p. 128, author's translation]

Benedikt also advocated the "faradization of the vulva" [WIE 88, p. 130] for hysteroid vaginismus, including vaginal ether irrigation and the use of coolants. Electrical treatment complemented a classic pharmacopoeia and highlighted images of diseases whose singularity was hard to find. While at the beginning, the general action of electricity had the advantage that it could be widely applied, the methods of galvanization, faradization or franklinization enabled differential associations to be made. For example, certain general pathologies, such as hysteria, corresponded to a series of particular reactions that allowed the pathology to be divided into several others, while emphasizing different causes:

> Rosenthal examined electrical reactions in two patients: in one, reactions to both currents were normal; in the other, there was an exaggeration of electro-muscular contractility and galvanic excitability of the plexuses and nerve trunks; [...]. [AXE 83, p. 911, author's translation]

Dynamic electricity was thus mainly used to combat hysteria spasms, convulsions, apathies and attacks of great hysteria (hystero-epilepsy). It enabled two types of action:

– in the event of a decrease in the activity of the spinal cord and the brain, it enabled a catlectrotonizing action (exciting, invigorating);

– in case of increased excitability of nerves and muscles (neuralgia, hyperesthesia), then it triggered an anelectrotonizing action (weakening, antinevralgic, calming).

The end of the 19th Century was marked by the desire to shed light on the mechanisms of action of electrotherapy:

> The application of electricity to living beings must not be left to empiricism if science and the sick are to benefit from it. It is absolutely necessary, and this will be my conclusion, that the electro-therapist must be an electrician as well as being familiar with physiology. [ERB 84, p. 729, author's translation]

We will discuss Jackson's work on epilepsy, with a view to showing how a disease considered to be the result of electrical disturbances contributed to the delineation between neurology and psychology. Thus, the heuristic value of epilepsy lies partly in its contribution to this typological difference. Very attached to the doctrine of psychophysiological parallelism, Jackson made a dichotomy between the anatomophysiology of the brain and psychological events. Therefore, knowledge of brain mechanisms does not allow us to understand normal or pathological mental processes because:

> There is no physiology of the mind as there is no psychology of the nervous system. [HUG 31–32, v. 1, p. 417, author's translation]

The neurologist needed to treat the brain centers in a materialistic way, since epileptic seizures do not depend on psychological facts, even though these may be the consequence. While the analysis of the role of the emotional and psychological event can allow us to understand the links between the general change in the body and the discharge of unstable nerve cells, it should not be treated by the doctor. Taken in a psychosomatic and dualistic model, moral causes were conceived as general disturbance factors but need not erase the possibility of delimiting psychogenic diseases and neurological processes even if psychological symptoms were involved in the symptomatology of brain diseases. Epilepsy represented a distortion of the electrical activity of the brain, dominating convulsive diseases and contributing, in connection with electrotherapy, to differentiating and empowering neurology and psychology. The challenge was to show that it was related to brain functions that could not be correlated with hysteria. While Jackson was developing his studies on partial epilepsy, Charcot, who was appointed to the Salpêtrière in 1862, was dealing with hysteria in the 1870s. He described it, initially, on the model of epilepsy by seeking to find in its process, the tonic, clonic and stupor phases. This was based on the definition given by Briquet (1796–1881) in 1859. Hysteria is a "neurosis of the brain whose apparent phenomena consist mainly in the disturbance of vital acts that serve to manifest emotional sensations and passions" [BRI 59b, p. 3, author's translation]. This twinning was classical and was described by Jean-Baptiste Louyer-Villermay (1776–1837) in 1818 as "epileptiform hysteria" [DSM p. 228]. In this context of the superimposition of evils, Charcot created the clinical entity of hystero-epilepsy. In believing that epilepsy was the primitive disease to which the hysterical manifestations were grafted, he also believed that the opposite process was taking place. Initially unaware of the simulation phenomena on the part of the hysterical, he therefore confused the two syndromes into one. Convinced of the existence of hysterogenous zones, he used an ovarian compressor which proved to have no effect on epileptic seizures. It was in 1872 that he differentiated between the two diseases despite the similarity in the convulsions. Paradoxically, it is thanks to his fictitious entity of hystero-epilepsy that he helped to delimit the neurological and nervous fields. His thirteenth lesson, published in 1878 and entitled *De l'hystéro-épilepsie*, is a key work in understanding this delineation:

> Hysteria is shown to be combined with epilepsy to form the mixed form, a kind of hybrid composed partly of hysteria and partly of epilepsy. The point here is that many authors do not deny that epilepsy and hysteria can occur in the same individual. Here, the intelligence becomes obsessed in the long run unquestionably by the fact of epilepsy, other combinations are mentioned but always the two affections exist simultaneously and work without acting on one another in a serious way, each of them preserving the appearances and the prognoses which are appropriate to him. Seizure accidents remain separate, distinct. [CHA 78, p. 372, author's translation]

A series of observations allowed him to invalidate his concept:

– the presence of a slight feeling of illness or vertigo epilepsy was not noted in so-called mixed seizures;

– compression of the ovary stopped the hysterical attack but had no effect on seizures;

– the integrity of intelligence was not called into question by hysterical manifestations.

Thus, Charcot's theory of hysteria developed from the dismemberment of the concept of hystero-epilepsy and its psychogenic origin became a logical counterpart to Jackson's neurological theory of epilepsy. Charcot used the term Jacksonian epilepsy, referred to as Bravais-Jacksonian epilepsy from 1894 onwards. This epilepsy covered the field of clonic, tonic, but also spasmic and vibratory motor focal seizures and corresponded to a disturbance in the rolandic region. It was manifested by isolated tonic or clonic movements or by a well-ordered sequence (thumb, fingers, arm, face, leg, foot) according to the somatotopic representation of the motor cortex and could generalize into a "great evil" seizure after loss of consciousness. Epilepsy research offered two issues in relation to medical electricity:

– to show that epilepsy is a disorder of brain electricity;

– to use this imitation disturbance in electrical therapy to calm other mental ills.

While epilepsy was not seen as a form of hysteria, the understanding of the first led to the cure of the second. Charcot took up the correlations between natural electricity and physiology, taking as the object of study, lightning, which he considered to be one of the causes triggering hysteria. But while natural electricity was the cause, then artificial electricity became a potential cure:

> We know that Mr. Charcot has established that nervous accidents caused by lightning are of two kinds. Sometimes the fulguration directly produces nervous accidents, particularly paralysis, and sometimes it causes the appearance of hysterical nervous disorders. [DAN 92, p. 27, author's translation]

Charcot studied the result of galvanic currents [CHA 82, pp. 63–64] on hysteria during hypnosis sessions. His use of electricity was combined with the use of drugs such as bromide or suggestibility techniques such as hypnosis. Thus the influence of atmospheric electricity on nervous disorders came as a diagnostic tool to differentiate hysteria from the effects of brain damage:

> The characteristic of hysterical neurosis is obvious in this case: this paralysis, which recurs during thunderstorms, is obviously due to hysterical neurosis. Rather than stop at this interpretation, however natural, he prefers to imagine I don't know what kind of brain lesion in a home, localized I don't know how, and enjoying the singular property of being rekindled, with each storm, under the influence of atmospheric electricity! [CHA 89, p. 460, author's translation]

In 1853, Brochin [BRO 53] described the resources that electricity could provide for the diagnosis of paralysis independent of brain damage as in the case of general paralysis of the insane, and in 1859, Brierre de Boismont (1797–1881) [BRI 59a] used localized galvanization to confirm a differential diagnosis of the various types of general paralysis. These therapeutic approaches made it possible to isolate hysterical paralysis, paralysis caused by a brain lesion:

> But the truly heroic remedy is electrification. Depending on the case, either faradization of the skin or muscles, nerves or nerve trunks, or galvanization is used. I have to say that, in my opinion, the galvanization of the marrow by downdrafts is the most successful method in rebellious hysterical paralysis. [BIA 68, p. 30, author's translation]

Electrical treatment was inseparable from the improvement of machines, their progressive miniaturization and a medicine that claimed to be interventionist and a symbol of progress. This was one of the decisive factors in the appropriation of electricity by medicine in the 18th Century. This is why it is interesting to highlight these profound links between electrotherapies, medicine and society. The "electricity fairy" [BEL 91] continued to spread its wings for the relief of mental ills until it entered private medical practices and homes.

5.1.2. *When treatment depends on techniques*

The very fact that these treatments depended on machines but used a natural force present everywhere, even in organic structures, had an advantage over a more traditional pharmacopoeia, especially with substances that could cause drug addiction[3]. Cases of addiction were a nuisance to society even when the addiction

3 The term originated in the 1880s and refers to "A product introduced into the bloodstream through the digestive or cutaneous tract (,) intoxicates the organs and brain and causes addictive behavior (mania)" [DUP 16, p. 95, author's translation].

started with treatment. The use of morphine posed many problems during the 19th Century. As with electricity the patient did not become addicted to it, its use as a therapeutic agent could be discontinued at any time. In the 19th Century, all progress in electricity was immediately followed by medical applications and many specialized services equipped with *healing machines* opened in Paris and in the provinces. Franklinization, often applied in the treatment of nervous disorders, was practiced by means of the static current produced by a plate machine, one end of which went into the earth and the other to an insulating stool where the patient sat for a static bath. The effects, especially sedative, seemed favorable in painful states or in cases of neurasthenia. It was also in this diversification phase, that alongside galvanization, the application of "faradic" currents was developed, which consisted of applying induction currents. The concept was proposed by Guillaume Duchenne de Boulogne. As early as 1837, coils for medical "faradization" were available [PAY 65]. The latter therefore uses the faradic current or induced current, deriving from the direct current supplied by one or two batteries and transformed by the faradic device, thanks to an inductive coil, an induced coil and a switch. The faradic current had a strong effect on muscle contractions. Duchenne de Boulogne emphasized the advantages of medical applications of relatively low intensity alternating induction electricity compared to direct currents. Faradization was initially applied through the break-up of two electrodes: one located at the feet and the other at the target area. Duchenne designed his stimulation tool for more localized electrotherapy. Faradization was spreading in the medical field, particularly in the context of asthenia. In 1892, Danion published the case of a 45-year-old woman suffering from severe intellectual and physical overwork, resulting in an increase in the excitement of her nervous state, which was considered hereditary by the medical profession. The most noticeable symptoms were persistent frontal cephalalgia, decreased memory and intelligence:

> At first, light galvanization of the brain was used, but the patient's nervousness prevented him from following this treatment method. Skin faradization was then used with the brush, applied to the nape of the neck, shoulders and arms. The session was ten minutes long and was repeated every two days. After three sessions, the headache was gone. Sleep returned little by little, so that the patient could sleep three to four hours a night. Melancholy and moral affection also disappeared, so that as a symptom there was only anorexia. Such a result had been obtained as early as two weeks of treatment. From that moment on, galvanic currents could be passed through the brain. [DAN 92, p. 52, author's translation]

In view of the history of its failures, it was in the diversification of techniques and machines used that one of the reasons for the recurrent success of electricity in medicine was to be found. Indeed, while galvanization failed, faradization could be tried, all with different devices. Moreover, it became a coercive instrument of treatment, which was not of little advantage in an asylum environment subjected to a significant empiricism and to rare or complicated treatments to be applied. In *De l'application de l'électricité au traitement de l'aliénation mentale* (1859), Dr. Teilleux [TEI 59], chief medical officer of the Maréville Asylum, wrote that electricity offers: "The immense advantage of being able to be used as an agent of coercion, and, if necessary, to often substitute for the shower, the corset, etc., all of which have serious disadvantages, whereas the application of electricity can never have any" [TEI 59, p. 63, author's translation]. The asylum atmosphere in turn became subject to the appliances providing electrical treatment: they began spreading pretty fast. In 1884, the Cochin Hospital acquired an electrotherapy department. The most famous of these services was undoubtedly that of the Salpêtrière. Romain Vigouroux and Charcot proceeded with collective electric baths or individual application of static electricity or induced currents (faradization). In Pau, Dr. Auzouy described his method of applying electric baths, a technique used since 1770. The patients were seated, free or tied up, in an armchair in front of a small table on which the electromagnetic machine rested. It was then easy to subject them to the action of the current by means of metal exciters, with or without wet sponges and held by the operator by means of insulating sleeves:

> M. Renaudin, rightly thinking that mental pathology could benefit from the intelligent and judicious use of electric fluid, began, in 1857, in Maréville, some trials which, as soon as I was set up in my department, he invited me to continue. Since that time, induction devices have become our usual auxiliaries in our asylum. May something useful come out of our persevering efforts! [AUZ 59, p. 23, author's translation]

The interdependence of device-based treatment required constant attention to their disadvantages and side effects. As we have seen in previous chapters, the Leyden jar or electric batteries often had serious disadvantages, such as difficulties in controlling shock intensity. A sample of batteries was developed from the voltaic type to the trough battery. The latter did not have very great disadvantages as long as it was not very strong. In the latter case and because of the continuous energy currents it produced, it could hyposthenize the muscular or nervous systems a little too strongly.

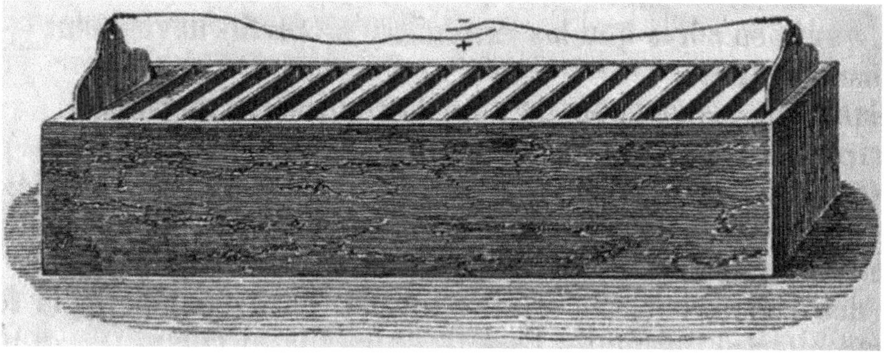

Figure 5.2. *Trough battery developed by the British chemist William Cruickshank (?–1810 or 1811) in 1802*

For example, induction devices could be harmful because of their power. For example, de Ruhmkorff's apparatus (1803–1877) had the feature of transforming an electric current of low voltage and high intensity flowing in an "inductor" winding into a current of very high voltage, produced in a second "induced" winding. Coils could generate shocks and sparks. Galvanizing, in the asylum context, also paved the way to explorations of the effects of electricity on psychosis. Within the framework of an organicist and materialist psychiatry, electrotherapy proposed to act on the organic center sought in the central nervous system and whose functions were disturbed:

> In the first line, your research will focus on recent and relatively easy to treat cases, particularly on the vaguer and still incomplete psychopathic states, on morbid anxieties, with insomnia, [...]; to these cases the simple longitudinal direction through the head, more or less following Neftel's method, is well suited. In this way you can choose the direction of the current, [...]; you can also carry out the simultaneous galvanization of the sympathetic and, possibly, of the cervical marrow. [ERB 84, p. 321, author's translation]

Neftel's method refers to this author's research on the association between hysteria and neurasthenia [SCI 02, p. 15]. The use of high-frequency currents on tissues and then on individuals understood as psychophysiological groups renewed and broadened the applications of electrotherapy. Thus at the end of the 19th Century, Nikola Tesla (1856–1943) presented a series of astonishing and highly theatrical experiments. Beyond its status as a spectacle, he initiated research on high-frequency and high-voltage currents and presented it at a series of conferences in the United States, England and France:

> One of the earliest observations I made with these new machines was that electrical oscillations of an extremely high rate act in an extraordinary manner upon the human organism. [TES 00, p. 175]

Here began talk of *Tesla currents*. Tesla had already written an article in the *Electrical Engineer* in which he mentioned the increase in body temperature under the influence of these currents, suggesting that a naked person at the North Pole could keep warm thanks to them. Furthermore, the theme of improving mental capacities emerged from his considerations on electricity applied to humans. Thus, Tesla seemed driven by the hope contained in the concept of life energy that it could be increased:

> Other sources of energy may be opened up, and new methods of deriving energy from the sun discovered, but none of these or similar achievements would equal in importance the transmission of power to any distance through the medium. I can conceive of no technical advance which would tend to unite the various elements of humanity more effectively than this one, or of one which would add to and more economize human energy. It would be the best means of increasing the force accelerating the human mass. The mere moral influence of such a radical departure would be incalculable. [...] Human performance will be increased, [...]. [TES 00, p. 175]

In an undated text entitled *Mechanical Therapy*[4], he described how the application of his inventions, particularly his oscillators, could ensure the regularity of organic functions without further medication. This point is important to understand the phantasmagorical appropriation of electricity as a potential benefit for humanity:

> It is not to be forgotten that the elimination of countless drugs, patent medicines and specific remedies of all kinds taken internally, by which millions of people doom themselves to an early grave, will be of untold good to humanity. [Tesla, see footnote 4]

Althaus (1833–1900) also advocated the fact that by regenerating brain functions, the deterioration of which in terms of neurons could be one of the causes

4 Tesla, N., *Mechanical Therapy,* http://www.tfcbooks.com/tesla/0000-00-00.htm.

of aging, electricity would be able to restore vital energy, which made it symbolically the key to life:

> Careful use of the constant current to the brain, especially to the vasomotor centre of the bulb, can significantly delay the involution of central neurons. [ALT 11, p. 170, author's translation]

On several levels, there was a gradual shift from electrotherapy in the treatment of behavioral disorders and nervous diseases to the idea that electrical therapies were a vehicle for improving the human condition. Althaus' research on brain rejuvenation and its positive consequences on physical and mental capacities was in line with the late 19th Century's perspective of slowing down the aging process. Electrotherapy was part of a problem of longevity and improvement of the human condition that complemented psychiatric care.

Intrigued by the experiences of Tesla, d'Arsonval, first assistant to Claude Bernard and then to Charles Brown-Séquard at the Collège de France, began research on the application of high frequencies in electrotherapy. He became professor of experimental medicine in 1894 at the Collège de France and in 1896 described the production of magnetosphenes[5] when a subject's head was introduced into a powerful magnetic coil. The founder of high-frequency electricity applied in medicine, he first studied the effects of high-frequency currents on animals. To detect weak currents in muscle contractions, he designed the ballistic galvanometer. Developed from the study of muscle contractions in frogs, it works with a very low current and was comparable with animal electricity. D'Arsonval was at the origin of the darsonvalization method also known as diathermy. This method is an application of high-frequency currents obtained either by means of high-voltage current or by means of a Ruhmkorff coil and two Leyden jars. His taste for experimentation led him to place his own head between two electric coils powered by an alternating current of 110 volts at 30 amps. He then saw flickering lights appear in his visual field, phosphenes, which he attributed to the direct stimulation of retinal cells. More than a century later, the technique born from this experiment, Transcranial Magnetic Stimulation, was developed both as a way of achieving brain exploration and as a therapeutic means for various disorders such as depression. It is likely that phosphenes were produced more by stimulation of the retina than by stimulation of the occipital cortex:

> The final apparatus described by d'Arsonval in 1893 consisted of two Leyden jars whose inner frameworks were connected to a large Ruhmkorff coil (or transformer) and a spark gap, while the outer

5 These correspond to a stimulation of the retina which gives a sensation of light points, due to small currents, induced by the magnetic field.

> frameworks were connected to a solenoid of about ten turns of large copper wire. In addition, to produce long sparks, d'Arsonval inserted into the solenoid, as Tesla had already done, a coil with a large number of turns of fine wire. [BRE 10, p. 58, author's translation]

D'Arsonval used these currents in an electrotherapy perspective and showed that currents above 10,000 Hz no longer caused muscle stimulation. High-frequency currents, generated by the discharge of a capacitor, would prove fundamental in several fields of physics and electricity:

> This discharge of the capacitor was studied by Hermann von Helmholtz in 1847, its frequency calculated by William Thomson in 1853, and finally its oscillatory character demonstrated experimentally in 1858 by the German physicist Behrend W. Feddersen. [BRE 10, p. 55, author's translation]

The first medical application took place in 1896 at the Hôtel-Dieu and the term "darsonvalization" was created in 1899 [MIN 09]. The current flowed through the patient between an active electrode (whose more or less thin shape increased the thermal effect) and a neutral plate applied to the patient. D'Arsonval was part of the paradigm of a new human nature whose nervous and intellectual properties could be influenced, studied or modified by electrical stimulation. The mental domain was therefore an object of study for the electrifying physician not only in pathological settings but also in terms of normal functioning:

> When penetrated by these ideas, as a physicist I examine certain phenomena that occur in the living being, I cannot help thinking that these so complicated syndromes that we call nervous fluid, thought, life force, etc., are perhaps at the bottom only modalities of energy that are still unknown to us. [ARS 81–82b, p. 555, author's translation]

He described three modes of therapeutic action of high-frequency currents: direct application, self-conduction and the capacitor bed. In the first case, the patient was directly connected to the terminals of the device via electrodes. In self-conduction, also known as induced faradization, he was immersed inside a large solenoid, a metal coil, which was part of the oscillating circuit. The patient was then not in direct communication with the device, but induced currents flowed inside his body. Finally, in the third treatment protocol, the subject held an electrode connected to a solenoid terminal which constituted a capacitor framework whose insulating mattress served as a dielectric. The metal bed, in communication with the other terminal of the solenoid, was the second capacitor framework:

> As these currents can pass through the living man with impunity, it is useless to insist on the hopes that such a method raises. It is therefore

> not reckless to say that high-frequency currents are opening up a new avenue for therapy. [ARS 96a, p. 264, author's translation]

Moreover, the craze for electrotherapy was also spreading quite quickly to the private sector, or even what we would call, today, the paramedical field. Thomas Courant's work [COU 81] was an example of the privatization of electrotherapy that accompanied a resurgence of magnetism and the belief that doctors could manipulate, even without machines, the forces of nature to rebalance nerves:

> Yesterday, the Les Mousquetaires performance was, for a moment, interrupted at the Théâtre de l'Ambigu: one of the spectators was caught unexpectedly by a strong attack of epilepsy, and his heart-rending cries, at the moment of one of the most pathetic scenes, stopped the actors and frightened the whole audience. Fortunately, a young man immediately rushed to the patient's aid, and through magnetism brought him to consciousness. This scene moved all the spectators, and the one who had given help if by chance evaded the recognition of the sick man: but the theatre manager asked his name. It was Mr. Th. Courant who is said to have often rendered such services by the same process. [COU 81, p. 14, author's translation]

In this single quotation, we find several points from the history of epilepsy: confusion with hysteria; the very common theatricalization of convulsions in the 19th Century, particularly during shows that conveyed emotions; and the importance of suggestibility. Courant's works were very often quoted in the popular press and marked a syncretism of Mesmerian-type magnetism with electric force. Thus, in the 1890s, many private practices were equipped with electrotherapy equipment. In Paris, the number of medical electricians continued to grow, reaching 10% in 1888 and 30% in 1900. Doctors such as Dr. Cachet, a friend of Van Gogh's and a former student of the Salpêtrière, proposed "applications of electricity to the treatment of chronic and nervous diseases" [BLO 10, p. 45, author's translation]. We can see the importance given to machines as a criterion of the scientificity of healing. Miniaturized electrotherapy sets were marketed and were very successful despite the opposition of some doctors, such as Danion in 1892:

> Is there an electro-specialist who has been able to seriously apply static electricity to a patient? For many reasons, the practice of electrotherapy is made impossible in these conditions; one can amuse the sick, but not do useful work. [DAN 92, intro., author's translation]

Industrialization and the expansion of cities were accompanied by massive advertising and the popularization of electrotherapy, a symptom of a trend towards home medicalization. The electric belt was one of these electric objects, a symbol of popularized medical electricity. Also known as the *galvanic belt*, it was invented by the London jeweler Richard Teed in 1802. Aldini, during his trip to London, contributed to its development. In Canada, this accessory was offered for a range of ailments, from depression and dejection to impotence:

> Therefore, it was considered a supreme remedy for the depressed and weakened of all kinds. In the panoply of treatments that he offered to rather wealthy clientele, Dr. DeBlois specifically recommended a treatment exposing men to galvanic and faradic currents in order to stimulate their genital functions. [GOU 02, p. 34, author's translation][6]

Electric hairbrushes for women were also made, as well as electric rocking chairs (1893), which were made from a dynamo and sent a small current of electricity to the person sitting on it.

Specifically intended for the treatment of nervous diseases, the machine praised in the advertisement in Figure 5.3 had considerable success in the USA as well as in Europe. Not only did private medical practices have them, but individuals also had access to them. Some physicians prescribed the home use of small devices with a dosage that was intended to be as precise as the dosage of pharmacological preparations. The role of such home "treatment" was also linked to the notion of control over morals that affected societies where cities were increasingly populated. Sexual behaviors were "therapeutic" targets, symbols of a way of understanding society and its norms:

> One way in which Lobb's [LOB 59] view of electrotherapy did stand out from the norm was in his insistence that it was a useful remedy for male sexual misbehaviour. He shared with others the more or less conventional view that female hysteria as an affection of the nervous system [...] frequently the result of excitement of the generative organs was amenable to electrical treatment. More unusual was his emphasis on electricity as a cure for the diseases emanating from the solitary vice. He described the symptoms that betrayed a young man's addiction to masturbation. [RHY 11, p. 127]

6 The author refers to the *Notes pratiques sur l'hydrothérapie, l'électricité et les rayons X* published in 1902 by Dr. C. DeBlois.

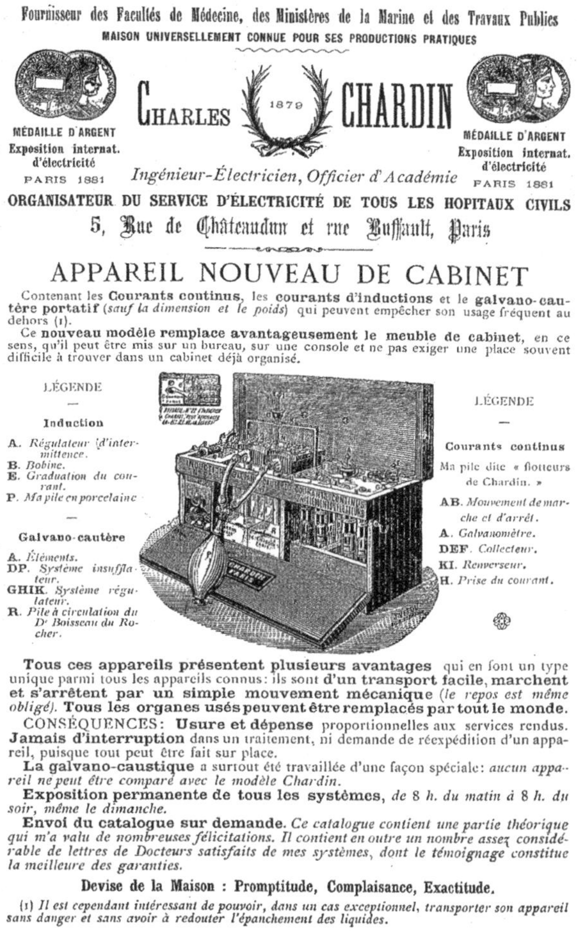

Figure 5.3. *This device illustrates the miniutarization of electrotherapy. Considered more practical, it still allows applications of direct currents as well as induction currents [DAN 92]*

This expansion of moral therapies came in place of the control of conscience by religion. It was by using expressions referring to sexual habits suspected of ruining the body's vital energy and repressed by 19th Century morality that the promoters of the electric belt believed they could provoke the act of purchasing their product. One of the main targets was onanism, suspected of provoking either convulsive crises or impotence and exhaustion:

> Extremely faithful to the pathological representations of the time, they painted the abnormality or talked to those more knowledgeable about

> the miseries of sexual appetite. This allowed them to finally have the last word: healthy sexuality was trendy sexuality! [...] The daily use of electrical therapy kept the body under the beneficial power of the rectifier current, which imposed its standard and functionality. [ROU 91, p. 202, author's translation]

The obstinate belief in electricity to combat diseases of the mind, dissolute morals and behavioral disorders was unfailing. Cases of more or less proven melancholy and of improved abatement by electrical treatment were reported by Auzouy:

> When he arrives, he remains motionless and mute in the place where we put him; his face expresses stupor, his pupil was dilated, his eye dull, his movements numb. He is completely insensitive; he was jabbed, pinched, subjected to the influence of a strong induction current, he did not even suspect what was being done to him and did not complain. He was a real automaton. One took the side of etherizing him and a slight improvement was manifested; affusions and the faradization to which he is frequently subjected were employed concurrently. Little by little he regained spontaneity, he answered questions and accepted a role as an assistant in the asylum's pharmacy. Still not very communicative and very focused, he was more and more sensitive to the electro-magnetic shock, and he ended up reporting his impressions. One day, he approached us with a smile on his face, and he told us that he feels his healing had taken place. [AUZ 59, p. 30, author's translation]

The doctor cautiously concluded that his intervention must have benefited from happy circumstances, not always objectified. This case was paradigmatic of the symbolic power of electrotherapy exerted on impressionable patients. In the rationalist environment of science, the electrotherapy machinery represented the action of humans on their fellow human beings and referred to the fact that humanity possessed the means to intervene on its physiological and mental interiority. This was how the technique developed by the neurologist Clovis Vincent (1879–1947), a pupil of Babinski (1857–1932), can be interpreted for the psychologically wounded from the First World War, who returned traumatized with equally traumatic physical deformities:

> In this conflict at ground level, writes Pierre Darmon, many of the buried soldiers, once they were freed, found themselves frozen in strange postures, bent over or crouching. Some still had twisted hands or feet, others were deaf or mute. But clinical examination revealed no lesions. They were therefore declared influenced or even treated in

> simulators. For them, Babinsky coined the term pithiatism (from the Greek I persuade and I heal). [DAR 01, p. 49, author's translation]

Thus, Clovis Vincent recommended the application of a very high intensity faradic or galvanic current, quickly nicknamed *torpedo* by the soldiers:

> The word torpedo was not coined by me. It was the first infantrymen I treated and healed with the help of the strong galvanic current who said, 'It turns you over like a torpedo'. The word they threw out was then passed from infantryman to infantryman. It did not have, moreover, in the minds of the first men who pronounced the word, a pejorative meaning. It was a cheeky, and basically French, way of referring to a treatment that transforms in a few moments. [VIN 16, p. 405, author's translation]

While the electric shock therapy used to send them back to the frontline seemed violent, it corresponded to emergency medicine aimed at freeing them from their disability as quickly as possible. In a conception that was basically very Pinelian, it also tended to emphasize the reversibility of psychological trauma. Locked in shackles or treated with electric shocks, soldiers sometimes regained temporary mobility.

The quarrels about the necessity, usefulness and dangers of electric shock treatments had already existed for a rather long time since they were defended by Ledru, Romain Vigouroux but strongly rejected by Paul Vigouroux. The technique of "torpedoing" became a symbol at the end of the First World War of torture by electricity, reviving the specter of electric martyrdom and provoking many controversies:

> The Austrian Commission into Military Misconduct, set up by the Austrian Parliament after the war, accused psychiatrists – including the eminent Viennese professor Wagner-Jauregg, future Nobel Prize winner for his work on malariatherapy – of having subjected soldiers suffering from neurosis and suspected of simulation to violent electrification sessions under duress. [RAS 10, p. 74, author's translation]

Freud (1856–1939) participated as an expert in this commission.

We generally have a dark perception of war electrotherapy, but what exactly was it? Generalist uses of electricity developed during this conflict, concerning the human body as a whole and encompassing the psyche, continuing the wave that medical electricity experienced from the 19th Century. The shortage of medicines and the growing number of patients were all factors that help to understand why torpedoes were considered a worthwhile treatment:

> During the conflict, war was analyzed by contemporary medical and hygienic thought, which perceived it as an object, as an environment, in the etiological sense of the term, as a set of favorable circumstances, engendering its own pathologies, leading to wounds and diseases specifically born of the conflict, posing singular questions, in unprecedented terms, to medicine, in particular to electricity, dealing with the living. [RAS 10, p. 75, author's translation]

The period of 1914–1918 did not give rise to the appearance of new electrotherapeutic techniques *ex nihilo*, but it did systematize their application as a matter of urgency, which implicitly included the use of electric shock. At a time of great political unrest, the numerous electrotherapy rooms and the already highly developed techniques were finding an important field of application. The survey, conducted in 1915 by the Royal Society of Medicine [SAY 15–16] in London on the organization of electrotherapy in French military hospitals, testifies to British interest in the development and effects of these techniques. The extent of the equipment, the diversity of electrical treatments such as the extension of hospital electrotherapy services or the large number of patients treated were described. This can be correlated with a strong clinical and empirical tradition in France. In a straight line from the end of the 18th Century, this generalization of hospital electrical practice encompassed a very wide spectrum of pathologies and appropriated the application field of traumas already studied by Silas Mitchell (1829–1914) [MIT 72] during and after the Civil War. During this conflict, Mitchell was in charge of nervous diseases and injuries at Turners Lane Hospital in Philadelphia. A specialist in neurology, he was known for having conceived the notion of the rest cure in the treatment of nervous diseases, for having tackled the notion of male hysteria and he "[...] described precisely in the aftermath of the Civil War what is called the 'phantom limb[7]'" [GEN 16, p. 171, author's translation]. Faced with these unknown disorders, with psychosomatic appearances that could not be explained by a cerebral lesion, he tackled on the one hand the theme of simulation; on the other hand, that of war neuroses, medicalized in the form of traumas or mental fractures.

Thus, far from representing a sudden eruption, shock electrification methods were colored by the urgency of the conflict and deployed a *continuum* of electrical, polymorphic practices from which medicine drew hope for healing its traumatized victims. The torpedo method was part of the therapeutic framework of physical cures for neurosis, whose success accompanied the decline of "moral therapy" in the

7 Refers to the sometimes painful perception of amputated limbs.

second half of the 19th Century. Electrical cures for hysteria were based in part on the patient's cerebral impressionability and suggestibility. In this perspective, they aimed to rebalance and reconnect nerve functions, seeking to reconstitute the physical and mental energies of the human machine by restoring the energy necessary for brain operations. While a sedative electricity developed in the 19th Century with electrostatic bath treatments, an exciting electricity triggering strong local sensations, by tingling, burning and tonic action of sparks, came to support the care of diseases such as melancholy, catalepsy or psychological paralysis:

> Nervous disorders resulting from war traumas (trembling, paralysis, aphasia, blindness, mutism), in the absence of visible or proven organic lesions, were explained either by a direct psychological cause (it was then a real morbid entity, caused by external factors), or by phenomena of suggestion or autosuggestion, as described and theorized by Joseph Babinski, Charcot's pupil, as 'pithiatic disorders'. [RAS 10, p. 86]

Simulation was considered to be subtly different from hysteria. Babinski had already differentiated between the ills that stemmed from a desire to escape from difficult working conditions and the blockages of thought. The first subject may have decided to come out of his torpor while the second needed to be convinced to remember that he could do so. For Babinski, the cure for hysteria was largely based on persuasion and suggestion. Thus, in the first instance, hysterical phenomena were considered likely to be relieved by the patient–physician relationship. High-intensity galvanization was used for patients who were resistant to mild therapies.

Clovis Vincent also pointed out that:

> It is important to make an absolute distinction between the hysterical and the simulator. The hysterical one, also called the pithiatic one, is a man who can, but does not know that he can. He even feels he can't. He won't want to until we force him to. The simulator knows and feels he can, but it doesn't want to. [VIN 16, p. 407, author's translation]

In 1920, in response to the evaluation of the treatment provided by Vincent, Freud proposed another way of interpreting psychic phenomena, positing simulation as unconscious and as already being a neurotic symptom:

> The school of psychiatry called psychoanalysis, which I created, taught for twenty-five years that neuroses of peace must be reduced to affective disorders. The same explanation was then applied in a very general way to war neuroses. We had further indicated that the

> nervous ones suffered from animal conflicts and that the wishes and tendencies that were expressed in the manifestations of the disease were unknown to the patients themselves, therefore unconscious. [FRE 84, p. 228, author's translation]

Faced with the porosity of the concepts of simulation, neurosis, trauma and hysteria, electricity played a therapeutic and diagnostic role. The existence of abnormal electrical reactions made it possible to differentiate between organic paralysis, simulation and hysterical or functional paralysis. Depending on the complete or incomplete degeneration[8] of the motor neuron, paralysis was considered differently. In contrast to those suffering from organic nerve damage, pithiatics were not sent for discharge but for treatment.

The treatment was carried out in three steps:

– torpedoing, carried out with faradic or galvanic currents, taking from a few minutes to several hours;

– this was followed by a time of consolidation of movements that returned to normal in order to fix the return of natural automatism;

– a third stage was marked by the use of moral treatment in a dialogical form in order to restore the patient's confidence in his physical and mental possibilities.

In just a few months, Vincent treated up to 700 patients. The torpedoes were applied through plugs, very close together, to keep the electrical circuit as short as possible:

> The patient did not give in until the current reached a certain intensity. It was during the painful treatment phase that the man, sometimes convinced that we were acting wrongly because he felt that he could not be cured, since four, five, ten doctors had considered him incurable, shouted, cursed at us, gave insults and struggled. This period was quite long. Then after a slightly stronger reaction than the previous ones, the man gave in. He's starting to perform the act that we wanted him to do and he's getting better. [VIN 16, p. 410, author's translation]

Electrical treatment during the First World War depended on structures and sets of devices that had already been developed. The history of shock therapies, whose beginnings can be traced back to Ledru's work, alternates between periods when

8 Already in 1884, Erb was trying to link the histological alterations of nerves and muscles to both quantitative and qualitative variability in electrical excitability. Abnormal electrical responses were referred to as degeneration reactions.

these techniques were fiercely criticized or even condemned, and then others during which they returned, often in crisis situations. The advent of electroshock therapy did not negate this idea. Electroconvulsive therapy (ECT) appeared in April 1938, with the work of Ugo Cerletti (1877–1963) [CER 50] and Lucio Bini (1908–1964) during the decline of electrical therapy in psychiatry that followed the end of the First World War. Although Cerletti applied an electric current, he did not refer to the classic works on the therapeutic application of electricity by Franklin, Nollet or Aldini. His treatment was guided, on the one hand, by the analogy with the chemical seizure drugs, camphor and metrazol, used to induce epileptic seizures; on the other hand, by the clinical idea that epileptic seizures protected the brain from psychiatric disorders, particularly schizophrenia. Brain numbness caused by epilepsy could therefore be artificially reproducible. Cerletti, in April 1938: "[...] The patient was schizophrenic, who had been found wandering around the railway station suffering from delusions, hallucinations and confusion" [ASH 13, p. 293]. This subject received a series of shocks without anesthesia. Calmer, he was then considered cured. Hundreds of patients were treated by Cerletti. However, his treatments were controversial, on the one hand because of the pain they caused, and on the other hand because of the ensuing effects such as short-term memory loss[9]. The introduction of electroshock therapy in psychiatric hospitals in France followed two logics: that of implanted clinical electrotherapy and the return of the war context. In the summer of 1940, when the first applications were made by the Germans in the Saint-Anne hospital on their soldiers, the mentally ill were suffering from hunger and the effects of occupation. In December 1940, Jacques Rondepierre, head doctor at the Ville-Evrard psychiatric hospital (Seine) and the radiologist Marcel Lapipe developed the first French "*sistonothère*".

In the face of the difficulties of moral treatment, these treatments seemed once again to be a cure for the incurability of madness and revived the hope of being able to return these patients to their families in a context where hospitals were struggling to care for them. This therapeutic hope was spreading to many countries:

> This did not prevent electroshock from spreading rapidly thanks to the launch of a small number of machines manufactured by Arcioni in Milan. Between the end of 1939 and the spring of 1940, electroshock therapy began to be used in Switzerland, Germany, Holland, England, Sweden, Denmark, the United States, but also in Argentina and some African countries under colonial rule. [BUE 10, p. 95, author's translation]

9 Hemingway testified in 1961: "What is the sense of ruining my head and erasing my memory, which is my capital, and putting me out of business? It was a brilliant cure, but we lost the patient" [ASH 13, p. 295].

The craze was quite immediate. In a period that can be described as euphoric, many psychiatrists liked to think that electroconvulsive therapy was a universal treatment and that its indications could be extended to all mental illnesses, including general paralysis and epilepsy:

> J. Rondepierre and J. Vié concluded from their trials that electroshock had no indication in comitials with spaced seizures, but that, in epileptics with frequent seizures, it improved mental disorders and social performance. [RON 43, p. VII, author's translation]

But in a second period, they admitted that while provoked epileptic seizures had dramatic effects on states of severe melancholy and some forms of mental confusion, other psychiatric conditions, particularly schizophrenia, responded poorly to treatment:

> The author focused on his very important personal experience of 76 electroshock treatments and reported a summary of his observations, and then specified the indications of electroshock according to the results obtained. The major indications were shock psychoses, exhaustion psychoses, first melancholic and manic attacks of unknown origin, mental contusions. The results were less consistent during manic-depressive psychosis, chronic hallucinatory delirium. They were more disappointing and especially transient in schizophrenic states, where they seemed to give nothing more than Cardiazol or insulin therapy. [DEL 43, p. 180, author's translation]

Yet, it was a matter of treating and curing patients quickly in times of shortage. While the mechanisms underlying electroshock therapy were uncertain and unknown, massive shock treatment empirically affected the electrical activity of nerve cells, causing an electrical storm to mimic epileptic episodes. For example, electroshock therapy used artificial electricity to blast brain electricity. The cultural and medical importance of seizure disorders was not without influence on the use of seizures. Initially provoked by Cardiazol treatments, electricity took over from chemistry, thus rediscovering a medical tradition of shock treatment and its effects on the subject. Electroshock therapy contributed to the renewal of interventionism and technicism in psychiatry, while reducing the mortality rate of patients sectioned due to the various shortages:

> In the context of an acute food crisis, the challenge was also to demonstrate that, contrary to deeply rooted prejudices, the insane were curable and therefore deserved to survive, since they could eventually regain their place and usefulness in society. It was in the name of the

> curability of mental illness that doctors in psychiatric hospitals were calling for an improvement in their patients' diets. [BUE 10, p. 100, author's translation]

Electricity appeared to be a "quasi-pharmacological" agent that was easier to obtain, quicker to implement and more immediate in its effects. Thus, electroconvulsive therapy has an ambiguous history, which follows the history of electrotherapy as classically understood, but also linked to the history of medical galvanism and of an electricity conceived as a force present in the organism on which medicine could intervene. Thus, electrical treatment is intrinsically dependent on the techniques and machines used and seems to develop differently by virtue of the approach to the use of this force, which is more clinical or more exploratory. Can electricity be considered a diagnostic tool for mental illness? Has it contributed to the demarcation of boundaries in mental illness and neurological diseases?

5.1.3. *Electricity: a diagnostic tool for mental illness?*

When did electrodiagnosis come into existence? While this treatment allows the assessment of muscular symptoms resulting from a nerve lesion or pathology, this was not until the end of the 19th Century and in particular Duchenne de Boulogne's research, which we will study at greater length in the second part of this chapter. Nevertheless, can electricity be a differentiated diagnostic tool for mental illness without brain or muscle damage? Several elements lead us to answer this question affirmatively.

In 1859, Auzouy in *Des troubles fonctionnels de la peau et de l'action de l'électricité chez les aliénés* used electricity as an indicator of the intensity of the effect of mental illness on clarity of mind. The deeper the obliteration of the faculties, the less patients responded to stimulation. On the contrary, the less the sphere of the intellect was affected, the more strongly the person reacted:

> The electrical agent measures, with a degree of accuracy that could never be achieved without its help, each subject's physical sensitivity. What is most striking in our experiments is that the accessibility to the suffering provoked coincides with moral aptitude, and that the electrification acts more or less energetically on sensitivity, depending on whether the individual is more or less gifted with the faculty of forming more or less clear-cut ideas, depending on whether he is more degraded in the intellectual scale, or more advanced in his mental aptitudes. [AUZ 59, p. 28, author's translation]

In 1884, Erb (1840–1921) cited research by Arndt (1835–1900) [ARN 69] and Tigges [TIG 77], both psychiatrists, on large groups of depressed and psychotic patients. For Arndt, electrotherapy was a unifying tool in the treatment of mental and nervous diseases. While Tigges recommended electrical stimulation of the brain in cases of deep and chronic depression; Arndt applied electrical stimulation to the brain in cases of severe psychoses with depressive symptoms of catatonia, hypochondriac delirium and melancholy. His research was proving to be heuristic for the development of the current tDCS method. Arndt recommended the use of faradic or alternating current as a stimulant against passivity, stupor, weakness and manic-depressive disorders. On the other hand, galvanization was recommended for disorders such as psychoses and psychotic symptoms. Vertical, horizontal and diagonal galvanization of the head, with two electrodes attached to the cranial bone, sometimes supplemented by simultaneous galvanization of the sympathetic system and the cervical spinal cord, were described as particularly effective in psychoses, recent anxiety and sensory hallucinations:

> Faradic current acts simply as an excitement, as an irritation; if one wants to obtain only this result, one can give it preference; it succeeds especially in states of simple depression, either primary, or coming as a result of tempestuous processes. [...] On the other hand galvanic current has, besides its exciting action, yet more actions (modifying, changing the moral state, calming, catalytic); its sedative and sleep-inducing actions especially, appear with great clarity; it is therefore suitable for almost all other psychoses, accessible to electrical treatment. But for success, the method used (the chosen direction of the current, the pole used for the action) is not at all indifferent; but very often one must first determine it empirically. [ERB 84, p. 317, author's translation]

It is worth noting the therapeutic modeling according to which an electric current corresponded to a disease. This conception, rooted in an organicist psychiatry, was a variation of the equivalence between a disease and a singular localization. For example, to psychoses corresponded galvanization. This meant varying the points of stimulation, the passage of electricity, the types of current and the intensities not only to treat the patient, but also to identify what the patient was suffering from. The similarity of physical and mental symptoms presented in many psychiatric disorders found here a way of objectification through therapeutic success. Intuitive questions were asked regarding the refinement of diagnoses but also the action of electricity on the brain centers. Indeed, brain stimulation techniques needed not only to identify the appropriate locations but also find the most favorable points for the passage of this energy into the brain structures:

> The faradization of the brain also has no influence on paralysis. In this connection, it is questionable whether electricity actually penetrates to the cortical centers that hold the arm movements under their control.

> There's no proof of that. But I am merely reporting here without comment on the experimental facts. [RIC 85, p. 745, author's translation]

This quotation from Paul Richer can be interpreted in two ways: on the one hand, under the theory of brain localization and the emblematic place taken by the brain in the functioning of the body, questions arise about the ability of the protocols for applying electricity to conduct it to the intracranial parts; on the other hand, the fact that electrical treatment had no effect on a paralysis, described in the context of hystero-epilepsy, may prove to be an opportunity to dismantle this entity, to differentiate the two pathologies and, more generally, to make it possible to delimit the fields of psychology and neurology.

Duchenne de Boulogne used alternating current to precisely stimulate and isolate one muscle group at a time. The induction device enabled him to measure the intensity of the current delivered to the organs and to adapt it continuously to the response obtained. Two exciters fitted on insulating sleeves allowed the condition of a muscle to be assessed.

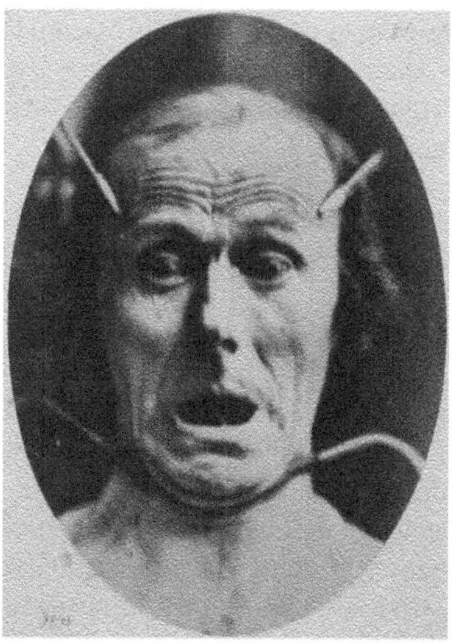

Figure 5.4. *Electrical contraction of the skin, of the frontal muscles, with lowering of the upper eyelid and looking down: expression of fear [DUC 62, fig. 61]*[10]

10 Duchenne de Boulogne's photographic collection available at: ENSBA Beaux-arts de Paris, l'école nationale supérieure, www.ensba.fr/.

Thus he carried out diagnostic evaluations on the state of the muscles and nerves of the wounded in Paris in June 1848. His electrical experiments also allowed him to detail each emotional expression by showing their muscular mechanics. He meticulously recorded all possible facial expressions from a man with paralyzed features which he captured through photography (see Figure 5.4).

In addition to showing the possible independence of psychological content and emotional mechanics, his technique allowed him to isolate pathological muscle circuits. He located several disorders such as *Duchenne muscular dystrophy* and discriminated the nerves actually damaged from those whose function was only suspended, by reflex or because of psychogenic disorders. Thus, the absence of reaction to the current could be interpreted as a sign of organic damage. Duchenne referred to the use of localized electrification to establish differential diagnoses:

> Unfortunately, this author (Landry) has not been able to distinguish the functional disorders resulting from the loss of feeling in the contracting muscle, as Ch. Bell said, from those caused by injury to the mental faculty of motor coordination; [...] It is because these two diseases, which are easy to confuse, differ essentially from each other by their symptoms, their operation, their prognosis and their treatment, that I reserve the right to explain at length the differential diagnosis. [DUC 72, p. 669, author's translation]

However, these differential diagnoses were not without overcoming the technical obstacles associated with diseases with symptoms that were as variable as they were invasive:

> Having taken care to gradually accustom the sick to the electric sensation, starting with very low doses, and thus dissipating their prevention, I no longer saw the nervous crises that were obviously produced by a moral influence. I therefore advise you to devote the first session to getting the hysterical ones used to the electric sensations in this way. By doing so, it is not long before the highest doses of the induction current can be administered. The decrease in muscle sensitivity, which complicates hysterical paralysis, would indicate the use of the first helix current and rapid intermittencies which act energetically on muscle sensitivity, at the same time as they cause the return of movement; but muscle faradisation with rapid intermittencies is not always applicable in hysterics. [DUC 72, p. 718, author's translation]

But when patients were prepared for faradization, thus during muscle contractions, seizures stopped. Depending on the degree reached by the hysterical attacks, Duchenne chose between fast intermittent muscle faradization and spaced intermittent muscle faradization. He thus outlined the boundaries of the pathological descriptions within hysteria itself and established clinical correlations between the intensity of emotions and hysterical disturbances of limb mobility and posture. He thus dealt with the cases of patients who suffered as a result of deep sorrow, a series of "hysterical accidents" [DUC 72, p. 816]. Vivid emotions such as fear and anger did not produce atrophic paralysis, but acted more specifically on the brain, "producing intellectual disorders, hemiplegia and generalized paralysis with contractures, without damage to muscle tissue" [DUC 72, p. 417, author's translation]. The electro-cutaneous excitation allowed him to recognize, *a posteriori* and in case of curative success, the cases of muscular hyperesthesia caused by hysterical disorders:

> Usually patients experience a certain amount of relief after the first electro-cutaneous excitation, the disruptive action of which continues, to the point where the muscular hypersensitivity gradually diminishes and may even disappear completely. [DUC 72, p. 822, author's translation]

In an article published in 1903 entitled "*Guérison d'un cas de Mélancolie à la suite d'un accès provoqué de Vertige Voltaïque*" [BAB 03], Babinski described the case of a patient in her forties who was suffering from melancholy that was resistant to all treatments. Opium, belladonna, hydrotherapy, balneotherapy and static electricity were tried without success. Babinski then decided to subject her to an electric voltaization of the head for a few minutes until she became dizzy. Almost immediately, this operation relieved her of her symptoms. After several sessions of increasing intensity, the patient declared herself cured. In his article, Babinski seems to doubt the truly curative role of *voltaic vertigo,* because, by his own admission, electrification had never proven itself in cases of melancholy. By causing experimental disorientation, he detected lesions of the cerebellum or vestibular apparatus. Manifestations of abnormal electrical reactions made it possible to differentiate organic paralysis from hysterical paralysis. This work, in which electricity was used to differentiate neurotic states from lesions of the nervous system, led to a reading grid for the physical and mental domains. Within the mental sphere itself, simulation pathologies and high emotional intensity disorders stood out, which needed to be differentiated from each other. Pithiatism and simulation raised the question of the differentiation between a voluntary crisis and one of psychogenic origin, often unconscious:

> I would reply that a thorough investigation often shows that this spontaneity is only apparent, that it is indeed the reproduction of a

> crisis observed in others, and that sometimes, what is called a hysterical crisis is nothing more than an association of cries, gesticulations, voluntary, conscious contortions, whose origin, in the person who is the author, is not emotion, but the desire to move those around him. It should be noted, moreover, that a mythomaniac who would like to feign convulsions, without ever having witnessed a hysterical seizure, could be spontaneously led to execute movements similar to those by which the legitimate attack manifests itself. [DUC 72, p. 18, author's translation]

A diagnostic tool and an empirical treatment since the second half of the 18th Century, can electricity be considered as a medicine? In 1765, Venel (1723–1775) defined a medicine as "...any material which is capable of producing in a living animal useful changes; that is, of restoring health, or of preventing disturbances, either by internal intake or external application" [VEN 65, p. 295, author's translation].

Electricity, in its different forms, softer or more tonic, was applied in the field of paralysis and then for nerve and brain diseases. Unpleasant effects, therapeutic effects, dosages, measurements, it seemed to bring together the conceptual framework of the medicine concept:

> When he combined the effectiveness and danger of electricity, Mauduyt reminded us that this is the nature of remedies and that this is the basis for the need to use concomitant remedies to prevent or correct its negative effects. Here the theme of risk and danger was a major element in the integration of electricity into all medicines. [ZAN 17, p. 97, author's translation]

While in the 18th Century, electricity was available somewhere between a remedy and therapeutic effects and the potential dangers of its applications, it became more precise through its various techniques (galvanism, faradization, arsonvalization) and was separated between therapies for the ills of the mind and body and an exploratory instrument for the nervous and muscular systems. The medicalization of electricity corresponded to the fact that the issues it addressed preceded it. The issues of materialism, localization of faculties and the definition of human nature motivated the integration of this exogenous force in pharmacopoeia. This inscription of a natural force in medicine represented the medical turning point of technician and interventionist practices. While the first applications to paralysis failed, its appropriation by the field of mental affections reappeared with the question of an application on the shape and mobility of the body. Thus, lines of convergence were drawn between mental and physical ailments that could be the cause.

While electrotherapies developed throughout the first half of the 20th Century, in the same period, an exploratory electricity, extremely technical in nature, deployed its instruments towards electrophysiology and knowledge of the electrical dimension of organisms.

5.2. Exploration and recording of nervous system activities

Measurements and explorations of brain electricity are all techniques that shed light on its physiological mechanisms, modes of conductivity and transmission from the brain to the nerves and vice versa. Transcranial or cortical stimuli locate, map and indicate the places where these mechanisms take place, while recordings of brain electricity advocate a new alphabet of thought, offering new images to understand how the mind demonstrates action of matter. Diagnostic tools, measuring instruments, cartographic references and electrophysiology are found in experimental laboratories couched in biophysical terms. A mixed modeling, localist and holistic, emerges from these specializations in the electrical explorations of life. Sometimes regionalized, sometimes functioning as a whole, the central and peripheral nervous systems constantly interact with normal, pathological, natural and artificial patterns of electricity.

5.2.1. *Electrophysiology: measuring and exploring from 1848 onwards*

Measuring the electricity flowing along the nerves provided a view of these pathways through living organisms. In fact, while the nervous system exercises control over the whole body, control which translates into voluntary or involuntary acts, sensations, whether conscious or not, it incessantly transmits messages, coordinates muscular movements, regulates emotions and makes you think via electrical force. In any case, this was the vision of the organism and its mechanisms that was developing in laboratories, particularly in Germany. Electrical medicine was continually reinventing itself, trying to understand and represent how the knowledge of animal electricity, inherited from Galvani, was able to offer leads for diagnosis and treatment. Electrophysiology, a fundamental discipline of contemporary science, is one of the main foundations of modern neuroscience as it allows the study of the functioning of brain circuits. These circuits, designed in analogy with metallic circuits, reinforce the perspective of the human machine. The concept implies recurrent paths along which an electric force, a nervous signal, circulates. However, while they underlie our movements, perceptions and even mental activities, it means that if we can measure them and measure the force that circulates through them, it is possible to regulate the malfunctioning of these natural circuits and influence their mechanisms. As Sherrington (1857–1952) pointed out in 1949, "Every sensation, every perception, even that which arises in us from the observation of a beautiful landscape, a beautiful painting, or from seeing a friend's face, is due to the flow of tiny electrical signals" [SHE 49, p. 113]. As part of a permanent electrical organic environment, these signals are constantly circulating along the fibers of the peripheral nerves or the central circuits of the brain. While their normal functioning conditions cognitive acts such as speaking, writing or thinking, electrophysiology corresponds to a program of research on the causal links between the anatomophysiological functioning of the brain and physical and mental behavior. These signals are triggered as electricity is constantly moving through the body and is moved by internal or external stimuli. Following the end of the polemics between Volta and Galvani, the key to understanding life is electric:

> The Bologna doctor and the Pavia physicist each believed they had found the true cause of the electric disequilibrium responsible for nerve excitation and muscle contraction. For Galvani, this disequilibrium was intrinsic to the animal organism, whereas for Volta, it originated from the metallic arc used to connect the excitable tissues of the frog preparation. [PIC 13 p. 269]

Developments in electrochemical experiments were important for understanding the mechanisms of electricity in tissues. For example, the experiments in 1800 by William Nicholson (1753–1815) and Anthony Carlisle (1768–1840) on the dissociation of water led to the discovery of many new chemical elements, including processes responsible for generating electricity in batteries. In this way, they uncovered the mechanisms of water electrolysis by passing an electric current through it, responsible for its decomposition into hydrogen and oxygen.

Matteucci and du Bois-Reymond [BOI 48–49, p. 638] perfected the voltaic pile and stimulation electrodes, allowing better control of the intensity, duration and location of the stimulus, thus opening the door to modern electrophysiological brain experiments. Between 1838 and 1841, thanks to galvanometers capable of measuring low electrical currents, animal electricity could be accurately detected. Improvements in instrumentation to understand and objectify electrical mechanisms could never be detached from medical, clinical or experimental practices. The convergence of the improvements in the electrodes, with those in the battery, played an undeniable role in understanding brain mechanisms, whether in the field of measurement or recording:

> In spite of the difficulties of the physiology of the elementary nervous signal, the electrodes enabled the physiology of the central nervous system to develop when they were inserted at the end of the 19th Century into the deep structures of the animal's brain. [DUP 99b, p. 252, author's translation]

Matteucci showed that a device, provided it was connected between the surface of a muscle and a damaged part of the same muscle, indicated the passage of a current, later called the linking current:

> In wanting to bring the origin of the proper current closer to that which we have admitted for the muscular current, we should suppose that, by a connection which is completely unknown to us, and which it may be for anatomy to discover, the tendinous surface which makes up the greater part of the frog's leg, represents the interior of the muscle; [...]. [MAT 44, p. 130, author's translation]

Medicine continued to get to grips with the progress of physics and the latter, to medicalize itself:

> The impact of a biological scheme on a physical discovery is illustrated by the discovery of the electric battery. Whatever the mental elaboration guiding Volta to the invention of battery, it is certain that in constructing this epoch making device, Volta's attitude was in fact that of creating an apparatus which, [...] was clearly much more related to complex machines of the "organic" type, derived from the realm of biology (the electrical organ of the fish), than to the simple physical devices of his time. [PIC 98, p. 405]

Chairs in medical physics were created, particularly in France, and bioelectricity was developed, reviving the skinned anatomical figures of the Renaissance, since it was often necessary to bare nerves, muscles and brain areas to access these natural currents:

> The first success with the instrumental recording of animal electricity was thus the achievement of a physicist. This shows how difficult it was to separate, even in that period, physics from physiology in this field of studies. As a matter of fact, the investigation of animal electricity was then one of the favorite research fields of many apparently genuine physicists and chemists. [PIC 13, p. 271]

du Bois-Reymond confirmed Matteucci's discovery of the linking current. In addition, he suggested that muscles and nerves were made up of molecules that were positively charged on one side and negatively charged on the other. These molecules could therefore orient themselves naturally like the particles of a magnet with a pole at each end. This idea, although later re-evaluated, was heuristic because it anticipated or introduced the idea of an electrical polarization of the nerve or muscle cell membrane. Thus, through a series of physiological events, this electrical movement, including at the level of molecules, caused the production of short-duration electrical signals, called nerve impulses or action potentials. These signals, propagated along the membranes of nerve fibers, were the basis of our nervous and cerebral functioning. Similar electrical impulses were responsible for the phenomena leading to the excitation of muscle fiber and played a major role in the rhythmic activity of the heart. They were therefore essential to life, to its definition, and our understanding of vital phenomena depends on their knowledge. du Bois-Reymond demonstrated the existence of currents specific to nerves and muscles, notably thanks to his improvements to the galvanometer. In an electrical anatomy approach, he removed a nerve segment and found a current that he called electromotive power or electromotive force. It also allowed him to recognize the electro-tonic force, a property of the nerve according to which, provided that a direct voltaic current was passed through the free part of the nerve, above and below the circuit closed by the galvanometer, an energetic deviation was observed. The current applied excited the muscle and stopped at the very moment the contraction took place. It was then possible to determine the time it took for the nerve impulse to travel a given length along the nerve. Helmholtz, to whom du Bois-Reymond entrusted this research, excited the nerve connected to a muscle of a hind leg and determined the time that elapsed between the excitation and the resulting contraction, then he activated the nerve at a point further away from the muscle and measured the increase. Thus, he attributed the difference to the time required for the nerve impulse to propagate between the two stimulated points. It had a conduction speed of only a few tens of meters instead of several hundred kilometers per second. In 1851, he was able to graphically record muscle contraction and calculate the speed of nerve signal propagation. These experiments represent an important milestone in the history of nervous system science. Helmholtz succeeded in bringing into the world of measurement, an elusive event in animal organization, other than through observation. He made graphically visible, print-worthy and precise, a manifestation of nervous phenomena related for a long time to immaterial entities:

> When an instantaneous electrical discharge has passed through a muscle of animal life, or the nerve branching there, there is first a period of time during which no appreciable effect is produced. After this time, the tension of the muscle increases by degree, reaches a maximum and finally declines to return to its starting point, corresponding to the muscle's state of rest. In frogs, I found 0.01 for the duration of the time remaining between irritation and the first manifestation of the mechanical effects of the muscle. From there to the maximum, there is 0.08; finally the decline in muscle tension, until it is completely relaxed, lasts from 0.3 to a full second. [HEL 51, p. 262, author's translation]

du Bois-Reymond first assimilated the *vis nervons* to the nerve impulse. The latter represented a step in the investigation of its identity with the electrical impulse. Did the impulse concept retain some relationship with liquids and flows? Because a dry brain receives brain electricity, compared to a wet brain dominated by chemistry:

> du Bois-Reymond was at the origin of 'mechanical physiology', one of the sources of contemporary biophysics. This dealt a definitive blow to the vitalist conceptions of the 'nervous fluid' which became the 'nerve impulse'. [DUP 99a, p. 32, author's translation]

This electrical model of the brain inaugurated a long line of brain-mechanical circuits. Action potential experiments[11] represented an electrical event along the nerve fiber:

> The nerve impulse (or action potential) is – as already mentioned – the mechanism for encoding and transmitting information that the nervous system has developed along the evolutionary process. It is characterized by a stereotyped amplitude and time course. Indeed, none of these parameters depends on the amplitude of the stimulus or of the physiological cause that had produced the impulse. As to the dependency on the stimulus intensity, nerve impulses obey the 'all-or-none' law, being produced in its full amplitude when the stimulus exceeds a certain threshold. [PIC 13, p. 310]

This raises the question of the specificity of the information transmitted to the different parts of the brain. Can we understand, in electrical terms or in terms of propagation speed, the transmission of different information through the optical or acoustic nerves? Johannes Müller (1801–1858) suggested in 1833 that the brain is the interpreter of these messages that arrive through the nerves. He spoke about specific nerve energy. His theory, contemporary with electrotherapy and phrenology, reinforced and clarified the conception of a theory of psychological functions located in distinct cerebral zones. Different parts of the nervous system could perform different functions under the influence of electricity. In 1857, Alfred Becquerel, brother of Edmond Becquerel, highlighted an autonomous action of the spinal cord which could not be considered as only conductive:

> Instead of applying the two poles to the upper part, if they are applied to two opposite points on the lower end, the muscles of the trunk and all four limbs contract in the same convulsive manner. This

11 An action potential is a unidirectional electrical signal that travels through neuronal axons and causes the release of neurotransmitters at synapses.

> contraction of the trunk muscles and upper limbs proves that the marrow is not a simple conductor of nerve fibers and that it has its own action. Indeed, in both cases there should have been only a simple convulsion of the lower limbs, while the more general convulsions that occur in the second case, prove the existence of a special action of the marrow, an action that here is ascending. These contractions of the trunk muscles and upper limbs are not produced by derived currents, but by the action of the spinal cord without the intervention of electricity. [BEC 57, p. 111, author's translation]

Ludmer Hermann (1838–1914), a pupil of du Bois-Reymond, drew attention to a fundamental aspect of the electrical phenomena involved in nerve conduction: signal propagation along a nerve fiber consisting of a negativity of the outer surface of the fiber. To stimulate the nerve, it was necessary for the outer surface of the fiber to become negative. He assumed that under normal conditions, the negativity of the outer surface of the fiber caused by the action current would act as a stimulus for the upcoming nerve segment. In this way, a circle could be closed, capable of taking into account the propagation of the nervous signal in a natural circuit. But it is to Julius Bernstein (1839–1917) that we owe the first measurement of the temporal evolution of the electric nerve impulse. Although Helmholtz was prompted to conduct his experiment on the electric model of nerve conduction proposed by du Bois-Reymond, his results did not account for an elusive velocity close to that of the electric field. Was nerve fluid therefore electrical in nature? In order to take the identity between nerve activity and electrical phenomena further, it was necessary to show that the negative variation propagated along the nerve at a speed corresponding to that of the nerve signal. This problem was thus taken up by Bernstein [BER 68]. He was a pupil of Helmholtz and du Bois-Reymond and in 1868 he began to produce this measurement using his differential rhizotome, an apparatus designed to study the depolarization current of a nerve or muscle by sending extremely short and close electrical impulses. Thus, by means of a mechanical device, this ingenious instrument enabled an electrical recording, based on a timing procedure of a sampling of the electrical impulse's path. It appeared at a given point on a nerve and within a defined time. It turns out that the speed of propagation of the negative variation and that of the nervous signal matched, which provided evidence in favor of the identity of the two events, nervous and electrical.

Thus, after 1850, discoveries on the physiological and biophysical foundations of nerve tissue function were made, as it was shown that nerve impulses were electrical waves whose speed of propagation along the nerves could be measured. This research aimed to shed light on the nature of life, in a perspective already opened up by Galvani, and made the electrophysiological properties of living organisms intelligible. Princely concepts for the understanding of the brain and peripheral nervous system were emerging. At a time when du Bois-Reymond was helping to show the existence of a potential for action, he uncovered a key element in understanding communication and transmission within the nervous system:

> I am not mistaken in thinking that I have fully realized the centuries-old dream of physicists and physiologists to equate the nervous principle with electricity. [IMB 06, p. 54, author's translation]

The nervous system entered the physics world, the world of the measurable, but also of psychophysiological investigation, where questions such as the speed of nerve impulses and the speed of thought were instantaneous. How could they be determined?

The contributions of electrophysiology were part of the phenomenotechnical paradigm enabling imitation, measurement and tracing of the body's electrical activities. It linked its destiny to the electrical theory of transmission. Imbued with the spirit of physics, equipped with its instruments, it enabled the theorization of the laws of nerve conduction and the development of these principles towards medical electrodiagnosis. Thus, alongside electrotherapy, regional electrodiagnostic practices, chronaximetry in France, electroencephalography in Germany, or electromyography in the Netherlands were developing.

At the beginning of the 20th Century, the physiologists who supported the existence of the neuron leaned towards the fact that the transfers of information between nerve cells or to a muscle were of an electrical nature or involved electrical phenomena. Santiago Ramón y Cajal (1852–1934) [RAM 09–11] demonstrated the structure of independent nerve cells outside of a reticular-type mesh. His histological preparations revealed cerebral morphological unity: in limited number, the complexity of neuron mechanisms was found in the proliferation of neuronal arborizations and in the large number of points of contact, called synapses. The neuron, a new unit of the nervous system, the term for which was coined in 1891 by Heinrich Wilhelm Waldeyer (1836–1921), opened up a new cerebral landscape.

Ramón y Cajal, thanks to the coloring technique developed by Golgi, made visible these "... cells with delicate and elegant forms, the mysterious butterflies of the soul, the beating of whose wings may some day – who knows? – clarify the secret of mental life..." [RAM 96, p. 363].

His concern for the intelligibility and the biological and anatomical visibility of the cognitive field through neuronal structures was found in the electric-alphabet analogy of thought. Lines of convergence could be drawn between studies of the fine structures of the brain, electrophysiology and the electroclinical approach. But while anatomists could, thanks to Ramón y Cajal's research on brain tissue, see its elementary parts, physiologists were struggling to demonstrate the existence of brain electricity, which had been suspected for more than a century. In 1875, Richard Caton (1842–1926) described how his galvanometer showed unexpected variations in the absence of muscle activity. He deduced that brain structures secreted electricity.

From 1934 onwards, electrophysiology made further advances. Alan Hodgkin (1914–1998), then a student at Cambridge, began a series of studies on the effect of local nerve blockage in the neuromuscular preparation of the frog. He essentially followed animal preparation, a true paradigm for this discipline, developed by Galvani. A student of Edgar Adrian (1889–1977), he was responsible for analyzing the changes in excitability induced in the apparatus' state of excitation by stimulating the tip of the sciatic nerve away from its insertion into the muscle, using a localized cooling technique.

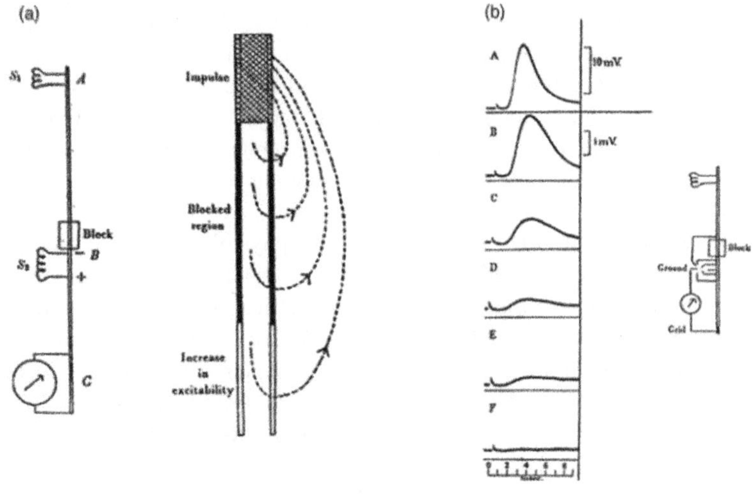

Figure 5.5. *Hodgkin's initial experiment on blocking nerve conduction by localized cooling in the preparation of frog nerve muscle [PIC 13, p. 290, Fig. 9.14]*

In 1919, Lapicque (1866–1952) gave a very successful form to the electrical theory of transmission based on the concept of action potential [LAP 43]. Thus he developed the theory of chronaxy studied by Jean-Claude Dupont in his *Histoire de la neurotransmission*:

> It underwent considerable development in French-speaking electrophysiological circles. Its power of seduction was essentially in the fact that it was part of an ambitious synthesis, aiming at a purely physical conception of the functioning of the nervous system. [DUP 99a, p. 37, author's translation]

Chronaxy, in the context of a physical and mathematical description of electrical events in the nervous system, refers to the length of time required to stimulate a muscle fiber or nerve cell with an electrical current of twice the minimum DC electrical current required to achieve a response to the excitation of a given organic element. Experimentally, Lapicque tested the effects of substances and poisons on chronaxy. An overview of the effects of pharmacology on cerebral and nervous electricity was thus outlined, highlighting in particular the conditions under which the transmission of excitation from the nerves to the muscles was interrupted as soon as there was a marked difference in chronaxy between them:

> Generally, the action of these substances for which pilocarpine[12] can be taken as a type is not simple; they also act on the nerve to reduce, in a first phase, the chronaxy which quickly returns to normal; on the muscle, after the first phase of chronaxy reduction, they present for a slightly high dose a second phase where the chronaxy is increased. [LAP 43, p. 95, author's translation]

We could say that Lapicque represented a culmination of the neuroanatomist movement who, from the last third of the 18th Century, refused the separation of soul and body or a negative definition of materialism, which would not be a reductionism but an extensionism towards a matter whose mechanisms were constantly being discovered. So Lapicque:

> [...] rejected the dualistic separation between life and soul (Descartes, 1664), mental experience and brain events (Sherrington,1906), or mind and brain (Popper and Eccles, 1978). In this, he was merely formulating a biological and psychological paradigm common in the scientific circles of the time, and now commonplace in ours, which sees in the complexity of the nervous system the conditions for the emergence of mental life, that is to say, according to Changeux's expression, a matter of thought. [DUP 99a, p. 38, author's translation]

The concept of chronaxy symbolized the harmony of a nervous machine based on electrical transmission and organic conductivity that was self-sufficient. The idea of electricity as the guarantor of the balance of the body was based on the fact that it participated in diagnosis, in the treatment of mental illnesses or in the definition of life and death[13].

---

12 Pilocarpine is an alkaloid extracted from jaborandi leaves, of the rutaceae family, having a direct parasympathomimetic action.

13 What was dying in the middle of the 20th Century? Developments in electrophysiology, coupled with the recording of brain electricity, enabled new explorations to understand the limits of life and death:

Molecular biology eventually showed that there are all the biological conditions necessary for the production of electrical phenomena in living organisms:

> The crucial element in the genesis of electrical potentials in living organisms is represented by the plasma membrane (or cell membrane), the thin structure that encloses the internal (or intracellular) compartment of the cell, thus separating it from the extracellular part. In the cell membrane there is indeed the complex molecular machinery responsible for the electrical phenomena of living beings, and particularly for the processes underlying the excitation and conduction of electrical signals in nerve and muscle. [PIC 13, p. 300]

Electrophysiology made it possible to establish that although the nerve impulse is indeed of an electrical nature, it is not transmitted as a current along a conducting wire; it is a changing condition, a wave that moves with a constant duration and amplitude and at a speed particular to each type of nerve and to each animal species. But where does this nervous electricity come from? Hodgkin (1914–1998) and Huxley (1917–2012) in 1952 [HOD 52] defined the action current as an electrochemical process based on changes in cell membrane permeability to sodium and potassium ions.

Thus, electrophysiology of the late 19th and early 20th Centuries painted a rather holistic picture of the functioning of the living, subject to perpetual physical movement. It brought the measurement and objective animal electricity from Galvani. The power of the physical and mathematical images that framed these explorations of biological mechanisms contributed to the development of this laboratory discipline that made the boundaries between physics, medicine and biology so porous. We will now approach electricity from the point of view of the location of faculties and brain functions and ask ourselves how electrical techniques, developed at the beginning of the 20th Century, allow us to renew philosophical and medical questions about the materiality of mental content, both normal and pathological.

Forward to 1954, when everything seems to have changed compared with the case in 1941, despite a persistent heartbeat, death was found because of loss of brain electricity and physical signs of brain functioning. In 1954, that persistence lost the significance it previously had. There are three ideas, though, that unite these two accounts even in their differences. First is an 'electricocentric' account of biology: that electrical cellular activity either could replace or more precisely reveal functional attributes of an organ and/or the organism – attributes critical to the reliable determination of it as living. Second is the suspicion that loss of brain electrical activity was the final event of human life. Third, and more implicit, is the emerging understanding that death was a slippery construct without clear lines or technical or clinical signs. Death did not close a door on an array of other continuing biological events that were a part of dying. [BEL 14, p. 127]

5.2.2. *Stimulations on animal and human brains: localist perspectives*

At the beginning of the 19th Century, animal brains were experimented *in vivo* and allowed direct intracranial galvanic stimulation. While Aldini performed his first transcranial stimulations on human subjects, animal experimentation played a fundamental role in direct brain stimulation. We will see that the surgery on certain electrical pathologies, such as epilepsy, proved to be extremely heuristic in applying the correlations between stimulation and localization in humans. In 1809, Rolando [ROL 09] described the effects of galvanic stimulation on the animal cortex. He was known for his research on the physiology and anatomy of the nervous system, and in the field of experimental physiology he was considered one of the first to study the effects of electricity on the brain. In 1822 the treatise *Inductions physiologiques et pathologiques sur les différentes espèces d'excitabilité et d'excitement, sur l'irritation, et sur les puissances excitantes, débilitantes et irritantes* [ROL 22] was published. Rolando discussed the theme of cerebral electricity in the context of his *in vivo* experiments on electric fish. He was one of the authors of the analogy between the galvanic cell and brain morphology. He conceived of the brain as a source of animal electricity and applied electrophysiological techniques to animal brain organs. He demonstrated that the central nervous system was electrically excitable:

> With the idea of observing the effects of a galvanic fluid current directed from the brain to different parts of the body, I pierced the skull of a pig and then inserted a conductor from the Volta electromotor into the hemispheres of the brain, testing either one part or the other, while the wire was applied to different parts of the body. From these repeated experiments in various quadrupeds and birds, I obtained violent contractions, and observed that these were very visible, when the metallic conductor penetrated the cerebellum. However, the hemispheres of the brain in the pigs had been quite torn by the repeated introductions of the tip of the conductor, so that the striated bodies, the ventricles were damaged: but the animal nevertheless lived for 12 hours in a dormant state, and would have lived longer, had it not been shocked more. [ROL 09, p. 18, author's translation]

It wasn't until the 1870s that the excitability of the cortex was commonly accepted. Rolando also showed, through his experiments, that the voltaic cell was a relevant stimulation tool to explore brain functions while following Gall's localist perspective. Yet, despite Rolando's work during the 19th Century, it seemed clear that the cerebral hemispheres were non-excitable. The doctrine was so firmly established that Gustav Fritsch (1837–1927) and Eduard Hitzig (1839–1907) had difficulty in conducting experiments to prove the contrary:

> This dogma was refuted in 1870 by the important experiments of Fritsch and Hitzig, who showed that although electricity can be applied to certain portions of the exterior without producing movement, there are others whose excitation invariably causes movement on the opposite side and that certain movements can be uniformly produced by the excitation of certain defined regions. These facts have since been extended and verified by several experimenters on various animals and on man himself. [FER 79, p. 25, author's translation]

Their experimental observations provided evidence of the excitability of the cerebral cortex and helped confirm the hypothesis of the localization of functions in the brain. The cerebral functioning linked to these experiments was based on an electrical anatomy that was conducive to locating the stimulation points and their consequences on the body:

> For example, it cannot be said that, however localized they may be, electrical excitations do not involve the white fiber brushes penetrating the cortical apparatus, nor the nerve cells of this apparatus. To argue the case for cortical electrical excitability, one is obliged to focus on the differences between the reactions obtained by simultaneous excitation of the exterior and white matter and those produced by excitation limited to white matter alone. [FRA 87, p. 309, author's translation]

Fritsch and Hitzig exposed cortical hyperexcitability, first in dogs, through systematic stimulation of the hemispheres and identified the regions in which it resulted in movement of the contralateral limb. They thus began by describing the organization of the hemispheres in relation to motor functions. Gradually, from the most elementary to the most complex functions, the stimulations made in different regions of the brain made it possible to locate each one of them. There were numerous controversies despite clinical studies of pathological cases, particularly aphasia, which, after 1861, supported the theory of cerebral localization by electrical stimulation. By applying electrodes to specific points in the cortex, they showed that stimulation of the convexity of the brain in its anterior part triggered muscle contractions on the opposite side. On the contrary, the posterior regions of the cortex remained inexcitable. They completed their first results by removing specific areas of the cortex [CAR 09, pp. 131–132], resulting in paralysis of the corresponding limbs. Their experimental protocol followed a mechanical logic: if the stimulation of a given area caused a movement on the opposite side, then the ablation of this same area must produce the opposite effect, that is, the loss of movement:

> In 1870, Gustave Theodor Fritsch (1838–1927) and Eduard Hitzig (1838–1907) used 'voltaic' (or 'galvanic') DC stimulation tools, perfected by du Bois-Reymond and Matteucci, to explore the functioning of the dog's motor cortex. The localization of the stimulation allowed them to draw the first map of motor functions in the brain. However, the current used strongly polarized the neurons and could create lesions, which limited the possibilities of functional analysis. [MIC 13, p. 320, author's translation]

Exploratory neurosurgery links with electrical stimulation techniques were performed directly on the brain. Faradization, which allows a more functional and less injurious type of stimulation, was used in the context of diseases or operations that left the brain bare. Ferrier, by using a Faradic type alternating current, could consider switching from the animal model to the human model while limiting the risks. He first transposed the mapping of the motor areas of the monkey to the human brain, which guided the first neurosurgeries. By exposing the monkey's brain, he applied electricity to it, which was used as a stimulant for the nerve centers, to show that a specific stimulation corresponded to a specific movement. He also performed ablation to make manifest the causal relationship between electricity and movement or its obliteration:

> It has been established by experimenting on monkeys – and at present it is so generally accepted that it is useless to go into great detail – that the destruction of centers whose excitation produces definite movements, produces paralysis of the same movements on the opposite side of the body varying in degree, intensity and duration with the extent of the destruction of these centers. [FER 91, p. 113, author's translation]

Electricity made it possible to locate the centers of movement, to work in voluntary and involuntary conditions. Stimulation of the nerve machine caused movements that required no reflection. Nevertheless, the passage from animal to human demonstrated the correlation between will, stimulation and location. Ferrier is considered a pioneer in functional brain stimulation. He used a Stöhrer battery, made from zinc and coal elements and the induced current from the second coil of du Bois-Reymond's magneto-electric device. Principles of successful mechanical excitation of the cortex were established as methodological principles:

> Luciani and Tamburini consider the following conditions to be necessary for cortical excitations to be effective, even to produce

> simple movements: 1. the animal, when the brain is exposed, must have lost little blood and become little excited; 2. it must be vigorous and adult; 3. the morphine anaesthesia must not be very deep, that of chloroform must be about to disappear; 4. electrical excitability must exist, an absolute condition. [ENG 87, p. 307, author's translation]

His work was confirmed in the early 20th Century by William Cushing (1869–1939) and Sir Charles Scott Sherrington. In this modeling of brain function, experimental conditions on healthy brains played an important role. But pathological conditions of the central nervous system were heuristic, especially when they resulted from epilepsy, which was considered a disorder of the brain's electrical activity. The work of John Huglings Jackson was not without influence on Ferrier's research. Jackson marked the advent of knowledge about epilepsies, which he declined in the plural sense, particularly in terms of pathological locations. Indeed, the epileptogenic foci differ according to the form of epilepsy (major, partial, etc.). His research was exploratory and contributed to our understanding of how the brain works. In 1866, he challenged the very concept of epilepsy by proposing that the term be downgraded to better describe the condition of nerve tissue in a sudden and temporary loss of function. He also saw partial epilepsy as a localized phenomenon due to occasional, excessive and localized discharge of gray matter. The unity of the disease resulted in many epileptic symptoms, as even a blink of an eye could be taken as a sign. In 1884 and 1890, *The Croonian Lectures* [JAC 84] and *The Lumleian Lectures* [JAC 90] were published, in which Jackson, based on the work of his master, Thomas Laycock (1812–1876), took up the idea that reflex action can occur within the cerebral hemispheres. He developed the concept of the sensory-motor function of the nervous system and addressed a recurring medical and philosophical issue in the history of electricity in the cerebral organ: the relationship between the brain and thought in the pathological context of epilepsy. His work was characterized by its orientation towards neuro-physiopathology as well as his interest in the theory of brain localization. He continued to question the pathophysiological influence of the higher nervous centers on these pathological processes and *The Lumleian Lectures* begins as follows:

> There is not rarely after epileptiform seizures local temporary paralysis, and sometimes aphasia with it. [JAC 90, p. 821]

The heuristic role of epilepsy in the understanding of brain mechanics should be emphasized. Thus, studying epilepsy provided an experimental framework for exploring normal and pathological brain functions, the former interacting with the latter. Jackson also initiated an exciting reflection, based on Darwin's work and, in the field of psychophysiology, on evolutionary perspectives[14] applied to nervous physiology. He thus developed the theory that nervous system disorders can only be understood with reference to the evolutionary levels of nerve functions. These evolutionary levels are locatable from the deepest brain layers to the outermost, which in this model were also the most recent. His theory implied a progressive phenomenon from the most automatic and primary function to the most advanced and least automatic function. Epileptogenic foci, depending on their location, affect different levels of function, more automatic or more cognitive. This contributed to an evolutionary mapping of human functions. Jackson gave different brain activities a dynamic and localizing pattern. His localized understanding of epilepsy informed the work in which Ferrier experimentally showed that electrical stimulation of certain brain areas in animals exposed the brain to constantly produce the same motor response. Jackson's analysis of cases of unilateral epilepsy taught him that a certain order is predominant in the onset and spread of seizures. He thus became convinced of the cerebral location of epileptogenic centers. By proposing the theory of linking convulsions with discharges occurring in the cortex region to explain partial epilepsy, he directed neurology towards a localization based on functionalist models. His model was strictly neurological: although his conception of the three evolutionary levels of the brain was not supported by anatomical data, he designated three sensory-motor brain states where epileptic seizures occurred differently. During discharges, it was considered that nerve cells panicked and released excess nerve energy. At the end of the 19th Century, his work was an important step towards a modern understanding of epilepsy in which seizures were considered caused by sudden electrochemical discharges occurring at the level corresponding to the location of the epileptogenic focus. According to Oswei Temkin (1902–2002) [TEM 71][15], Jackson broke with the *falling sickness* model, marking the entry of epilepsy into the field of neurology. The concept of *falling sickness* refers to the disease conceived through the clinical prism of generalized crisis. Although it caused an explosion of the very concept of epilepsy, Huglings Jackson enabled a

14 Jackson conceived the nervous system according to three levels of evolution, all of which are involved in pathological processes, particularly epilepsy: the median, the superior and the inferior. Three principles also frame his evolutionary perspectives on brain and nerve development:

- The evolutionary phenomenon is consistent with an organization that goes from the least organized to the most complex, marked by the transition from the lower to the higher centers.
- Therefore, the evolution is accomplished by going from the simplest to the most complex that prevails in the higher centers. If, for example, a nerve center has two sensory roots whose synapses articulate directly to two motor roots, Jackson described it as simple and well-organized. On the contrary, if an area of the nervous system of four sensory roots combines several synaptic facilitations of a motor nature, it will be more complex but less well organized because it is more recent.
- The evolution progresses from the most automatic to the most conscious level.

15 Temkin used the term *falling sickness* as a prescientific concept for secular epilepsy.

focus on the amplitude of the symptoms, their diversity and the possibility that each one depends on a different brain area. His ideas, crossed with Charcot's analyses of hysterical disorders, made it possible in particular to more skillfully delimit the fields of neurology and psychology. Epilepsy has played a dual role in the history of medical electricity: the disturbances it causes interact with normal brain function and are responded to in the experimental setting of an exploratory clinic. Thus, epileptic disturbances are as much a subject of medical research as they are an opportunity to understand the links between cognitive functions and brain electricity. In paroxysmal seizures, it helps to model the normal functioning of the brain but also to consider the stakes of its electrical exploration. It makes visible what needs to be measured and localized, and makes it possible to consider the benefits of electrical stimulation as a means of updating brain functions in their organic core. Thus, Ferrier studied monkeys [FER 74b] in order to provide evidence for Jackson's ideas about the cortical, focal and "irritative" origin of epileptic seizures:

> The paper contains the chief results of a research commenced with a view to test the accuracy of the views entertained by Dr. Huglings Jackson on the pathology of Epilepsy and Chorea [...]. In order to put this theory to the proof the author determined to expose the brain in various animals, and apply irritation to the surface. The method of irritation was suggested by the experiments of Fritsch and Hitzig who had shown that contractions of definite groups of muscles could be caused in dogs by passing galvanic currents through certain portions of the anterior regions of the brain. [FER 74b, p. 2]

By using faradization to excite circumscribed parts of the brain rather than massively irritating all of the cerebral hemispheres, he inferred that the gray matter did not behave as an inert layer that would simply transmit electrical current to the spinal cord fibers. On the contrary, like other nerve centers, it stored and transformed excitations into a force of its own:

> After this haemorrhage, when the brain stops beating, and often long before the animal dies, the current does not produce any excitation or, if it does, the phenomena are so complicated that no conclusions can be drawn about the location of brain function. That is why it is better to discover only one part of the brain at a time, the part whose functions you want to study, rather than to expose an entire hemisphere at once. In order to discover the anterior and inferior part of the hemispheres, it is necessary to extirpate the globe of the eye, to remove the arch of the eye socket and the zygomatic region and the temporal muscle; the shock and hemorrhage have then so depressed

> the excitability of the brain that any experimentation becomes impossible. [FER 74a, pp. 3–4, author's translation]

Thus, thanks to applications on the cortex of the faradic current, which produced longer contractions than galvanic current, he drew up the first map of the monkey's motor cortex, which was the basis of Sherrington's work on primates and Penfield's (1891–1976) work on humans. Ferrier started with experiments on guinea pigs or cats from which he removed a portion of the skull sufficient to uncover the left hemisphere, indicated in Figure 5.6 below by the shaded area.

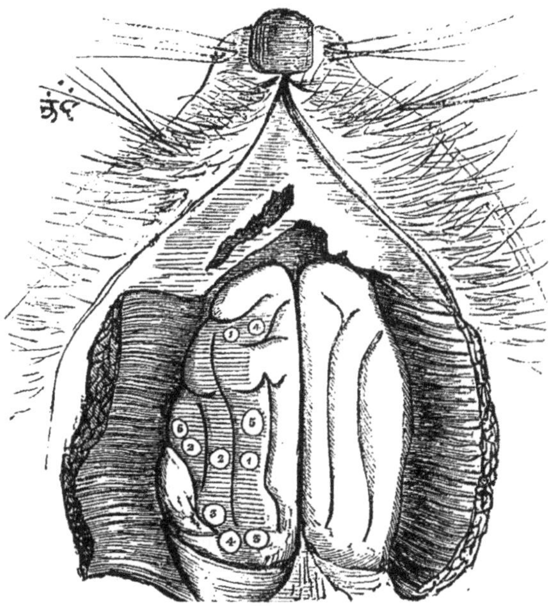

Figure 5.6. *Drawn from a cat, we can see that he sketched the convolutions discovered and carefully marked the points where the electrodes were applied [FER 74a, pl. 2]*

He drew conclusions of comparative electrophysiology on humans and animals:

> If we take into account the more extensive development of the anterior parts of the brain in humans, we will see that they correspond to the centers of movement of the forelimb in cats and dogs: this is a very satisfactory confirmation of the results and conclusions of my personal experiences. [FER 74a, p. 65, author's translation]

Several points were concluded from Ferrier's electrical stimulation experiments:

– the anterior parts of the cerebral hemispheres contain the centers that preside over the voluntary movements and external manifestations of intelligence;

– each convolution forms a separate and distinct center. Certain groups of convolutions are the anatomical core of the centers that preside over the various movements of the eyelids, face, etc. Differences in animal behavior are matched by anatomical differences. For example, the centers that direct the movements of tails in dogs, legs in cats, mouths in rabbits are developed and differ greatly from each other;

– the action of the hemispheres is generally crossed even though some movements of the mouth, tongue and neck are coordinated for both sides in each of the cerebral hemispheres;

– since the proximate cause of different types of epilepsy depends on lesions in different centers of the cerebral hemispheres, it is possible to artificially produce an attack in animals in order to explore and clarify the correlations between its effects and the brain center from which it originates;

– the optic lobes or quadrilateral bodies, in addition to their role in terms of vision and iris movement, are centers for the extensor muscles of the head, trunk and limbs. Irritation of these centers determines an opisthotonos[16];

– the integrity of the centers that make up the cerebellum depends on the preservation of the body's equilibrium.

Ferrier's electrical stimulation of the animal brain not only uncovered normal brain mechanisms, but also provided a better understanding of the symptoms of neurological and cerebral diseases by provoking them mechanically. In this way, cores could be targeted, a fundamental point for neurosurgery as well as for stimulation. In 1884, Erb emphasized that electrophysiology can indicate new therapeutic pathways by making visible the effect of organic electrical stimulation on physiology:

> Arguments are provided by the data we have on the electrophysiological influence experienced by the brain vessels; in the first line, their direct modification (narrowing and widening), as Löwenfeld later proved. Thus I open at least a clearer possibility to accelerate or slow down the circulation in the skull and brain, to act as a modifier on the conditions of nutrition, perhaps to erase pathological phenomena such as anaemia, hyperaemia both primary and secondary with their consequences. [ERB 84, p. 287, author's translation]

16 The body, struck by an opisthotonos, undergoes a contracture, is curved backwards and the limbs are involuntarily extended. This position is consistent with the symptoms of tetanus.

Conversely, operations on diseases such as tumors or epileptic foci became an opportunity, particularly in Penfield's research, to confirm the site of high cognitive functions. The combination of surgery, clinical and electrical stimulation provided an experimental framework for understanding how cognitive abilities are located in the brain and how they interact with brain disease. Turning to the human model, Ferrier found that by touching the exterior with the electrodes in an area that corresponded to the anterior portion of the pre-olandic convolution, he obtained wrist and finger movements: the subject's hand reaching out. Over the region in which these movements were obtained, the application of current produced a movement of the left elbow with an extension and flexion of the shoulder. Experimental passage in humans is difficult when the currents used are potentially injurious. In 1882, the Italian neuropsychiatrist Ezio Sciamanna (1850–1905) [SCI 82] conducted a series of systematic experiments with electrical stimulation on a trepanned patient with brain trauma [CAS 16]:

> Influenced by Ferrier's publication in 1873 documenting his initial experience with cortical stimulation mapping on several species, 19th Century experiments applying electrical current to the exposed human brain soon followed. [CAS 16, p. 186, author's translation]

In 1883, the Italian-Argentinian surgeon Alberto Alberti (1856–1913) conducted an experiment in brain stimulation. It lasted more than eight months, with experiments on a woman with a tumor eroding the skull and allowing direct access to the surface of the dura mater. Bartholow (1821–1904) also conducted studies on the excitability of the brain using electrodes. He strived to understand and locate the sections of the brain responsible for certain motor actions. In 1874, he published *Experimental investigations into the functions of the human brain* [BAR 74, p. 727], a short article in which he detailed the medical history of a patient named Marta Rafferty who had a deleterious epithelioma. These observations remain significant milestones in human mapping. Systematic observations on the topography of the brain were also made in 1887 by the British surgeon Victor Horsley (1857–1916) [HOR 87]:

> Horsley performed his first craniotomy on May 25, 1886, in Queen Square in London. The patient was a 22-year-old man who had experienced posttraumatic Jacksonian status epilepticus. The seizures were caused by a cortical scar, which was excised with a half centimeter of surrounding brain. The wound healed, and the patient had no further seizures. By the end of the year, Horsley had performed 10 craniotomies; the outcomes ranged from clinical improvement to complete recovery and were successful in nine cases. [TAN 02, p. 608]

At the beginning of the 20th Century, Fedor Krause (1857–1937) [KRA 34] made the first map of the motor cortex by faradic stimulation. It is based in part on the work of Sherrington.

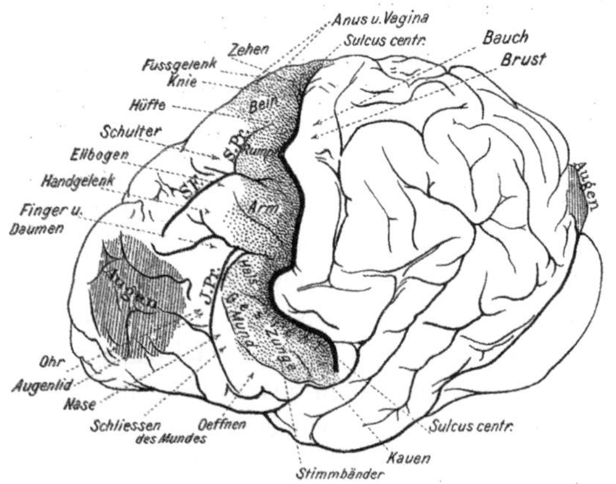

Fig. 64.

Hémisphère cérébral gauche d'un chimpanzé (troglodytes niger) avec les centres déterminés faradiquement d'après Sherrington et Grünbaum.

Figure 5.7. *Left cerebral hemisphere of a chimpanzee with centres determined faradically according to Sherrington and Grünbaum [KRA 34, p. 178]*

Wilder Penfield[17] initiated many technical advances, particularly in the exploration and electrical stimulation of the brain in the medical context of developments in neurosurgery. He was at the center of developments in neuroscience, epileptic medicine, electrical understanding of mental activity and the exploration of faculties through electricity. These themes were at the interface of fundamental and clinical brain research, as he used the clinical setting to research the localization of major cognitive functions such as language and memory. His work symbolizes the heritage of a culture where electricity was considered a tool for brain investigation and a means of recognizing the human being in their moral and physical dimensions [RHY 13, p. 9]. Penfield also frequently mentioned the Jacksonian localization model of epilepsy and developed analog links between the brain and the machine, particularly in his research on memory:

> The application of electrical stimulation to human's brains resulted in yet another replacement, the transformation of not only the body but

17 In Oxford, Penfield met Charles Sherrington in 1916. It sensitized him to the study of the human brain. He then entered the Johns Hopkins School of Medicine in Baltimore, Maryland, where he obtained his doctorate before practicing as a surgeon at Columbia University Presbyterian Hospital and spending seven years at the New York Neurological Institute where he was already devoted to nervous system pathologies. He gave a new direction to Canadian neurology in 1928, when he moved to Montreal where he initiated the opening of the Neurological Institute in 1934. This institute marked his dual commitment to relieving the illness and pain caused by neurological diseases, while advancing research in the fields of neurology, neurosurgery and psychology.

> also the human psyche into an electromechanical device, experiencing automatic sensations due to the electrical stimulation of the brain. This was to happen a generation later, when Wilder Penfield started to explore systematically the human cortex by electric stimulation. [BOR 18b, p. 189]

In the thought current of a localized electrification within gray matter, often compared to a galvanic battery, the fine electrical stimulation of the cortex was at the center of Penfield's work. The clinical context was designed as an experimental setting where electricity was the means to understand the relationship between body, mind and the outside world. Because Penfield used this neurosurgical and clinical framework, he could conduct his research on humans:

> The cell bodies are collected together, forming islands, or blankets of gray matter. The branching connections form the white matter. This whole system vibrates, one might say, with an energy that is normally held in disciplined control, like that of a vast symphony orchestra, while millions of messages flash back and forth, to as many functional targets. [PEN 75, p. 11]

How electricity relates to the invisible mechanisms that hold this symphony between mind, brain and body together is a continual subject of research in both the clinical and neurology research fields, "Neurosurgery must go hand-in-hand with neurology" [FEI 16, p. 18]. Penfield worked on the creation of the Montreal Neurological Institute. At the same time, he remained gripped by questions about the nature of epilepsy: what are its effects in the brain? How do we treat it? Epilepsy is at least as old as human history, if not older, because it affects animals such as *papio papio*[18] baboons. This gives it a role in understanding brain evolution, a point already made by Jackson. In his book *Mystery of the Mind* [PEN 75], Penfield points out that he suggested the existence of more complex brain structures corresponding to the functions of the mind and the most recent stages of brain development [PEN 75, p. 83]. The Montreal Institute quickly became a renowned institution for its treatments, especially for severe epilepsy. Penfield initiated a research program on the localized causes of this pathology and also worked with his collaborator, Cobb (1887–1968), on primates. The two scientists thus attempted to cross-reference basic animal research with human clinical practice. They faced the difficulty of locating cognitive functions in apes compared to humans, thus underlining the obstacle that the animal model may constitute in relation to the differentiated development of cognitive functions. Thus, by highlighting the techniques of direct and electrical brain stimulation, Penfield developed an innovative method that took place in the

18 Penfield helped pave the way for researchers such as Robert Naquet to study hereditary epilepsy in animals [KIL 66].

neurosurgical context of epilepsy and made it possible to map the areas of cognitive functions. This method consisted of locally anesthetizing the patient and then stimulating his or her brain areas during open-head surgery. In his memoirs, *No Man Alone* [PEN 77], he recounts how he conceived of what would later be called the *Montreal procedure*. Following craniotomy, an electrode was placed on the cortex, allowing an electric current of a few milliamps to be applied during the operation. During this stimulation, the patient was awake and able to describe his sensations, allowing the surgeon to identify, *in situ,* the part of the brain involved while not damaging the centers of functions such as speech:

> One touches the cortex with a stimulating electrode and, since the brain is not sensitive, the patient does not realize that this has made him aphasic until he tries to speak, or to understand speech, and is unable to do so. [PEN 75, p. 51]

The epileptogenic focus could thus be removed while maintaining cognitive integrity. The fundamental challenge was to overcome the lesional problems and remove the scar without damaging mental functions. The counterpart of this therapeutic challenge was the fundamental exploration made possible by this technique of exploring brain mechanisms in the waking state. This surgery of conscious patients quickly became known for its effectiveness. For example, Penfield performed lobectomies and sometimes removed more than a quarter of a frontal lobe from a patient after identifying the area of the brain that caused the seizure [PEN 37, 40]. Between 1935 and 1936 [PEN 35], he presented his findings at the London Neurological Congress and marked a victory in the history of neurosurgery. He developed his investigations by electrical stimulation by collaborating with Herbert Jasper (1906–1999), who helped to make the E.E.G. a device that made it possible to specify epileptogenic foci. The collaboration between the two men allowed the encounter between surgery and electrical recordings to be taken further [PEN 52]. Penfield was also at the origin of an annex of the Montreal Institute dedicated to clinical E.E.G. for the study of epilepsy and mental illness. In this way, he simultaneously developed electrical and surgical investigations of the human brain. Hundreds of exploratory craniotomies were performed and enabled the delineation of sensory-motor representations in the cerebral cortex and the localization of cerebral areas. This step represented one of his fundamental results: the mapping of the cerebral cortex and the representation of motor and somatosensory *homunculi*. The *homunculus* is a representation of the effect of the functions of the different parts of the human body on the surface of the motor cortex. This modeling leaves much room for hands and mouth, both of which are fundamental to communication [PEN 37, 50]. It therefore represents functions in proportion to their physiological importance to the human species. Thus, the space allocated on the cortex is proportional to the complexity of capacity of the movements.

These representations, resulting from the meeting of conscious surgery and electrical stimulation techniques, account for the interactive mapping of projections on the motor cortex of the different parts of the human body, the links between the sense organs and the location of stimuli on the cortical surface. These explorations of the motor cortex and the somatosensory cortex, centers of analysis of information from our bodies, are paradigmatic for understanding the links between our body and our brain. The *homunculus* gives an averaged representation of stimuli and reactions from one individual to another. While scientific precision remains dependent on an inductive method, it remains a metaphorical representation of cognitive functions, subject to a principle of probability and highlighting the prevalence of some of them over others. The electrical stimulation of the brain sometimes gives totally unexpected reactions which all refer to the links between the cerebral physiology and the mental sphere. Indeed, laughter and other exacerbated emotions, a feeling of *déjà vu*, mystical experiences, hallucinations and memories of real-life scenes known as flashbacks are all consequences of these stimuli. This last point stimulates Penfield's interest in the mechanisms of memory and the recording of memories often conceived in analogies to the machine: "The brain is like a computer, the mind is the programmer" [PEN 75, p. 57]. For example, surface cortical stimulation or direct cortical electrical stimulation (DCES), developed as part of neurosurgery, is used to develop neurophysiology and neuropsychology. During stimulation, recording the patient's performance on a series of cognitive tests allows the identification of those disturbed by the electrical examination. Thus, speech disturbance indicates that the area under the electrode is involved in this function. It thus gives individual behaviors an organic origin that can be stimulated and associates operations of consciousness with specific cerebral areas. These brain stimuli contributed to the development of neuropsychology at the Montreal Neurological Institute. In 1959 Penfield published *The interpretive cortex; the stream of consciousness in the human brain can be electrically reactivated* [PEN 59] as well as with Mullan, *Illusions of comparative interpretation and emotion production by electric discharge and by electric stimulation in the cerebral cortex* [PEN 59]. Here is what he wrote about his method:

> Since a gentle electrical current interferes with the patient's use of a convolution of the brain and sometimes produces involuntary expression of its function, a stimulating electrode could be used to map out the cortex and to identify the convolutions as the patient described his sensations and thoughts. [PEN 75, p. 13]

In 1954, Penfield published a treatise with Jasper entitled *Epilepsy and the functional anatomy of the brain* [PEN 54]. It introduced the concept of an integrated central nervous system. He defined it on the basis of two mechanisms: the mechanism whose action is essential to the existence of consciousness and the sensor-motor mechanism of coordination. He supposed that the neurological and psychological mechanics arose from the interaction between the diencephalon and the cortex of the two hemispheres. His research raised the question: is the brain the messenger of the mind[19]? This neuropsychological questioning is the subject of clinical, surgical and electrical investigations using electrophysiological methods:

> Throughout my own scientific career I, like other scientists, have struggled to prove that the brain accounts for the mind. But now, perhaps, the time has come when we may profitably consider the evidence as it stands, and ask the question: Do brain-mechanisms account for the mind? Can the mind be explained by what is known about the brain? [PEN 75, p. xiii]

His quest was not without marking the resurgence of a Hippocratic assertion that the brain is the messenger of consciousness. Penfield operated an epistemological shift from the question that the brain explains the mind to the questioning of the evaluation of knowledge about the links between the brain and the mind. The central question is whether the human being is essentially determined by the body alone or by mind and body as separate elements. An explorer of the *mind-body problem,* he was involved in brain matter throughout his career. If a scientist constructs problems to which experiments can be applied, Penfield materialized the question of the links between consciousness and the body, considering the former in relation to brain function. In an article published in 1952 [PEN 52] on the mechanisms of memory, he argued that without the integrated centrencephalic system there would be no conscious processes of the mind. He thus defended the role of the brainstem and the reticulated formation of the thalamus[20] in psychic mechanics. It is in the context of research on the brain's links with consciousness that he showed that, provided that it is operated on a conscious brain, ablative surgery on a pathological center can stop the negative interactions with healthy functions. Although in 1935, in the article "*The frontal lobe in man: a clinical study of maximum removal*" [PEN 35], Penfield and Evans stressed the fact that destroying parts of an organ is not the best way to understand its functions, Penfield disseminated his method of ablation of foci, particularly epileptogenic foci, while emphasizing the fact that as soon as the electrical disturbances cease, interactions between normal functions and pathological symptoms cease. In collaboration with Hebb (1904–1985), he published in 1940

19 In 1975, the second chapter of *Mystery of the Mind* was entitled, "To consciousness, the brain is messenger".

20 In the late 1940s, Magoun and Moruzzi observed that electrical stimulation of this area during sleep is able to awaken the subject. Their conclusion was that the tonic activity of the cross-linked formation is governed by sensory impulses. This fact is made visible by recording the activity of the nerve cells in this formation. These recordings reveal that some neurons receive information from the external environment (visual, auditory, somathetic), while others may be excited by variations in the internal environment (anoxia, hypoglycemia for example) [MOR 49].

[PEN 40] a study of the patient known as *K.M.* The 16-year-old *K.M.* received a blow to the head and remained unconscious for 10 days following the accident. According to Penfield, he changed, became infantile, irresponsible, excited and forgetful. Chronic epileptic seizures brought him to Montreal where he underwent lobectomies with the unexpected result of increasing his abilities. Hebb and Penfield derived the theory that while a neurological condition may interfere with conscious behaviors and actions, the removal of the pathological focus enabled the restoration of the brain norm. A disturbed electrical activity, therefore, disrupts cognitive functions. While artificial electricity is an exploratory tool, disturbances in brain electricity provide an experimental framework that crosses the issue of brain exploration. Penfield also worked to understand the links between brain location and memory [PEN 52]. In his experiments on temporal lobe surgery, he described memory as a continuous tape comparable to that of a cassette recorder:

> The left hemisphere of the brain is outlined, with the brain-stem and spinal cord shown beneath, to illustrate the results of electrode stimulation of the cortex in motor, sensory, and what may be called psychical areas for the recall of past experience. [PEN 75, p. 26][21]

If the brain is the messenger of the mind, then electricity is the pen. The brain and electricity are the physical basis of mind and consciousness. For Penfield, the organic substrate of the mind can be located within certain areas of the cortex and in the upper part of the brain stem:

> The indispensable substratum of consciousness lies outside the cerebral cortex, probably in the diencephalon (the higher brain-stem). The realization that the cerebral cortex, instead of being the 'top', the 'highest level' of integration, was an elaboration level, divided sharply into areas for distinct functions (sensory, motor, or psychical), came to me like a bracing wind. It blew the clouds away and I saw certain brain's mechanisms begin to emerge more clearly, and they included those of the mind. [PEN 75, p. 18]

In addition to his many discoveries about the localization of cognitive functions and his developments in neurosurgery, Penfield brought the notion of anatomical brain-mind boundaries back to the forefront of neuroscience. Through electrical stimulation, he tended to make such boundaries visible, permeable to the action of the mind, and highlighted the fact that different localized brain control centers participate in the genesis of consciousness. The latter is not a linear strip, these different dimensions, past and present, being able to operate simultaneously:

21 Penfield also noted that when the interpretative cortex is stimulated, it obtains recorded sequences of successive states of consciousness from the past that reactivate [PEN 75, p. 26].

> If, to the contrary, the truth is that the highest brain-mechanism is busy creating the mind by its own action, one might expect mental confusion when the neuronal record is activated by an electrode so that a stream of past consciousness is presented to the mind along with the presentation of the contemporary stream of consciousness. [PEN 75, pp. 55–56]

Thus, through his application of electrical stimulation during neurosurgical operations, he helped model mental activity on the basis of the localization of brain function. Penfield also conducted fundamental research, based on instrumentation and electrical exploration, to investigate the relationship between psychology and neurology in order to understand the human brain in its normal and pathological, physical and moral characteristics.

5.2.3. *Brain electricity recordings and the electric alphabet*

The use of milliampere meters to externally record brain activity via electrodes placed on the scalp led to the genesis of electroencephalography. Caton, by stimulating the bare cortex of rabbits and apes, visualized in 1878, low voltage variations between two electrodes. One was placed on the cortex and the other on the bone. He showed the recordable presence of spontaneous electrical activity in the brain. The Russian physiologist Vladimir Pravdich-Neminsky (1879–1952) [PRA 13] published a trace of electrical activity and coined the term *electrocerebrogram* in 1912. These recordings held out the hope of deciphering mental content thanks to developments in new methods of brain representation. Electroencephalography carried such hope until the 1960s. By giving a graphical representation of brain events, it opened up a perspective for interpreting the activity of thinking on an electrical trace:

> Today the brain writes in a secret language, tomorrow scientists will be able to decipher the neuropsychiatric conditions and the next day we will be able to write our own letters in brain language. [FIN 30, p. 7, author's translation]

In 1924, Berger [BER 69] succeeded in recording the first human *enkephalogramm*. In 1929, he explained the ways in which the overall electrical activity of the human brain was recorded. His aim was not so much to know the brain mechanisms as to understand the nature of man as a psychophysiological being. Thus, by his own admission, he saw these electrical phenomena as a new means of deciphering a human nature dependent on his cerebral physiology. Going from the activity of thinking about an electrical trace is a dream of unlocking the secret of the mind, but also the craze for electricity as the key to human nature. Certain theoretical principles developed in the 19th Century, coming from the fields of physics or electrophysiology, opened up perspectives on electroencephalographic techniques for the physiological knowledge of the brain. We can mention three principles:

– the notion of residual energy or electricity. This dynamic and mechanistic model of muscular and nervous physiology has a fundamental influence on new conceptions of vital properties and, by implication, on the relationship between body and mind;

– the second postulate used in the conception of E.E.G is the dependence and reduction of biological processes with physico-chemical mechanisms;

– finally, the postulate, derived from developments in anatomoclinical discourse in the 18th Century, that mental processes depend on the brain areas responsible for the various brain functions, is not without an influence on the early recordings of the electrical activity of the human brain.

Berger spoke about mental energy to characterize brain electricity and proposed to determine its power. His willingness to measure it was part of a search for tangible signs of this mental energy which represented the physical and measurable aspect of thought and experience, the product of a series of energetic transformations. The correlation between a mechanistic understanding of thought and cerebral localizations was at the basis of these recordings insofar as the transcription of thought activity into an electrical trace was part of the project to graphically represent its mental and cognitive contents. Since the end of the 19th Century, electrography has made it possible to read the signs of the body's dynamic activities. These activities were extended to the mental domain by Berger. This technique thus inherited the graphic method initiated in experimental physiology, notably in the work of Etienne-Jules Marey. The treatise *La machine animale* [MAR 73], published in 1873, deals in particular with human and animal locomotion and the graphic recordings of its mechanisms. Similarly, the four volumes of *La Méthode graphique dans les sciences expérimentales* [MAR 78] summarize the applications of his method as a research process applicable to human mechanics[22]. Thus, techniques for recording physiological mechanisms tended to provide a reified image of the latter, the plot being devoid of any animistic or metaphysical perspectives. The intention was indeed to give objective images, based on a co-evolution of the body and the machine, the beginning of a permanent meeting point between the fields of biology and engineering:

> Philosophically, the most significant implication of this conceptualization is its epistemological reversal: because of this technological externalization of biological principles, and only as its consequence, the body becomes accessible to experimental

22 The electrocardiogram, by allowing the internal changes of heart rhythms to be reduced to a graphic line, promises direct access to vital phenomena in their own language.

> exploration and scientific explanation. Technology is not the application of knowledge but its very source. Because of the recursive loop between technology materializing biological principles and scientific research investigating biological objects, the body becomes amenable to intervention. [BOR 18b, p. 191]

Peter Gloor, in his translation of Berger's treatise, recalls the problems linked to interpreting this first recording of human brain activity. Can electroencephalographic techniques claim mechanical objectivity as developed by Loraine Daston and Peter Galison? Images can carry with them a certain neutrality, which is part of their power of seduction and persuasion, but can they be dissociated from the culture, the disciplinary field and the theories in which they are inscribed? Images seem to be, in themselves, elements of language that contribute to the construction of a scientific theory while aiming to clarify it. Thus, the E.E.G. participates, in its application to psychology, in mechanizing and making visible the thought and its contents, while proposing to interpret its stages by means of cerebral waves whose variations have been made accessible both to analysis and interpretation:

> Depicting individual objects 'objectively' required a specific, procedural use of image technologies, some as old as the lithograph or camera lucida, others as freshly late-nineteenth-century as photomicrography. These protocols aimed to let the specimen appear without that distortion characteristic of the observer's personal tastes, commitments, or ambitions. [...] Mechanical objectivity required a certain kind of scientist, long on diligence and self-restraint, scant on brilliant interpretation. [DAS 07, p. 121]

The entry of the E.E.G. into brain studies formed part of a double culture, of the electric brain and the human machine, the two fields responding and echoing each other. The translation of mental events into electric waves has the power of scientific images, which are vectors of both the scientist's objectivity and the objectification of the field of study. The applications of electroencephalographic techniques to the understanding of cerebral mechanisms developed from 1930 onwards in the context of emerging neurosciences:

> The EEG shaped brainwaves into an electrical brain that could only partly be brought to coincide with the subjects of other branches of brain research. One important characteristic of this electrical brain was its medial and mediatory position between anatomical findings and psychological observations. Unlike neuroanatomical, electrophysiological, and psychological research, electroencephalography allowed observation of the brain 'at work'. [BOR 18a, p. 16]

The idea that the success of electroencephalographic techniques was involved in providing access to mental and cerebral activities *in vivo* is fundamental to understanding the development of tools for brain exploration [CHE 15] other than through the prism of technophilia. Thus, Grey Walter (1910–1977) spoke about the correlation of psychic states with tracings, that in 1934, Adrian and Matthews (1906–1986):

> [...] recently gave an elegant demonstration of these cortical potentials. [...] when the subject's eyes were open the line was irregular, but when his eyes were shut it showed a regular series of large waves occurring at about ten a second. [...] Then came the surprise. When the subject shut his eyes and was given a simple problem in mental arithmetic, as long as he was working it out the waves were absent and the line was irregular, as when his eyes were open. When he had solved the problem, the waves reappeared. [BOR 08, p. 368][23]

In 1935, Durup and Fessard (1900–1982) concluded that alpha rhythm was associated with the attentional state of the subject [DUR 35]:

> Apart from the action of sensory stimuli, it has been shown that mental work, in the case where a rather large effort of attention is necessary (for example a difficult arithmetic operation), abolished Berger's rhythm by a mechanism similar to the one just exposed. A simple conversation, an attention directed towards the outside (on non-visual objects) with no task to accomplish, does not in any way hinder Berger's rhythm. They even favor it. Emotions also seem to be capable of producing the same kind of action, but there is not yet much data on this. [DUR 35, p. 7, author's translation]

Thus, scientists were able, by simply reading these recordings, to determine when a person was engaged in an arithmetic problem and the duration of his or her concentration. The development of electroencephalographic techniques in the field of psychiatry and psychology were such that the brain was seen as an organ to be deciphered, within which neuropsychiatric specificities, whether normal or psychological, were within the reach of doctors. These recordings renewed the program of deciphering the psychological nature of humans through the reading of cerebral rhythms. Electricity found here an anthropological meaning relative to the human being inscribed in nature and determined, even in his consciousness, by the forces of physics. Thought could thus be reduced to the different states of cerebral electrical activity and to a series of specific signs. Adrian compared brain activity to the hustle and bustle of

23 Borck cites Grey Walter's treatise *Thought and Brain: A Cambridge Experiment* [WAL 34, p. 479].

the crowd and explored the problem of the psychophysiological significance of brain waves, about which he was skeptical. In 1936, he compared his brain pattern with that of a water beetle, thus highlighting their indistinguishability (see Figure 5.8).

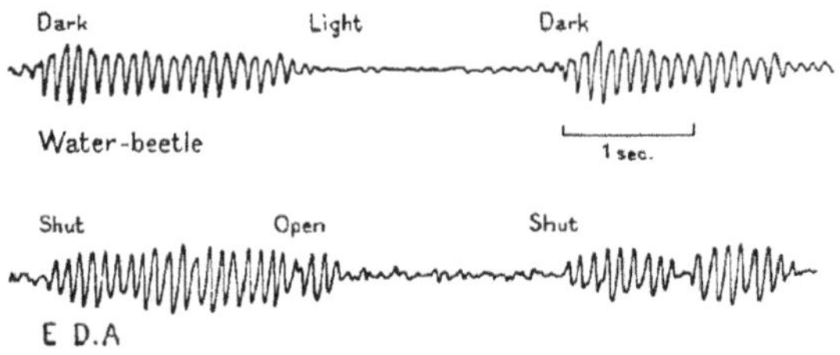

Figure 5.8. *Adrian's visual assimilation of human brain waves to potential swelling recordings [ADR 34, p. 373]*

Despite Adrian's skepticism, the hope of establishing an identity between cognitive functions and the electrical language of the brain persisted throughout the first half of the 20th Century. By postulating from the outset a community of signs, the identity of nature was deduced from them. Thought materialized in graphic form was perceived as the effect of brain activity. This is why psychology questions the possibilities allowed by E.E.G to determine the causal links between brain activity and psychological performance. This question was at the origin of many interpretations of brain patterns. For psychiatry, always in search of a neurophysiological basis for mental disorders, it soon became evident that the majority of diseases affecting the central nervous system have electroencephalographic correlates. But could its correlates be considered relevant? Henri Gastaut (1915–1995), an epileptologist in the second half of the 20th Century, saw in electroencephalography, in addition to its clinical applications, the means to explore mental activity and to establish a typology of characters based on electrical tracings. Since each individual possesses a singular electrical activity, he conceived of it as revealing the great traits of the personality, in its normal states as well as in mental pathology. The E.E.G., in addition to enabling the representation and recording of brain electricity, was used here as a "psychological trait developer". This theory was reminiscent of the notion that there are individuals whose nervousness is caused by an excess of electrical fluid [PET 87, p. 86]. Gastaut correlated certain electrical phenomena such as desynchronization with the simultaneous expression of mental states or emotions: his electroclinical studies of psychopathological cases made these correlations more obvious. In 1949, he published "*L'activité électrique cérébrale en relation avec les grands problèmes psychologiques*" [GAS 49], an article in which he already developed his desire to show: "How dynamic electroencephalography […] makes it possible to 'track' the electrical expression of the physiological processes underlying thought to its last haunts" [GAS 49, p. 64, author's translation].

While the E.E.G. did not keep its promises in terms of understanding mental activity, it was a question of methodology for Gastaut. And yet:

> In the face of such findings, psychiatric disorders captured by the EEG appeared to be barely distinguishable from the transient anxiety or emotional states of 'normal' subjects. The pathological was thus distinguished from the normal only by a slight difference in degree, intensity. Some EEG users, however, wished to define 'pathological thresholds', by 'stimulating' the experimental subject, according to the model adopted to detect non-symptomatic cases of epilepsy. [PID 10, p. 42, author's translation]

These problems of interpretation and of the relationship between psychological limited states and abnormal traces pushed Gastaut to differentiate the static E.E.G. from the dynamic recordings. Static E.E.G consisted of collecting brain electrical activity under well-defined and coded conditions, so as to make the results as comparable as possible. It appeared that the states of wakefulness and sleep, attention or indifference, consciousness or coma were accompanied by more or less characteristic cerebral electrical activities, while the patterns remained the same regardless of the subject's state of mind or degree of intelligence. Dynamic, also called functional, electroencephalography found its maximum expression in the so-called activation methods with which electroencephalographists strove to modify the electrical activity of the brain by the most diverse means with the intention of establishing archetypal psychological thresholds. In collaboration with the field of psychology, Gastaut practiced, for example, dynamic recordings of subjects submitted to Rorschach tests in order to highlight the influence of theta rhythm variation in neuropaths:

> There are other ways to induce theta rhythm in neuropaths. Thus Faure, from Bordeaux, and Bert, in my laboratory, chose to subject their subjects to the effect of Murray and Rorschach tests. In the majority of cases, most of these boards were ineffective, but in other cases, on the contrary, some of them can have a spectacular effect; an effect all the more striking when using a harmonic frequency analyzer that can highlight the slightest variations in theta rhythm proportion. Under these conditions we could see the proportion of this rhythm increase considerably when certain plates were subjected to the reflection of the subject, and I remember in this connection a case which had been reported to me by J. Bert. This was a perfectly normal young man whose life and that of his mother had been completely disrupted following his father's decision to abandon them a few years earlier; this subject showed a significant theta-wave discharge when he was subjected to the plate where a man appears behind a pensive

> child. Interestingly enough, this young man was waiting for the days following the visit of this offending father whom he had not seen since the abandonment. [GAS 49, p. 84, author's translation]

Gastaut carried out hundreds of recordings of the electrical activity of deep structures. He highlighted a possible involvement of the rhinencephalon in the expression of emotions and changes in consciousness. These studies, important for their localization and functional aspects, determined the development of E.E.G outside the clinic and diagnosis. The localization of this research was related to the localization of recorded electrical activity, as in the case of studies of the rhinencephalon, while the functional aspects were related to the overall electrical representation of brain function and thought. In 1953, during the symposium *Brain Mechanisms and Consciousness* [GAS 54], in a text entitled "*Le tronc cérébral et l'électrogenèse relative à la conscience*", Gastaut highlighted, through a study of the desynchronization processes of cortical electrogenesis, their influence on the mental state of subjects, particularly in psychopathologies marked by states of major anxiety. Based on the parallel reading of clinical symptoms and brain traces, he proposed an electroencephalographic typology based on two criteria:

– clinical behavioral characteristics such as anxiety, calmness or nervousness of the subjects studied;

– brain rhythms grouped into classes by the community of electrical signs were therefore associated with the subject's mental characteristics.

Gustaut's willingness to interpret the subject's character in relation to the brain's signals, to decode it in the same way as all physiological mechanisms, referred to the modeling of the body in terms of machines and even computers as far as the cerebral organ was concerned. Gastaut seemed to "copy" mental decoding on to computer coding. He discussed three syndromes associated with three types of electrocerebral disorders recorded in subjects with no mental pathology: hyperexcitability, hypoexcitability and cortical instability [GAS 54, p. 272]:

– hyperexcitability was associated with a rapid but small-amplitude alpha rhythm, accompanied by a high beta rhythm (15–20 cycles per second) and intermittent desynchronization. This recording was correlated with personalities marked by dynamic temperament, hypersensitivity, emotionality and impulsiveness;

– hypoexcitability was associated with alpha waves, which were slower than usual, with no phenomenon of desynchronization. It was associated with methodical and thoughtful personalities;

– cortical instability was associated with low and high frequency phenomena for the theta and beta rhythms with the alpha rhythm in the background. This last syndrome concerned impatient, aggressive or hostile personalities.

Gustaut therefore proceeded with a typology of patterns based on a causal analysis of electrical brain phenomena on behavior [GAS 51a, b]. These explorations were influenced by previous studies [SAU 49] on the relationship between E.E.G and mental activity and by the research of Mundy-Castle (1923–2015) [MUN 53a b, MUN 54] on the correlations of psychology with electroencephalograms, published in 1953–54. This classification of brain rhythms into three categories turned into electrical signs, without explaining them, the existing relationships between cortical rhythms and attentional behaviors or states. In addition, Gastaut showed the possibility of varying electrical and behavioral factors in parallel, in particular through the absorption of psychoactive substances. One of his studies, published in 1953 [GAS 53], was based on the E.E.G. examination of 12 subjects, considered to have normal psychological activity, to whom a team of researchers administered 60 micrograms of d-lysergic acid diethylamide[24]. He used this substance as a "mental scalpel" to better understand the integration of consciousness into matter and the neurological mechanisms underlying its contents. In the context of dynamic E.E.G., it was not a question of recording an artificially created phenomenon and seeing if it differed in the subjects, but of creating different psychological conditions and appreciating their electrographic translation, exactly as one does in neurophysiology to appreciate somatic conditions in cats or rabbits: it was finally a question of electrophysio-psychology. Thus, after one hour, 9 of the 12 subjects showed parallel modifications of their tracings and their psyche: the recordings showed a typical hyperexcitability marked by an acceleration of the alpha occipito-parietal rhythm in connection with an exacerbation of the subjects' perceptions accompanied by emotional instability. Gustaut highlighted three main effects that he intended to study in parallel:

– neurovegetative effects;

– psycho-affective effects;

– translations of these effects into electroencephalographic signs.

A line of convergence was found from these correlations: the most marked psychological phenomena appeared with the most important neuro-vegetative modifications and with the most modified patterns:

> Fig. 2: Same derivations, same medium, same layout as in Fig. 1. On the left, before the LSD, the intermittent light stimulation does not cause any appreciable 'practice' and allows the alpha rhythm to persist at 10 c/s. On the right, three hours after LSD, the same stimulation causes a remarkable 'practice' of the parietal-temporal-occipital rhythms. [GAS 53, p. 115, author's translation]

24 D-lysergic acid diethylamide, discovered in 1943 by Hoffman, was marketed for medical purposes between 1950 and 1960 by Sandoz, which distributed it to the medical community.

While he did not deduce from this the absolute interdependence of the traces and states of consciousness, he nevertheless noted the visible reciprocity of this influence [GAS 54]. Gastaut defined certain grapho-elements characteristic of the human electroencephalogram such as the occipital points at the opening of the eyes and studied their possible relations with certain psychological manifestations such as emotional immaturity and repressed aggressiveness.

Thus, while psychologists saw very early on, in the application of electroencephalographic recordings to thought, the means of solving the psychophysiological problem, by laying the electrical foundations of the link between the cognitive faculties and physiology, it also seems to have been established as early as 1960 that electroencephalography was not a tool for interpreting thought. Was there a possibility to talk about brain language? Moreover, new perspectives were opened for the application of these techniques insofar as the E.E.G., as a clinical and diagnostic tool, allowed the reduction of neurological symptoms to localized and identified pathologies such as in the case of tumors or different types of epilepsy. Thus, electroencephalographic examinations were prescribed in many clinical situations to check the overall and spontaneous electrical activity of the brain but were no longer, after 1960, perceived as a key to entry into the human psyche. As this exploration tool proved to be very effective in determining pathological foci, the neurological trail was explored as early as the 1930s. Thus, in 1936, Walter was already locating and recording slow potentials from a brain area suspected of being the site of brain tumors. Electroencephalography thus provided the model for a pathology based on a reading of electrical disorders that did not require further analysis. Thus, beyond the spectrum of symptoms that could be caused by a brain tumor, E.E.G allowed the reduction of its clinical manifestations. Thus, it generated representations whose images could be interpreted differently according to the more or less marked interest in neuronal or cognitive mechanisms or in the clinic. In the case of brain tumors, it appeared that without further neurophysiological investigations, the electroencephalographist's trained eye could detect the pathology that would otherwise be difficult to identify [GIB 41]. The electro-stereotactic techniques developed by Bancaud (1921–1993) and Talairach (1911–2007) in the 1960s were an interesting example of the application of electroencephalography to neurological disorders. They allowed not only the localization of pathological foci in deep structures but also precise interventions. Jean Bancaud's main research concerned the stereotactic functional exploration of epilepsy that was resistant to drug therapy. Some failures in conventional surgery seemed to him to be caused by the uncertainty of operative indications based on often ambiguous clinical, radiological and electroencephalographic criteria. By using a stereotactic technique allowing the placement of a high number of intracranial electrodes based on rigorous anatomical mapping, it became possible to explore the activity of the different brain structures involved during a seizure. In 1962, Bancaud and Talairach developed stereo-electro-encephalography. This technique made it possible to study the modalities of propagation of discharges in the vicinity of epileptogenic zones, thus giving a fairly accurate spatiotemporal representation of the paroxysmal dysfunction. It also made it possible to reconsider the etiophysiopathogenic problems of epilepsies and offered the possibility of developing functional epilepsy surgery using open skull pathogen extraction processes and stereotactic focal destruction techniques. Thus was born the association between electroencephalography and stereotaxis. In 2007, in his book *Souvenirs des études stéréotaxiques du cerveau humain,* Jean Talairach describes how these methods were used to locate deep epileptogenic foci:

> The aim here was to establish, using multiple electrodes, strict correlations between the supposed anatomical location and the starting point of the seizure, the EEG perturbations and the clinical symptoms, in other words, to allow triple anatomo-electroclinical correlation. In addition, the electrodes, between them, made it possible to highlight the dynamics of the seizure, i.e. to locate the epileptogenic zone, the place where the phenomenon that triggered the seizure occurred, to appreciate the irritative zone which was in its vicinity but could also extend to the opposite side, and finally, to highlight, especially if the seizure tended to (become) generalized, the propagation routes. [TAL 07, p. 35, author's translation]

Since the suspicion of the existence of brain electricity, thought of by analogy with natural electricity, neuro-anatomists and neurophysiologists have been exploring different ways of recording this energy. Then the need to attribute a meaning to it in relation to human nature was very quickly felt. Thus, the invention and developments of electroencephalography were linked to the fantasy of being able to represent and localize the different cerebral activities as well as their psychological correlates. In this perspective of reifications of thought content, the fascination for neuroimages came to be a persistence of the attraction exerted by the first techniques of visualization of mental processes.

While electrical stimulation and recording developed separately, both geographically and disciplinarily, the tendency is to couple these therapeutic techniques in psychiatry. This coupling could facilitate the neurophysiological effects of brain stimulation by controlling the patient's neurocognitive activities at the time of stimulation. Brain stimuli are indeed integrated into the framework of neural network activities. In other words, although it is most often always a specific brain region that is stimulated by rTMS (repetitive Transcranial Magnetic Stimulation)[25], the effect of this stimulation on one or more of the patient's cognitive

25 rTMS is a transcranial stimulation technique used at the diagnostic and therapeutic level for psychiatric and neurological illnesses, but it is also an instrument of scientific investigation aimed at understanding the interactions between cerebral activity and these pathologies.

and behavioral activities is due to its distributed effect on one or more of the neural networks underlying the cognitive activities, an effect that can only be understood by taking into account the state of activity of these networks at the time of stimulation. We will now look at the articulation between the notion of cerebral control and brain stimulation techniques in the field of psychiatry around 1950 in order to see how this articulation generated care practices newly applied in the 21st Century to improve psychiatric symptomatology.

6

Disorders and Resurgences of Electrical Neurostimulation Therapies: From Heath to Deep Brain Stimulation

In the history of deep brain stimulation, Professor Alim Louis Benabib (1942–) and his team from Grenoble are cited as the discoverers of a new technique to reduce the symptoms of Parkinson's disease[1]. This perspective, without denying the contributions of the French team, will be deconstructed on the condition that it gives a history to deep brain stimulation, which was rooted in the field of mental illness treatment at the heart of the 20th Century. On the one hand, the history of invasive stimulation in deep cerebral zones is long, complex and agitated by polemics and a scientific imagining which is sometimes close to the theme of "augmented man". On the other hand, this history conveys the misconception that neurostimulation primarily targeted movement disorders, whereas, psychiatric applications were the first to be made. It is not a question of knowing which ills of the body and of the mind were the first to be modulated or stimulated, but rather of highlighting the tensions that have coexisted between these two fields since the beginnings of medical physics. Thus, the narrative story of deep brain stimulation (DBS) could be summarized as follows: it was invented in 1987 at the same time as the benefits of high-frequency stimulation; neurosurgery benefited deep stimulation. Its non-motor effects on the subthalamic nucleus prompted, at a second stage, its use in psychiatric and behavioral illnesses. The aim here is to resituate French research on two levels: on the one hand, within a long history which has its roots in the first applications of medical electricity on the obliteration of faculties, convulsive diseases and psychic disorders; on the other hand, within a shorter history, in the 20th Century, which saw the development of electrical surgery with a stimulatory aim within the framework

1 The first studies on correlations between basal ganglia and movement disorders date back to 1975 [BEC 75].

of psychiatry. From the body to the brain, from mobility to the mind, internal electricity interacts with artificial electricity, explores brain mechanisms in the therapeutic framework, and merges its mechanics with organic structures. It is also not a question of reducing deep stimulation to a renewal of psychosurgery, against which it is a reaction. It is more relevant to approach deep stimulation from the angle of an analogical operation with the ablative lesion:

> While psychosurgery was a lesional unselective, irreversible manipulation of a brain area [...] the primary goal of DBS is to rebalance the damaged neuronal circuits through an electrical selective reversible manipulation (stimulation). [SIR 11, p. 5]

Electrical stimulation has been shown to be close to post-epileptic effects in its effects on neuropsychiatric disorders, which brings it closer to the effects of electrical convulsive therapy. As discussed in the previous chapter, on the relationship between mapping and electrical stimulation, in 1874, the American physician Robert Bartholow [BAR 82] reported the results of studies on the electrical stimulation of the cerebral cortex in a conscious human. A second publication on the electrical stimulation of the human brain was made by Horsley [HOR 84] ten years later. He determined the presence of functional tissue in an encephalocele[2] by electrical stimulation. In 1887, he performed [HOR 90] the first electrocorticography in epilepsy surgery and inserted electrodes through an invasive basal cell carcinoma of the skull of a conscious patient and observed the painless contraction of the contralateral limb caused by the electrode in the brain matter. These experiments, in the continuity of Fritsch and Hitzig's research, confirmed the excitable character of the human brain and the controlaterality of the motor cortex.

At the beginning of the 20th Century, Chardin's electrotherapy device promised mountains and wonders: a cure for paralysis and epilepsy, treatment of tumors, cancer, gangrene, anemia and even obesity. But how did it work? Chardin placed a first electrode on the patient's head and a second on the affected area. The current was continuous and the intensity was adjusted to the subject's sensitivity. The recommended duration of treatment was 6 to 10 hours per night. His technique was based on the idea that electricity is the immediate remedy for all ills that are circulatory in origin. He used what he called the E.C.V. method: electrokinetic-vascular or the action of electricity on the vessels. He therefore subjected patients to transcranial electrification in cases of insanity for an unlimited period of time:

> The fact is that in the last two months, the illness had worsened so much that the child, already nervous by temperament, was almost

2 A hernia of the brain outside the skull.

> mad, due to insomnia and food deprivation. From the first application, the patient became calm and slept. [CHA 17, p. 90, author's translation]

6.1. Stimulation, control and improvement of moral and cognitive capacities

In the context of the Second World War, electro-convulsive therapy was widely developed in the treatment of patients suffering from depression or psychological trauma. At the same time, electrophysiological studies and measurements of the electrical fluid in the nervous system were being mathematized and instrumented. These clinical and experimental laboratory and physiological practices came together in brain stimulation, a practice characterized by three criteria described by Morlacchi and Nelson [MOR 11] as crucial components of medical innovation:

> Improving in the ability to develop effective medical technologies, learning in medical practice and advances in biomedical scientific understanding of disease. [MOR 11, p. 511]

In the case of brain stimulation, these three principles interact strongly, in an anticipatory way, to understand and improve the physiological conditions of the brain:

> From the 1940s onwards, this specialty disseminated rapidly throughout the world, providing (or attempting to provide) therapeutic relief to patients with what were then otherwise untreatable neurological and psychiatric conditions. [GAR 13, p. 710]

In 1947, at Temple University in Philadelphia, the American neurologist Spiegel (1895–1985) and neurosurgeon Wycis (1911–1972) [SPI 47] designed a stereotactic device, designed to be used in humans for precise ablative procedures, which were less dangerous to the integrity of brain mechanics than lobotomies.

Brain stimulation thus has a dual disciplinary origin: neurostimulation techniques and stereotactic neurosurgery, also known as functional neurosurgery. Used in conjunction with imaging technologies:

> A stereotactic device is a movable metal frame that can be positioned on the head of a subject or animal, which constitutes both a skull restraint system and a reference space using a polar coordinate system to hold an electrode in position and then guide it by sliding it into the brain space at a specific location. A micromanipulator and a guide for holding the microelectrode fit on this frame and can be positioned in

> the reference space by choosing certain coordinates, so that they allow the microelectrode to descend into the brain at a precise point determined by its coordinates. [BAR 18, author's translation]

This equipment facilitates stereotactic neurosurgery with less lesional stereotactic neurosurgery in its post-operative consequences. This technique developed further at the end of the 1950s, particularly in Europe with Bancaud and Talairach, and in Canada with Penfield's research. Permitted operations such as cingulotomy or anterior capsulotomy were first used in psychiatric illnesses to replace the frontal leukotomies of Walter Freeman (1895–1972). Since 1930, animal brains have been studied in a conscious state and in depth when W.R. Hess (1881–1973) developed a procedure to implant very fine wires into the brain of an anesthetized cat. After the effects of anesthesia have worn off, the relatively free animal can then be electrically stimulated by connecting long electrodes to the terminals of the implanted electrodes. Very quickly, these deep implantations, which do not interfere with arousal, mobility or attention, are conceived as research avenues for understanding human and animal behavior. Delgado (1915–2011), in works that were often very theatrical and staged during public demonstrations, developed these techniques for their potential benefits to humans. Thus, in the context of comparative electrophysiology, he practised the implantation of electrodes in the area of the caudate nucleus, especially in the bull:

> […] Delgado was sufficiently confident of this idea that in 1963 that he stepped into a bullring in Cordoba in front of an aggressive fighting bull. As it charged towards him, Delgado stood his ground and calmly twiddled a button on a remote control device that sent a signal to a transceiver connected to an electrode implanted in the animal's brain. [ASH 13, p. 304]

Electrode implantations were not only very quickly considered from a treatment perspective, but also from a behavioral control and cognition improvement perspective. The boundaries between care and improvement were very porous. Where did the first stop and the second begin? The answers were not simple and called into question the definitions of therapy and disease. Beginning in the early 1950s, Heath (1915–1999) implanted electrodes into patients and stimulated their brains intermittently over long periods of time. In the context of psychiatric treatment, these interventions contributed to the physiological study of the mental sphere. Working at Tulane University, his research took place in a post-World War Two context in which a large percentage of hospitalized veterans, around 50% [FRA 18, p. 24], were hospitalized for psychiatric disorders. Lobotomy, which was still widely practiced in the United States, was limited and then entered into decline by implantation:

> Between 1936, when Portuguese neurosurgeon Antonio Egas Moniz reported that he had performed the first frontal lobotomy to treat schizophrenia, and 1949, when he received a Nobel Prize, twenty thousand Americans had been lobotomized. [FRA 18, p. 23]

The development of brain implants was conceptually situated between the irreversibility of psychosurgery and a post-Freudian context of the biologization of disorders. Heath's technique developed all the more easily because, from 1960 onwards, Chardack (1915–2006) and his colleagues [CHA 60] developed the first fully implantable pacemaker. A proponent of biological psychiatry, Heath was looking for an organic substrate, the site of madness, on which to intervene through deep electrical stimulation. He followed an organicist doctrine that organo-functional defects caused mental illness. As a result, mental problems needed be treated physically. A prolific scientist, he developed, in no less than 425 articles and three books, a type of neurosurgical operation with a therapeutic aim. In 1952, he created the Society for Biological Psychiatry and conducted his first experiments mainly on cats and monkeys. His studies [HEA 54b] showed that stimulation of the deeper cerebellar strata, located in the vermis up to and including the fastigia and amygdala, animated a pleasure circuit in the brain; a circuit he sought to activate in his patients. Thus he spoke of *currents of happiness* which the subjects themselves could stimulate. Thirty-eight patients were treated at Tulane between 1955 and 1978 with a chronically implanted cerebellar stimulator. All of these subjects suffered from schizophrenia, chronic depression or epilepsy, accompanied by behavioral problems. Heath considered that patients who responded best to treatment were those who suffered from depression, had behavioral pathology as a result of epilepsy or had psychotic behavior as a result of structural brain damage. Their follow-up period varied from a few months to 27 months:

> At the Society meeting a year ago, we gave a preliminary report on the first 11 patients in whom we had implanted a cerebellar pacemaker [...] as treatment for untractable behavioral pathology. [...] This series has now increased to 38 patients. Three of the 38 patients had second operations, making a total of 41 operations. [HEA 80, p. 243]

By implanting the electrodes in the cerebellar area, he correlated psychotic disorders with the mechanisms of the cerebellum to which movement management was usually assigned. He concluded that brain damage could induce psychotic behavior. Thus, diseases were subject to choice, with schizophrenic patients being set aside, and it was considered that not all psychiatric pathologies were equally involved in the same brain circuits. In addition, two conditions, still applicable today in the context of deep brain stimulation, were prerequisites:

– the disorder needed to have been thoroughly tested with all other forms of treatment without beneficial results;

– on the basis of these therapeutic trials, the medical team needed to be able to affirm the incurability of the disease.

In 29 patients, cerebellar electrodes were placed on the upper surface only (Figure 6.1). Current probes indicate that the propagation of stimulation is effective with these electrodes but limited to 4 mm beyond their implantation site. In nine other patients, the upper surface electrodes were placed as cathodes and the inferior surface electrodes were placed as anodes, so that the stimulus were more intense at the upper surface (Figure 6.2):

> Many parameters of stimulation have been tried. The most effective is at 100 Hz with a pulse width of 150 to 250 used at a voltage setting determined by sensory evoked potentials recorded from the scalp and the patient's clinical response. [HEA 80, p. 246]

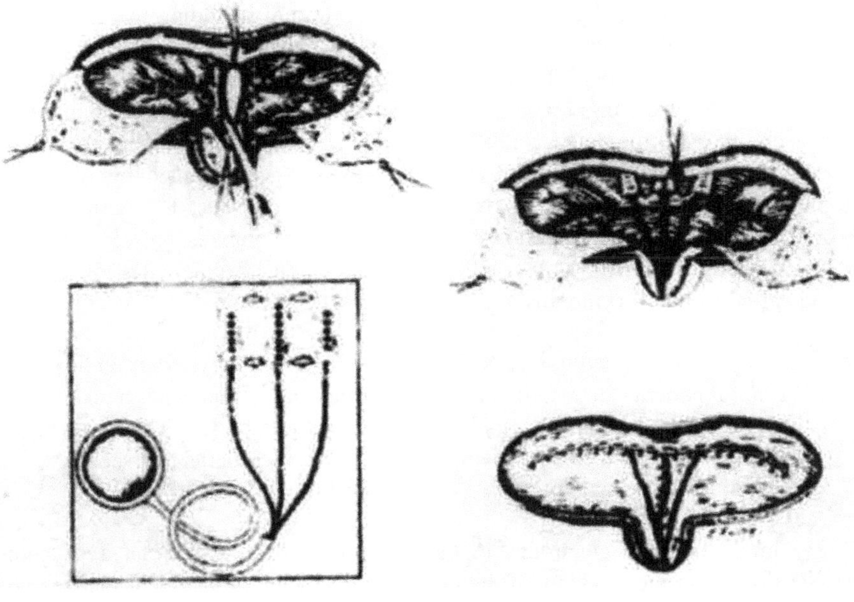

Figure 6.1. *Drawing illustrating the location of the superior surface electrodes. Top left: exposure of the cerebellum. Bottom left: the electrodes: the middle chain consists of seven platinum contacts, each 2 mm diameter. Top right: electrodes in place. Bottom right: Dural closure showing exiting electrode wires [HEA 80, p. 245]*

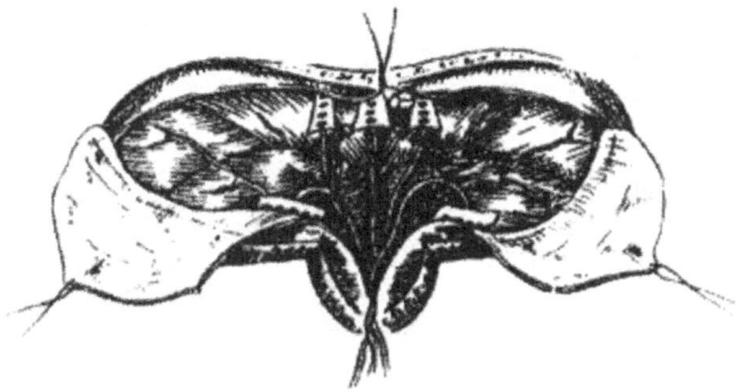

Figure 6.2. *Drawing of the superior-inferior electrode configuration with the inferior electrodes placed under the lateral hemisphere. On the superior surface, there are six contacts in the midline chain and five in each of the lateral chains. Inferior electrodes consist of two chains on each side, each made up of five contacts [HEA 80, p. 246]*

The results on schizophrenic patients were mixed. Still, in terms of neuroleptics, three were institutionalized and few used their stimulators regularly. While in the case of patients with depression, while improvements were noticeable, the device remained difficult to tolerate:

> Five of the six patients in this group are largely free of depressive symptoms and require no medication. With stimulation, the sixth patient, the paranoid-hypochondriacal man, experienced prompt relief of symptoms. He soon began to complain, however, that the stimulus was uncomfortable and that the wires in his neck were painful. He subsequently stopped wearing the pacemaker and insisted that it be removed and replaced with another. Because we had some evidence he felt the stimulus turning on, and because it was possible that a current spread to the tentorium was causing his discomfort, we replaced the electrodes and receiver. After prompt relief of depressive symptoms for a few days, he once again began to complain of pain from the "cord" (electrodes) in his neck, he ceased to stimulate himself, and he insisted that the wires be cut but not removed. Eventually the entire implanted device was removed. [HEA 80, p. 249]

Surprisingly, in a report published in 1980 [HEA 80, p. 250], the findings of "... a patient with hysteron-epilepsy in the original series of 11, continues to be essentially symptom-free". What did hystero-epileptic mean in the 20th Century? In its current meaning [AUX 12], this term refers to a more or less sudden change in behavior and contact with the environment, "compatible with an apparent alteration in consciousness and associated with abnormal, sometimes spectacular movements of variable duration" [RUT 16, p. 124, author's translation] but without neurological episodes of epilepsy.

In his series of observations, Heath noted the case of an implanted patient with mild mental delay, accompanied by aggressive and uncontrollable behavior. After the operation, he was considered to have become asymptomatic and, Heath pointed out, he was able to finish school and get a job. The heuristic aspect of these operations must be emphasized: in addition to relieving symptoms, they were likely to influence the entire cognitive and behavioral life of the subjects. This provided the catalyst for the idea that pathological disturbances disrupted the entire cognition. As Penfield also noted, both faculties and behavior could be relieved following implantation with brain stimulation. Treatment was combined with improvement, which became part of psychiatric care. Heath concluded that cerebellar stimulation produced its best therapeutic results in patients whose main symptoms were related to deep aversive emotion. The least significant results were shared between subjects who rejected the use of the stimulator and those who, despite their attention to using it, had only limited effects. This work helped to identify organic targets that could be improved by this technique. While no specific area can be assigned to insanity, it seems that singular brain areas can be modulated according to the disorder treated. Moreover, therapeutic failure is also a field of experimental knowledge. In the group where the results were less favorable, the recorded responses were characteristically inconsistent, with a drop in amplitude occurring after one or two stimulations to be followed by an increase. This phenomenon suggested intermittent interference in the neural network targeted by the stimulation. These failures were heuristic and pathophysiological principles could be deduced from them. In the case of schizophrenia, Heath suggested that the biochemical damage underlying the disease altered the response to stimulation. It must also be considered that the disease is not "pure" in its mechanics but that, as a pharmacoresistant pathology, it bears the traces of the neuroleptic treatments followed, which influences its expression and interactions with stimulation:

> In summary, results obtained with the cerebellar pacemaker continue to be encouraging. Most of the patients, all of whom were previously intractably ill, have benefited. We have gained more information about the types of patients who are most apt to respond. Those who are ill because of a preponderance of profound adversive emotion (depression, rage, violent behaviour) benefit significantly. The stimulation also works well with the clinical diagnostic entity of depression, including anhedonia[3]. Psychotic manifestations of

3 Anhedonia is a disorder characterized by the inability to feel positive emotions in situations that are considered pleasant.

> epilepsy have also been largely eliminated, and the therapeutic procedure has removed many symptoms in patients with organic brain syndrome. Whereas some schizophrenics have been notably helped when stimulation has been possible, they have been less responsive as a group for reasons that are still obscure. [HEA 80, p. 255]

Guided by experimentalism, this research depended on knowledge of the electrochemical principles at work in mental illness; knowledge which Heath also contributed to enriching. This knowledge also motivated technical improvements. While exploratory, these operations contributed to understanding human behavior in physiological terms. Convinced that it was possible to control uncontrollable impulses with an implanted microchip, Heath described the connections discovered in the brain and the communications between the outermost layer of the cerebellum and the emotional areas within the limbic system.

In 1952, Delgado [DEL 52] developed a technique for implanting electrodes for recording and chronic stimulation in psychotic patients, thus moving from an experimental animal model, the bull, to a therapeutic application on humans. The following year a symposium on "Intracerebral electrography" took place, including a paper on "Neurosurgical and neurological applications of deep electrography":

> An observation that may have some practical significance was that several of our psychotic patients seem to improve and become more accessible in the course of stimulation studies lasting several days. [BIC 53, p. 182]

In 1961, a book entitled *Electrical Stimulation of the Brain; An Interdisciplinary Survey of Neurobehavioral Sciences* [SHE 61] was published. It was devoted to comparative behavioral studies on the consequences of implants in animals and humans for recording and subcortical stimulation in epilepsy, obesity or aggressive behavior:

> To take a different and extreme case, Roberts and I have watched electrical stimulation cause a cat to lower his head and lap water from a dish, but when the dish was moved to one side the cat lowered his head and licked the floor. We were obviously observing a motor response instead of a general motivational effect. Another satiated cat began eating when stimulated [...]. [SHE 61, p. 387]

Considering the application of this type of electrical conditioning in humans refers to the boundaries between treatment and control. The context of application alters the intent and perspective of the transaction. Delgado, in the field of research on electrical stimulation techniques applied to the study of the behavior of primates and humans, developed a theory involving the interactions between *psychic energy* and external stimuli:

> Symbolically, we may speak about "psychic energy" as the level of intracerebral activity which could perhaps be identified in neurophysiological terms by electrical and chemical processes located at specific neural fields. This psychic energy may be considered a main determinant of the quantity of intellectual and behavioral manifestations. [DEL 69, pp. 9–10]

Initially, from an ESB[4] applied to the motor cortex, an inhibition reaction to abnormal hyperkinetic movements was noted throughout the application, "allowing patients to perform skilled acts which were otherwise impossible" [DEL 69, p. 87]. Thus, Delgado advocated the wearing of a small box, used by the patient to stimulate the brain to temporarily inhibit abnormal mobility and restore physical and mental skills. Citing Heath's research, he highlighted the correlations between these non-invasive practices and the reduction of violent behavior. As part of Heath's self-stimulation program, films showed a patient who stimulated his brain to suppress an aggressive mood. Delgado also noted the links between violent behavior and the inhibitory effect of repeated stimulation of the cerebellar tonsil. These perspectives went beyond treatment to ideas developed by Laycock [LAY 40b] about how electricity can help prevent emotional outbursts and violent demonstrations. Although the ideological context was different, the application of stimulation was a tool for controlling the subject's self-control:

> Because the brain controls the whole body and all mental activities, ESB could possibly become a master control of human behavior by means of man-made plans and instruments. [DEL 69, p. 88]

These moral–behavioral applications of electricity, which seemed to move away from the psychiatric field but remained in the mental sphere did not question free will as long as they were considered to be related to neural activities. On the basis of research on electrically inhibited behavior, researchers such as Heath and Delgado assumed a physiological identity between voluntary triggering and electrical stimulation in that they would activate brain mechanisms in a similar way. If both spontaneous and electrically evoked behavior involved the participation of the same set of brain areas, then the two types of behavior should be able to interact by reciprocally modifying the inhibitory and excitatory influences of the other.

4 ESB here stands for non-invasive electrical stimulation of the brain or non-invasive electrical stimulation.

Is the individual defenseless against direct electrical manipulation of the brain? In previous experiments, electrical stimulation of appropriate intensity accompanied voluntary acts provided that the subject practiced self-stimulation.

Recalling Penfield's work on the localization and stimulation of different types of memory, Delgado saw the role that ESB could play in memory disorders:

> Like spontaneous memories, the recollections induced by ESB could bring back the emotions felt at the time of the original experience, suggesting that neuronal mechanisms keep an integrated record of the past, including all the sensory inputs (visual, auditory, proprioceptive, etc.) and also the emotional significance of events. Electrical stimulation activated only one memory without reawakening any of the other records which must be stored in close proximity. This fact suggests the existence of cerebral mechanisms of reciprocal inhibition which allow the orderly recall of specific patterns of memory without a flood of unmanageable amounts of stored information. [DEL 69, p. 68]

Thus, brain modulation and stimulation techniques were developed as exploratory and therapeutic tools in the field of mental pathology as early as the 1950s. Heath and Delgado's links to control devices made these techniques morally questionable, giving them an ambiguous status in the history of neuroscience. We can nevertheless underline the proximity of the problems of morality control in the 19th Century to the themes evoked by these techniques. Nevertheless, brain stimulation has epistemological and historical roots. Thus, this research contributed to refining the cerebral, functional and anatomical targets in action in many pathologies. The dynamic localization of networks of brain activity was made possible by the development of stereotactic stimulation devices or implants. A history of deep brain stimulation can be extended beyond the late 1980s. In 1987, Alim Benabib published an important article [BEN 87] on the stimulation of intermediate ventral nuclei to treat symptoms resistant to any therapy in Parkinson's disease[5]. It highlights a great many developments and convincing effects on tremors. He and his colleagues used pre-operative electrostimulation to confirm the brain area involved in the disorder being treated. They quickly pointed out that it is high-frequency stimulation that reduces some of the motor symptoms of Parkinson's disease:

5 Natalia Petrovna Bekhtereva (1924–2008), a neuroscientist in Leningrad, is considered the first to have used chronic stimulation of deep brain structures as a therapy for motor disorders. In 1963, she published a book on the use of multiple electrodes implanted in subcortical structures for the treatment of hyperkinetic disorders. She spoke about electro-stimulation therapy. However, she remains little known, because she writes in Russian.

> Deep brain stimulation arose that way, somewhat by chance, when we realized the unexpected effects of electrical stimulation, which we used for something else when it exceeded a certain frequency. [BEN 17, p. 18, author's translation]

Thus, the Grenoble group was certainly the first to systematically study the therapeutic role of a high-frequency electric current in deep brain stimulation and established 130 Hz as the ideal frequency, now commonly used. This frequency can inactivate neurons in the target structure, either by blocking depolarization or by causing the release of an inhibitory neurotransmitter. It is also hypothesized that this frequency is responsible for the local and synchronous cellular activation of a large number of neurons, which may result in a "non-significant" message preventing the expression of disease-related abnormalities. Although treating Parkinson's disease and other pathological tremors are the most common indication for this technique, a number of patients with obsessive compulsive disorder, depression or epilepsy have access to it. The behavioral consequences of the interventions carried out by Benabib and his colleagues on Parkinson's have reopened the door to the application of deep stimulation in psychiatry and neuropsychiatry:

> Acute hypomania was observed in 4–15% of individuals, with a complete manic picture, even in some patients who had never previously suffered from bipolar disorder. Pathological laughter phenomena have also been triggered by DBS, with some spectacular cases on the operating table. It should be noted that laughter was often associated with a sense of bliss. The literature still reports cases of disinhibition with exhibitionism, hypersexuality and pathological play in the aftermath of the intervention, [...]. [GRO 09, p. 61, author's translation]

There are some hypotheses about the role of personality disorders and the thymic history of patients following these operations. These reactions can also be correlated with the fact that different behaviors can be activated in the case of open brain surgery, as we have seen with Penfield. The subthalamic nucleus is also involved because of its interactions with the limbic system and the frontal cortex. Inhibited by stimulation, it could lead to behavioral problems. There are more hypotheses than certainties, and how deep brain stimulation interacts with behavior or cognitive and affective abilities is still largely misunderstood. Here we find the trial-and-error structures of experimental medicine which accumulates reports of cases where the operation has or has not worked. A perfect knowledge of the mental mechanisms in their electro-chemical functioning is necessary to shed light on how deep stimulation works. The only certainties are that it causes behavioral variations and that its applications, present and future, are multiplying. This neurosurgical technique was the catalyst for the material movement of the view of all human capacities to within brain matter and inscribed behaviors from a body perspective whose central circuit is the brain. Future applications of neuromodulation range from Huntington's disease to obesity, and there are many stimulation targets.

Thus, it appears that the history of deep brain stimulation in the field of psychiatry has a temporality that goes back beyond the end of the 20th Century. Moreover, this technique is part of the paradigm, emerging between the late 18th and early 19th Centuries, of an organic mental sphere influenced by the action of artificial electricity and dependent on its natural electrical mechanisms.

6.2. Deep brain stimulation and psychiatry

Electroencephalography, fMRI and neurostimulation are all tools that contribute to understanding the normal and pathological mechanisms of thought. Forms of treatment or exploratory instruments, they contribute to the development of a brain model dependent on modulation, recording and stimulation techniques. Magnetoencephalography is an interesting contemporary example of the analogies made between the speed of thought and the speed of electrical impulses. This technique makes it possible to capture very weak magnetic fields, produced by the passage of electric current through the brain circuits, thanks to sensors placed on a subject's skull. These detect the electromagnetic activity of several thousand neurons, provided that the discharge of the electrical pulses is synchronous. The recorded speed is considered to be comparable with the speed of thought. At least the speed could correspond to the speed at which information, which helps to generate thought, is transmitted. By allowing the identification of functional neural networks and the synchronicity of different brain regions, this technique sheds light on the question of the temporality of functional neural networks. Moreover, as a diagnostic tool, it allows the correlation of dysinchronic brain rhythms with certain psychiatric disorders such as schizophrenia:

> Since the first experiments in the 1970s, a growing number of studies have used MEG to analyze the dynamics of brain activity associated with multiple forms of cognitive expression: perception, memory, learning, attention, motor planning and action, language, social interactions, decision making, and emotions. [CHE 15, pp. 141–142, author's translation]

Cerebral neuromodulation, of which deep brain stimulation is one technique, is experiencing significant growth with a number of emerging applications. In addition to treating movement disorders such as Parkinson's disease, dystonia or essential tremors since the late 1980s, it is experiencing a resurgence in the psychiatric field. Its success in more than 55,000 patients worldwide has facilitated the acceptance of an implanted brain stimulator. The emerging use of brain stimulation for the treatment of neurobehavioral disorders, such as obsessive-compulsive disorder and depression as well as Tourette's syndrome will be discussed. Other applications, for eating disorders, drug addiction, obesity, tinnitus or Alzheimer's disease are also being considered.

With regard to psychiatric indications, such as depression or obsessive disorders, it can be considered that the alteration of the targeted brain structures together with the corresponding neurological deficits can determine behavioral problems. This neuropsychiatric conception of mental, behavioral and emotional disorders makes it possible to understand the rebalancing of specific neurophysiological substrates made possible by deep brain stimulation and the subsequent improvement by harmonizing the physical and psychological dimensions of the treated subjects:

> Probably the most closely watched area of expansion for DBS is behavioral and psychiatric disorders, including Tourette's syndrome[6], major depression and obsessive-compulsive disorder (OCD). [TAR 08, p. 73]

The above statement connects the long history of this technique marked by its first applications in the mental field, with the results expected in the future. Chronologically, if we focus on brain stimulation techniques, movement pathologies have been targeted as a second step:

> The first use of DBS mirrors that of stereotactic ablative surgery: both were initially performed to treat psychiatric disease. Chronic subcortical stimulation was not originally introduced for the treatment of movement disorders. The chronological order of applications of old time DBS was first for psychiatry and behavior, then for pain, then for epilepsy and last for movement disorders. [HAR 10, p. 7]

In addition, the tension between therapies for mobility disorders and the treatment of the psychological sphere is part of the history of the application of medical electricity to humans. The concept of a brain target is extremely important. On the one hand, it is correlated to an anatomofunctional paradigm of brain functions; on the other hand, it brings into play the notions of anatomical regions of the mind. It refers to an integrated human nature that is dependent on the proper functioning of the physiological mechanisms of the brain. It is highly likely that deep brain stimulation exerts its effects through interdependent and correlated mechanisms at the site of stimulation. High-frequency stimulation, developed by the Grenoble team, has been proposed to modify transmission, using neural scrambling to produce a functional lesion, mimicking the effect of a therapeutic ablative lesion. The treatment applied through this technique tends to unify the disciplinary

---

6 See [MMT 04] for more information.

boundaries between psychiatry and neurology, with treated disorders becoming neuropsychiatric. As part of the management of language and behavioral tics, which are highly invasive in Tourette's syndrome[7], Visser-Vandewalle [VIS 99] and colleagues treated a 42-year-old patient with implantation in 1999. A noticeable reduction in the number of speech tics was noted four months after the operation and was followed by complete resolution after one year. These clinical benefits have been attributed to the deactivation of frontal cortical areas by stimulation of circuits involving the ventral striatum in the limbic and dorsal area. In neuropsychiatric pathologies, circuit stimulation of the limbic area is often performed. In the small number of implanted patients with Tourette's five years after surgery, thalamic stimulation was associated with a 72% to 90% [MUK 08, p. 1,124] reduction in tics with complete resolution of all vocal and major motor tics.

The small sample size of patients receiving deep stimulation makes it difficult to consider this therapy as potentially standard in the field of psychiatric illness and leads to recurrent and troublesome side effects being highlighted:

> For example, patients with TS in whom DBS systems have been implanted have experienced poor healing, infections, or hardware malfunctions due to compulsions to repeatedly touch their incisions or push on the subcutaneously-buried hardware. [MUK 08, p. 1, 124]

The disparity of studies and targets, including in this particular disease area, needs to be emphasized. Indeed, the stimulation sites, the techniques used and the evaluation methods vary:

– in the 1999 trials [VIS 99], bilateral high-frequency stimulation of the median and intralaminar nuclei of the thalamus, applied to three patients, resulted in an 80% improvement in tics;

– also in 1999, bilateral stimulation of the parafascicular mid-nucleus of the thalamus (CM-Pf) or of the internal globus pallidus (GPi), in its anterior and medial parts, or both, in a patient with severe Tourette's syndrome, produced the following results: in bilateral stimulation of the CM-Pf, a 64% improvement in tics and a decrease in impulsivity was noted. In bilateral stimulation of the GPi, an improvement in tics of 70% assessed with the YGTSS[8] and a disappearance of

7 This syndrome is marked by an extremely wide and variable range of symptoms. Indeed, all symptoms may appear or may not appear successively in the pathological process. The most commonly described are obsessive-compulsive behaviors and disorders, self-aggressive behaviors often associated with the previous disorders, attention deficit, hyperactivity, impulsivity, anxiety, depressive syndromes, etc.

8 The Yale Global Tic Severity Score is a measurement scale for rationally reporting and evaluating post-operative success.

self-harming was also noted. In addition, no significant changes in depression or anxiety, including concomitant stimulation of both sites, could be reported;

– bilateral stimulation of the sensory-motor territory of the GPi, allowed in one patient [DIE 05], caused a 73% decrease on average in the frequency of tics per minute, with an improvement in depressive and anxiety symptoms, but with a side effect of bradykinesia of the left extremities.

The case of deep stimulation applied to Tourette's syndrome is emblematic of the difficulties surrounding the evaluation of their clinical success and the importance of the criteria chosen to carry out this evaluation. In addition, disorders such as obsessive-compulsive disorders, long considered to be of psychogenic origin, have undergone paradigm shifts related to historical breakthroughs in the way they are considered and to the epistemological and societal shift from a Freudian psychoanalytical perspective to a cerebral and functional conception of these disorders:

> Thus far, work in obsessive-compulsive disorder (OCD), the first neuropsychiatric condition studied using modern DBS devices, has shown consistently positive results across multiple small-scale studies. [TAR 08, p. 511]

Breakthroughs were visible even in the changes in terms to characterize disorders that moved from the mental sphere to the brain areas. A few steps in the history of obsessive-compulsive disorder make this journey intelligible. These disorders are defined by the intrusive and incessant irruption in the thinking of an idea or representation. Compulsions are repetitive response behaviors that reflect the fight against obsessions, while having the effect of reducing the anxiety that results from their emergence. This definition leads us to take into account, at the heart of obsessive-compulsive symptomatology, this impression for the individual of being at fault and/or in a situation of error. Compulsive behaviors are thus intended to put an end to the error signals perceived by the subject. This is why they are then led to reproduce them on a loop on the basis of an internal emotional and motivational state, oriented towards obtaining lasting relief. These symptomatological aspects suggest the alteration of a number of psychological functions, from error detection to emotional, motivational and reward processes.

The determination of obsession has undergone several paradigm shifts and has a long history. Until the beginning of the 20th Century, these disorders, after having been part of theological medicine, became part of moral treatments and philanthropic medicine[9]. Then Janet (1859–1947) [JAN 08–11] proposed a clinical

9 These disorders were first described by exorcists and confessors, notably by Duguet (1649–1733) who, in his *traité des scrupules,* described their types, dangerous consequences and

breakdown and an explanatory model, according to which obsessive illness evolves in three stages: psychasthenia, forced agitation and obsessive-compulsive disorder. Finally, Freud described obsessive neurosis, the theory that obsession is the association of an idea that imposes on the patient an emotional state of "doubt, remorse and anger". The notions of guilt and superego play a crucial role in this mechanism. It was only in 1980 that the term obsessive-compulsive disorder began to be used: this shift in language is not insignificant and corresponds to changes in the technical, ideological and cultural framework.

Thus, Freud marked the treatment and the psychic description of the obsessive neurosis. In *Nouvelles remarques sur les psychonévroses de défense* [FRE 73], he devoted a chapter to obsession entitled *Essence et mécanisme de la névrose obsessionnelle* (essence and mechanism of obsessive neurosis), in which he details its psychological stages and attributes its cause to an infantile trauma. The importance of memory and the power of remembrance, which can act as if it were a repeated current event, are emphasized. In the context of a century that never ceased to extend and absorb a significant number of disorders in the hysterical entity, he brought the internal logic of obsessive neurosis closer to that of hysteria. Obsession was described as hysteria complicated by new mechanisms. In his introduction to *L'homme aux rats,* he wrote:

> The means by which the obsessive neurosis is used to express the most secret thoughts, the language of this neurosis is in a way only a dialect of hysterical language. [FRE 10, p. 200, author's translation]

Then he highlighted the fact that, in obsessive neurosis, the primary affect is marked by guilt and reproach attached to memory. In other words, the fundamental mismatch between memory and the pleasure it might have implied is part of the obsessive neurosis of guilt and blame that the subject feels. While guilt can be understood as a defense against obsession, obsessive thoughts are described as the subject's response and obsessive behaviors as the symptoms. It is this dynamic of guilt and moralization of the trauma that helps to break down the conception of the psychogenic origin of obsessive neurosis. In this case, the compulsion comes from the repressed affect, the reproach, which returns in the form of new symptoms. This first psychoanalytical interpretation of the obsession makes it possible to interpret, within a precise mechanism, the rituals that mark its development. Around 1920, in the midst of the "condemnation" of an interventionist psychiatry, Freud extended his

---

possible cures. Then Esquirol (1772–1840), in 1838, classified obsessions within the framework of affective or reasoning monomanias. He also related the first post-Pinean case of obsession: that of Mademoiselle F., suffering from "moral madness" and compulsive washing sessions. Carl Westphal (1833–1890), a German psychiatrist, in 1878, spoke of coercive neurosis. He described obsessions in an intact mind and intruding into the normal thought process against the subject's will [DUG 18; ESQ 38].

reflections on the obsessive trait, on the links with collective psychology, and highlighted the formation of a very/too strong superego at work in the regression and transfer of obsession onto objects devoid of affect. It is within his intellectualized construction of obsessive neurosis that a split occurred.

From obsessive neurosis to obsessive-compulsive disorder is more than just a change of terms. Therapeutic practices differ in each case. This change of name is, first of all, the effect of a conceptual and clinical transformation in the definition and therefore in the conditions of observability of obsessions and compulsions. The notion of OCD has its roots in the scientific framework of a naturalization of intentionality and *compulsive* action. The reasoning mobilized by cognitive behavioral therapy (CBT) rejects the moral and intellectual dimension of this disorder.

This epistemological change could only take shape socially, through a change in definition. Indeed, at the end of the 1970s, developments in pharmacology, the development of behavioral therapies, a reversal against psychoanalytical cures and the adoption of a biologization of disorders opened up the therapeutic space for both pharmacology and interventionist medicine. This definitional, therapeutic and methodological breakthrough was important insofar as the notions of guilt and patient responsibility were set aside.

In this context, the *Association Française des TOC* (French Association of Obsessive and Compulsive Disorders) was founded in 1992. Moreover, this double movement of de-semantization and de-moralization of obsessive symptoms was part of the re-medicalization of the field of mental pathologies, driven by the neurosciences and symptom of a naturalization movement of the psychological and subjective aspects of life. Thus, the descriptive and mechanical framework of Freud's obsessive neurosis may have been considered too moralizing and restrictive. The behavioral model for interpreting OCD has a structure in which the compulsive behavior and the anxiety that triggers it are disassembled. It was outlined in the 1960s by Isaac Marks (1935–) [MAR 88, 87]. In this reasoning, we start from the obsession with which an increase in anxiety is associated; from a more etiological point of view, it is anxiety that is primary and the obsessive representation is contingent as to its content. Acting, even in a form that does not go beyond thinking, has a power of relief: the anguish of obsession diminishes when one gives in, even "in thought", to the desire to check, to wash one's hands, etc. A paradoxical loop then begins: the reward circuit is disrupted by this decrease in anxiety and it gradually becomes impossible not to give in once again to the discharge of anxiety through action. The act has gone from obsessive to compulsive. This modeling reverses the therapeutic strategies: the compulsion, which is the false solution, must first be deconditioned by relieving temporary anxiety, and only then must an attempt be made to tackle the obsession. But the difficulties of this model, for cognitive behavioral theories, are numerous, and a significant number of patients remain resistant to this treatment as well as to pharmacology. Why are some representations so distressing? Can one speak about psychological addiction? are some of the many questions without real answers which motivate the testing of deep stimulation therapies. A final factor in the naturalization of OCD and the retreat of the subjective-moral model of neurosis can also be added: these disorders cannot be limited solely to the mental framework of neurosis insofar as they are present in neuropsychiatric conditions such as Tourette's syndrome. However, if, in a condition of this type, whose psychological etiology is difficult to target, there is a recurrence of obsessive disorders, then these can be apprehended as neuropsychological disorders or a brain illness. Not only is this interpretation in line with developments in neuroscience, but it is motivated by the dispersal of compulsive symptoms:

> This technique then confronts us in practice with the naturalization of a set of mental processes and allows an anthropological analysis of the transformations that neuroscience brings about in the definition of the individual in terms of cerebrality. [MOU 08, p. 175, author's translation]

What influence has the evolution of neurostimulation techniques had on the therapeutic management of obsessive-compulsive disorder? As we have seen, the development of deep brain stimulation for psychiatric illnesses is not a new idea[10]. Theoretical devices and models of neurocircuits have progressed as anatomical and physiological knowledge of the brain has progressed, but have also been shaped by the rise of theories about the materiality of the human psychological nature. The therapeutic success of deep brain stimulation for the treatment of movement disorders has led to a revival in the use of such procedures in the treatment of refractory neurological conditions, and in the field of psychiatry, a new frontier for its practice [DEN 12]. The choice of brain targets, by virtue of the disorder to be treated, involves the concept of brain localizations. In the case of OCD, limbic structures are preferred, as they are considered to be involved in mood and behavior modulation. Empirical aspects mark the encounters between deep stimulation and these disorders. In 1999, Nuttin's initial observations [NUT 99] highlighted the fact that deep brain stimulation can compensate for the consequences of damaging small amounts of tissue. Located in the areas that interconnect the large regions of the frontal lobes involved in mood and anxiety, these tissues cling to a network of interconnected regions that regulate human activity. Nuttin considered the inner capsule as a major brain center, linking various areas on the surface of the brain to the activity of the deeper limbic regions. Then, in 2002, two cases of patients with Parkinson's disease who also suffered from obsessive disorders were reported: implanted with a view to improving the effects of the neurodegenerative disease,

---

10 In 1948, Pool used the implantation of a silver electrode in the caudate nucleus in an attempt to treat a woman suffering from depression and anorexia [POO 54].

they also experienced a significant improvement in their OCD. From this, experimental research is being developed in direct contact with the surgical treatment of this behavioral disorder. Certain prerequisites are required, the same ones Heath already indicated: that patients be resistant to cognitive behavioral therapies as well as pharmacological treatments. Stimulation of deep brain areas is therefore only recommended as a last resort in 25% to 30% of cases, but it can be used as a substitute for neurosurgical operations:

> Various neurosurgical approaches were then used, including bilateral lesions of the anterior capsule, the area where the fibers connecting the thalamus to the limbic cortical areas pass through [...]. These neurosurgical techniques[11] have proven to be effective in 60% to 70% of cases, but they are much less practiced today because the lesions caused are irreversible and can lead to cognitive and emotional complications (euphoria, psychomotor agitation, emotional blunting, motor aspontaneity...). [AOU 05, p. 792, author's translation]

The use of high-frequency brain stimulation has opened up new perspectives [ABE 05] for intervention for the most severe OCDs and was effective in three of the four patients operated on [NUT 99]. In 2004, Aouizerate and his team, using a case of chronic obsessive compulsive disorder, showed "...a significant reduction in the severity of depressive and anxiety symptoms in the first three months of DBS, with remission after six months" [AOU 04, p. 486, author's translation]. The physiopathology of this syndrome allows a better understanding of the advantages of this technique. From an anatomical point of view, the functioning of the subcortical-frontal system is based on a system of neurons connecting various structures (striatum, internal and external globus pallidus, sub-thalamic nucleus) and extending from these to the frontal cortex via the thalamus. These basal ganglia are subdivided into sensory-motor, oculomotor, associative and limbic territories. They appear to act as a complex filter that makes a selection from multiple pieces of information and then feeds it back to the motor areas of the cortex. The functional organization is thus represented as a set of loops with a specific function and responding to particular cortical territories with the thalamus as a representative. The associative loop is involved in cognitive functioning and the limbic loop in emotional processes. Deep stimulation thus appears as a tool for exploring cognitive, behavioral and emotional circuits at the same time as a therapy:

> To be judged properly, the extent of benefits in multiple areas will need to be monitored over the long term. Our group has very recently reported on a series of articles on the percentage of patients with OCD who meet stringent criteria for severity and resistance to treatment and

11 See [JEN 98] for more information.

> have undergone DBS in a striated ventral/ventral internal capsule (VC/VS). These patients had quadripolar stimulator electrodes implanted bilaterally in this area. DBS was activated in open mode three weeks later. Eight were followed for at least 36 months. [...] Four out of eight subjects had at least a 35% reduction in severity according to YBOCS [...] in 36 months. In two other patients, the numbers decreased by 25–35%. Depression and anxiety have also improved. The overall assessment of data designating functioning improved from 36.6 at baseline to 53.8 at 36 months. This corresponded to improved self-management of health and independence, work, school and social functioning. [TAR 08, p. 517, author's translation]

The role played by this technique in the management of OCD is involved in the functional and organic modeling of the brain structures that determine voluntary actions and mood changes. Neuropsychiatric illness provides an experimental framework for neurophysiological exploration. Improvement of these symptoms by learning as much about the pathological process as about the electrical-chemical networks of the brain is described:

> The biomedical paradigm in psychiatry has rethought the patient as a dysfunctional organism and aims to identify and intervene in terms of intracerebral pathological mechanisms. [Tabb, 2014 in CHE 18, p. 17, author's translation]

This technique of brain exploration and stimulation must therefore be understood in the context of the neuroscientific turn[12] because:

> The burgeoning interest in the use of functional-imaging studies to interrogate the neural correlates of behavioral and social phenomena is due in part to the role functional imaging played in the development of cognitive neuroscience over the past two decades. [LIT 12, p. 188]

Symbolically, a dynamic extension of the stimulation techniques with neuroimaging can be made as the effects of deep brain stimulation on the brain make visible the cognitive aspects of the brain which will then be imaged. Indeed, once the brain is placed at the center of knowledge on cognition and behavior, stimulation seems to be a key to understanding how electrical mechanisms influence chemical substances and all physiological operations in the individual:

---

12 The multiplication of new hybrid disciplines beginning with the prefix "neuro" (neuroeconomics, neuroethics, neuroeducation, neurolaw, etc.) attests to the extent of this phenomenon and constitutes what some call a neuroscientific turning point.

> Psychiatric disorders are typically characterized by a combination of affective, behavioural, cognitive and perceptual traits that affect how individuals think, feel and behave. Increasing evidence has accrued in recent years regarding the impact of psychiatric disease on the structural and functional processes occurring in the brain. [DEN 12, p. 21]

If we consider that the cerebral organ became central at the end of the 18th Century, then the path of electricity in the medical sciences, applied to nervous and mental disorders, became structuring for the neuropsychiatric understanding of the latter. The development of deep brain stimulation towards the treatment of obsessive-compulsive disorders is situated in this context of the mechanization of brain exploration. This technique makes it possible to reopen a field of research in the context of the management of neurological and psychiatric disorders, not only with a view to treating them, but also, and perhaps above all, with a view to shedding light on the functional foundations of the human psyche:

> Greenberg became fascinated by obsessive-compulsive disorder. In 2000, Rasmussen recruited him to Brown University and Butler Hospital in Rhode Island to head deep brain stimulation studies, first in obsessive-compulsive disorder and then, three years later, in depression. [TAL 09, p. 69]

These applications of deep stimulation are at the crossroads of clinical [GRE 03] and research:

> In addition, three patients, operated on by high frequency bilateral electrical stimulation of the anterior arm of the inner capsule, with improvement in two patients, one of whom showed a disappearance of symptoms of obsession and anxiety when the stimulation was initiated, showed an improvement in the disorders that persisted after more than 32 months. [MER 08, p. 240, author's translation]

Thus, from a neurosis of psychogenic origin, obsessive disorders gradually became psychiatric and then neuropsychiatric disorders. Their emotional components are reinterpreted and thought of in terms of brain function disorders. The practice of deep stimulation is developed at the intersection of different scientific cultures and aligns neurological and psychiatric disorders on the same level. Its application to these diseases participates in human biological materialization while returning to a tradition of the tool of electricity as an exploratory instrument. Nevertheless, deep brain stimulation for psychiatric disorders still represents an ongoing body of research that requires careful selection of targeted diseases in order to determine the effects, both short- and long-term, of inserting electrodes into patients' brains. In an additional dimension [MOU 14, 18a, b], which is added to the organic dimension, the specificity of the psychiatric disorder in relation to the neurodegenerative disorder is at stake. By focusing on emotional and behavioral symptoms, instead of going in the direction of a total reduction of psychiatric pathology to its neurobiological substrate, this technique proposes to disarticulate, in as many networks, the mechanisms that preside over the singularities of human faculties. Baptiste Moutaud developed the idea that the distinction between neurological and psychiatric illnesses makes sense provided that the persistence of a specificity of the psychiatric disorder in relation to its apprehension by neurology is emphasized. However, how obsessive compulsive disorder is viewed in relation to newer techniques (MRI, DBS) affects the functional and organic causality attributed to it.

Deep brain stimulation opens up a neuropsychiatric action program in which explanatory models of disorders are aligned at the brain level and merge around neurology, psychiatry and neuroanatomy. Although the mechanisms of action are not all known, this technique has the potential to precisely target regions and deep brain circuits that are thought to be involved in the formation of several diseases that have become neuropsychiatric again. Reversible and adjustable, it generates great hopes. Neuroimaging data supported the hypothesis of a predominant role of the orbitofrontal and anterior cingulate circuits in the emergence of obsessive-compulsive symptoms. Functional research has shown that several brain regions have abnormally high activity in patients with this disorder.

All of the above experimental arguments suggest the causal role of brain circuit dysfunction. Thus, this disorder is, today, considered to be the very example of a neuropsychiatric pathology in which many anatomical structures are involved. In May 2016, an interview[13] was conducted with Professor Paul Lespérance[14] in order to better understand the challenges of deep brain stimulation applications in psychiatry. He immediately underlined the importance of personalized neuromodulation in psychiatry. In the singular context of deep brain stimulation, he differentiated pathologies without visible brain markers from diseases such as OCD. The question of patient choice, the identification of primary symptoms, markers and the definition of a disease with visible organic or functional substrates are all questions which, according to him, guide the choice of diseases that can be electrically stimulated internally. These criteria make depression, for example, a problematic therapeutic target, since depression has very broad symptoms. The concept of refractory depression would, again according to Professor Lespérance, be more a problem of diagnosis and choice of treatment to be regulated individually than the framework of

13 This interview was conducted during a research visit to Canada in May 2016 as part of the ANR Normastim; des neurosciences à l'expérimentation clinique, led by S. Desmoulin-Canselier, M. Gaille and B. Moutaud.

14 Dr. Paul Lespérance is head of the Department of Psychiatry at the CHUM in Montreal. Available at https://docplayer.fr/46080493-Le-toc-paul-lesperance-md-unite-de-neuromodulation-psychiatrique.html.

a recourse to deep brain stimulation. While, in the context of the management of obsessive compulsive disorder, even if the targets remain complex to refine locally, the diagnosis seems less risky, the placebo response is low and the number of cases resistant to treatment is quite high, around 30%. The highly deteriorated quality of life of the subjects affected is a further reason to consider this technique. Moreover, the neurobiological basis is solid and ancient irreversible neurosurgery has made it possible to build up an important body of knowledge about the regions involved. Thus, choosing one pathology rather than another, or between two patients, depends on objective factors, such as the extent of the symptoms and subjective factors relating to the psychological state or the individual patient's ability to comply with the care protocol (frequency of appointments, etc.). In fact, there are pre-assessments performed by nurses prior to the psychiatrist's assessment involved.

Another field of application that is becoming paradigmatic for understanding the clinical and experimental issues of stimulation is that of depression. As we have just seen, the polymorphism of its symptomatology makes it a problematic disease, at the primary or secondary turn. However, between 1970 and 1990, many patients with depression underwent neurosurgery, particularly in the United Kingdom. Since 1990, several techniques to directly or indirectly modify the electrical activity of the brain have been used clinically or in development for depression. In 2001, Yves Agid (1940–) reported, in the *New England Journal of Medicine* [BEJ 99], a case in which electrical stimulation of the subthalamic nucleus accidentally activated neighboring nerve fibers that connect to the limbic system, the network of brain areas on which mood depends. In one patient with Parkinson's disease, it triggered an overwhelming sense of sadness and despair that dissipated when the stimulator was turned off. This episode led to the idea of a therapeutic link between deep brain stimulation and depression. Meanwhile, Helen Mayberg (1956–), a specialist in the study of the brain circuits underlying this disease and a pioneer in the functional mapping of the depressed brain, says she uses depression "to find normal systems in the brain" [MAY 03, 08]. She believes that depression provides a heuristic clinical framework for understanding the internal workings of the human brain. For this reason, her work focuses on the range of symptoms, stages and treatments for patients with chronic depression. With its many physical symptoms, such as sleep, appetite and cognitive disorders, along with disturbances in attention or mood, this disease allows us to focus on the links between brain mechanics, its effects on the body and faculties. Very quickly, Mayberg was led to focus on the so-called limbic structures connected to the cortex, striatum, thalamus and brain stem. These areas make good candidates for deep stimulation targets. In 1999, she left to conduct research at the University of Toronto, ready to design and test deep brain stimulation in severely depressed and drug-resistant patients. In 2002, her team obtained authorization to operate on six patients. The first operation was carried out in 2003.

In 2008, deep brain stimulation for depression was still very experimental and the mechanisms underlying clinical changes in patients remain a mystery.

Other candidate target regions have recently been proposed [KOP 04] as the result of empirical work in the field of psychiatric neurosurgery. Determining the interactions between new targets and similar neural networks is a fundamental research question:

> Currently, the practice of psychiatric neurosurgery is much more refined, restricted and regulated. Candidates must meet stringent criteria for severity and for resistance to conventional multimodal therapies. DBS is an invasive procedure and, although it is non-ablative in nature and theoretically reversible when stimulation is interrupted, the evidence that it may be useful in psychiatric disorders is limited to approximately 50 patients worldwide for OCD and to fewer cases of major depression. [TAR 08, p. 522]

Neurology

- Abnormal movements
  - Parkinson's disease
  - Trembling
  - Dystonia
  - Progressive supranuclear palsy
  - Chorea
  - Wilson's disease
- Cognition
  - Alzheimer's disease
  - Parkinson's disease
  - Minimal level of conciousness
- Epilepsy

Psychiatry

- Tourette's syndrome
- Obsessive compulsive disorder
- Major depression
- Bipolar disorder
- Autism
- Eating disorders
  - Obesity
  - Anorexia
- Addiction
  - Alcohol
  - Nicotine
  - Substances
- Aggression due to an intellectual disability
- Lesch-Nyhan syndrome and other self-harming disorders

Otorhinolaryngology

- Tinnitus

It is interesting to note the abundance of psychiatric indications which express a return of deep brain stimulation to its roots, the first operations involving the implantation of electrodes into the brain in the 1950s being carried out for psychiatric indications.

Figure 6.3. *Indications for deep brain stimulation in the dual field of neurological and psychiatric diseases [BAL 15, p. 978]*

It is important to remember that, unlike the Parkinson's disease model, the benefits of developing animal models to study psychopathology are very limited. Deep brain stimulation generates considerable interest in neuropsychiatry, offering the benefits of adjustability to improve the effectiveness of the treatment or reduce side effects. Focusing on obsessive-compulsive disorder, Tourette's syndrome or depression, the results, even on small numbers of patients, remain encouraging. This technique is now moving towards aggression disorders, resistant addictions, bipolarity or eating disorders and diversifies its potential applications in the wide field of deviant behavior.

Deep stimulation involves a double implication: on the one hand, it enables us to target cerebral networks whose activity works on a given pathological process; on the other hand, it contributes to understanding the integration of our behaviors and thought content in these networks. Can it be thought of as a tool for moral neuro-augmentation [FAU 18]? This is a question posed by the philosopher Luc Faucher in an article published in 2018. If stimulation techniques help to modulate individual moods, impulses and inclinations, are they likely to make individuals better? We still need to know what we mean by "better". In the same way that in the 19th Century, galvanic doctors advocated galvanizing for the control of behaviors, this technique could improve certain skills. Are moral competences part of it?

> On the other hand, is there any reason to think that DBS alone could produce "behavioral" changes? In some cases, the acquisition of a new moral capacity should be accompanied by learning: for example, if empathy is necessary for moral learning, as in the case of psychopaths, shouldn't they learn it once their capacity for empathy has become functional? If a person who was addicted to cocaine no longer feels anything for the drug and what is associated with it through DBS, will he or she then be motivated by the "good" things (family, work, relationships with others)? [FAU 18, p. 237, author's translation]

Beyond the theme of behavioral improvement, is stimulation a technique for self-improvement? How much room does it leave for free will? Does it represent a new frontier in the understanding of human nature? Whether we are dualistic or monistic, the brain is in constant dialog with the body and the surrounding world. When an apparatus is introduced into the brain structures to modulate moods and emotions, fields of the deep personality, a third element is introduced: the mechanical element. If we like to describe our loved ones as angry, shy or extroverted, what happens when the exacerbation or inhibition of these traits depends on good technical functioning?

6.3. Man, brain and machine

Is deep brain stimulation capable of becoming a cognitive enhancement technique? This question is part of the perspective that this technique is a key to understanding the neural mechanisms underlying mental acts. A significant amount of research in cognitive psychology or neuropsychology tends to show functional correlations between the application of stimulation, including transcranial stimulation, and the training of cognitive operations, particularly memory:

> Recently, training combined with transcranial direct current stimulation (tDCS)) has shown promise in improving cognitive performance. [JON 67, p. 2]

While it seems premature to conclude that it is possible that an individual's faculties can be improved by these techniques, it should be pointed out that, since the second half of the 18th Century, human mental faculties, progressively inscribed within brain matter, have been doubly correlated with research on electricity. On the one hand, electrical therapies appear as a means of understanding the behavioral and intellectual dimensions through the mechanization of faculties and thought contents. On the other hand, the discovery of brain electricity has reinforced the hypothesis that electricity plays a determining role in our cognition. From exploration techniques to stimulation techniques, the emotional and intellectual dimensions are thought of in terms of intervention and materiality, both in their normal and pathological forms. The interest of deep stimulation is also played out in a fundamental research perspective, where one of the challenges is the differentiation of anatomo-functional structures involved in the processing of perceptual, emotional, cognitive and motor information. We have seen that, around 1960, psychologists and neuropsychologists questioned the fact that electroencephalography could become an instrument for interpreting a cerebral language. However, the applications of deep brain stimulation contribute to the renewal of such empirical questioning on the links between mental faculties and the brain's electrical activity, starting from psychiatric and neuropsychiatric disorders:

> With many psychiatric comorbidities, including anorexia nervosa and obsessive-compulsive disorder, obsessive-compulsive disorder and depression, depression and addictions, it is highly likely that new targets tested in one type of condition may ultimately prove effective in the treatment of a comorbidity. The miniaturization of electrodes and tomorrow's nanotechnologies will make it possible, thanks to less invasiveness and the recording possibilities of the electrodes, to considerably increase our knowledge of the dysfunctions of cerebral activity in mental pathologies. [LEV 14, p. 29, author's translation]

In this sense, this technique is part of a new relationship with the human being, enhanced by the link to the machine within the framework of an interventionist medicine. In 2013, in an article entitled *Brain, Mind and Machine* [LIP 13], Nir Lipsman and Walter Glannon question the possibility that the apparatus can be considered as a new intermediary between the subject and his free will:

> In this respect, the stimulating device can enable one to be an agent in the fullest sense and thus something with which one would want to identify as integral rather than alien to one's psyche. [LIP 13, p. 469]

While such an investigation seems to be recurrent for many psychiatric, pharmacological or surgical treatments, it is a matter here of questioning the singularity of the intervention of a mechanical assembly (electrodes, batteries, case) in relation to the persistence of mental health through its proper functioning. Is a new individuality being built in its connection to the machine? Deep brain stimulation can be seen both as a third player in mind–brain linkage studies and as a key to activating and deactivating various mental functions. It is, according to these considerations, at the intersection of treatment and research and has opened up one of the most recent avenues for describing the mechanisms of cognitive abilities. Within a continuous evolution of neuromodulation techniques, it represents a means to study *in vivo* cognitive functions in all their forms:

> In the continuation of the identification of links between perception, cognition and emotion, the major challenge will be to understand the link between these functions and the motor system. Embodiment theory[15] offers an innovative and original framework for investigation from this perspective. Indeed, the basal ganglia could constitute the ideal neuroanatomical medium to explain the behavioral phenomena highlighted by these theories. [MER 08, p. 261, author's translation]

Can we talk about human–machine fusion? Does an individual merge with his prosthesis if he loses a limb? If the case of deep brain stimulation may seem singular in that it affects the organ that symbolically represents the heart of the person, it soon becomes apparent that a subject receiving a mechanism for better living either forgets its presence or appropriates it completely. In the case of a healthy subject, this technique appears as a transhumanistic way to improve individual faculties. The application of external artificial electricity was seen as a tool for improvement as early as the 19th Century. Let us recall that, after 1850, a parallel medicine of

15 This term takes into account the registration of thought operations, emotional contents and psychic movements within the body in relation to the interpretations made by the brain and in relation to the stimuli of our environment.

electrical accessories, including electrotherapy boxes, was a great commercial success. Now:

> More recently, transcranial electrical stimulation has been tested to increase the cognitive abilities of healthy subjects. Some manufacturers, such as the British Foc.us or the North American Thync, already sell direct current transcranial stimulation devices to private individuals, which they can use at home at their leisure to, according to them, think faster, memorize longer or stay focused for hours on end. [LLE 17, p. 39, author's translation]

These external stimulation devices, seen as the resurgence of electric para-medicine, have no more scientific basis than electric belts or stimulating brushes. Yet this does not stop a renewed interest in the theme of potential cognitive improvement. The philosophical and technical challenge is that of creating accessories whose use shapes human nature in return. In the absence of mythological thought, the technique ensures, at least in the collective imagination, a continuous evolution of the human species. The idea that external electricity, both natural and artificial, can modulate brain function is an old one. Brain stimulation has continued to evolve, from the application of electric fish to the use of modern anatomical and functional imaging methods to implant chronic stimulation electrodes in the deep centers of the brain, to treat a variety of psychiatric disorders:

> Methods of electrical or magnetic stimulation of the brain provide access to the relationships between brain structure and function on the one hand, and behavior and cognitive functions on the other. [FAU 12, p. 88, author's translation]

Advances in nanotechnology, biology, informatics and cognitive sciences, which together make up the "NBIC convergence", provide functional psychosurgery with relevant tools for clinical and basic research. The case of Kevin Warwick (1954–) [WAR 02, 12] is emblematic for improvement and exploration:

> Kevin Warwick, professor of computer science at the University of Reading (Great Britain), had a microprocessor implanted in the median nerves of his left arm in 2002. This microprocessor, connected to a radio transmitter, allows him to connect his nervous system – or rather, the motor subsystem controlled by the nerves of his left arm – to one or more computers, whether or not they are connected to the Internet, or to another human being. [PRO 09, 127, author's translation]

Neuroinformatics works for an increase in cognitive or perceptive capacities that are not limited to the therapeutic field. The challenge of this project is to increase the natural capacities or to correct dangerous tendencies of humans through the use of appropriate techniques, whether they are invasive and penetrate, or not, the nervous system. The Brazilian scientist Miguel Nicolelis (1961–) [CAR 03] has been working, since the beginning of the 21st Century, on the implantation of electrodes, particularly on primates, and, in addition to the potential for improvement, is seeking to deepen the classical analogy between the body and the machine. He perfected two types of techniques:

– an invasive technique in which silicone electrodes are permanently implanted in the motor and premotor cortex and send the "volitions" to a brain-machine interface. It transmits them to a robotic member or a computer screen. This technique has allowed about thirty patients, often struck by locked-in syndrome, to find a way to communicate with their loved ones:

> Neuromotor supplementation techniques seem to fall unambiguously into the category of unquestionable progress. [PRO 09, p. 132, author's translation]

– the second technique is non-invasive. Mental information is transmitted through a headset that is able to detect the differential neuronal activity that the patient is producing at any given time. The most advanced work in this field uses E.E.G. helmets, with the number of electrodes placed on the surface of the skull varying between 64 and 250 to extract the brain signals composing the various known frequencies.

These techniques can therefore be used for cognitive enhancement.

How can we evaluate scientific progress? Can deep brain stimulation claim the role of progress? From psychiatry to neurodegenerative diseases, it has the hallmarks of innovation. Guided by "clinician-researchers" [ARA in DES 18, p. 101], its close history is marked by a set of interrelationships between research, clinical and the project of an augmented human. It is therefore an object of transrelational research that does not only involve therapeutic objectives and requires collaboration between many disciplines: cognitive psychology, neuroscience, surgery, robotics and artificial intelligence.

Medical electricity is a technique that has won over all sectors of society and has been able to renew the forms of its applications in each of its singular contexts. The machines used by Nollet are not Aldini's stimulations, which themselves are not the later stages of cerebral, transcranial or deep brain stimulation. Technical evolution is not just a historical relationship between objects. The history of internal or external brain electricity is not continuous. The different fields of application modulate and shape electrical treatment, giving it purposes that meet different needs: the urgency of a traumatized soldier, a subject's obstructed capacities or the control of behaviors deemed decadent. The practices are modified by answering a philosophical, now medical, question about human nature and its ills:

> The concrete technical invention has multiple determinants, but some are external and others internal [...]. The technical object that emerges in society is made up of a technological core and a layer of social overdeterminations, implicit valuations, tenacious myths, etc. [GUC 05, p. 351, author's translation]

The perfection of methods is not sufficient to qualify different electrical therapies. The latter participate in constructing the problems to which they respond, such as irreversible lesion surgery or a profound questioning of the vital properties of matter. Warwick's research in robotics or Conelolis's research on brain implants opens up perspectives on electrical applications in correlation with the theme of the augmented man. They are also part of an imagining of electricity whose lines of convergence have been found since the 19th Century.

Conclusion

At the end of this study, we hope to have shown, and made intelligible, the presence of an electricity movement within the sciences of the brain, in its relationship with the psyche as well as in its relationship with the body. This journey was initiated by physicists studying natural and atmospheric electricity in the mid-18th Century, and is part of a long history. In the course of experiments on the application of electricity to the body and then on the effects of convulsive and nervous diseases, this force was found to be both internal and external. Although the electrical effects of electric fish have been well known since Antiquity, it has become part of the history of human biology by making the body a conductor and being found even in its organic structures. For example, Galvani's research has uncovered the presence of a neuro-electric fluid, an animal electricity whose source can be located in the brain. This electrical structure of the brain was confirmed by du Bois-Reymond in the mid-19th Century. It appears that medical electricity applied to the brain was in constant tension between the internal manipulation of electricity and the external application of this force. The interactions between these two levels of treatment were decisive in understanding the polymorphic developments in therapies and knowledge about biological electricity. Indeed, electrotherapy could not do without the knowledge, devices and techniques developed to understand, measure and objectify the action of this force in the central and peripheral nervous system. Moreover, from the 18th Century onwards, addressing the links between electricity and cerebral activity has been part of the questioning of the human being's dualistic, monistic or materialized nature. This is far from insignificant for understanding the intersecting interests of electrical therapies, organic psychiatry and, more recently, neuroscience. These philosophical questions form the basis for future developments in electricity, as part of the modeling of brain activity. Thus epistemology, history and neuroscience intersect to build a framework in which it becomes possible to understand how research on humankind, its place in nature and the electrical treatment of the ills of the mind have been intertwined:

> Philosophy and neuroscience benefit from observing each other, establishing and maintaining contact to constrain each other. They benefit from questioning each other, not superficially, but knowing each other well enough to ask the right questions. [FAU 08, p. 5, author's translation]

From localizationist models to dynamic representations of brain activity, cognitive functions and human behavior are naturalizing and electrifying. If the mechanisms and properties of matter are electrocentric, if humans are a part of it, then localized electrification, initially non-invasive, is potentially able to ensure the proper functioning and relationships between the bodily and mental dimensions. Electricity is seen as a remedy that improves and helps normalize behavior and is also considered a treatment for illnesses such as depression or schizophrenia. Starting in the second half of the 20th Century, implant techniques were developed to stimulate the deep structures of the brain. These developments were taking place at the same time in two fields: that of behavior in Delgado's research and that of psychiatric illnesses with the techniques developed by Heath. The applications of electricity to humankind had, very early on, at least two intentions: the regulation of the links between the nervous system and the mental domain as well as the treatment of mental illnesses. These two intentions complemented each other and are enriched to this day by the themes and discussions on the differences between therapeutic medicine and medicine for improvement.

This study also raises the theme of an epistemology of data and outcomes. Indeed, how can we understand the renewed craze since 1740 for electrical techniques that often fail to alleviate the ailments it targets? Electricity aims at a naturalistic objectification of human mechanisms, understanding them in relation to each other in physical and moral dimensions. These searches for objective data on human organic nature are sometimes disjointed from their results. We had not to judge this medical technique by their standards, but rather to reconstruct the steps leading to its involvement in medical, cognitive and psychiatric sciences. The objectivity of the data, while they are based on the interpretations that are made of them, opens up to other applications and feeds other research:

> However, historically, research aimed at discovering a mechanism often begins with the localization of complete tasks or activities in one of the components of the brain [...]. Even if they do not directly generate an understanding of the mechanism, such efforts often play an important heuristic role. [BEC in FAU 08, p. 95, author's translation]

We approached this theme by exploring the interpretations of the electroencephalographic recordings made in relation to the recorded subjects' mental content. The fact that this approach did not allow the systematic correlation of an electric alphabet with thought content, on the one hand, allowed the E.E.G to be inscribed in a clinical and neurological field and, on the other hand, to raise new questions about electrical brain mechanics. Thus, the correlation between an electrical recording and a specific stimulus given to the subject has been shown to be heuristic in the context of inferring a measure, and useful for understanding the temporal pattern of a neural process. This made it possible to objectivize and measure the response potentials mentioned. The data are placed in a context that makes them audible or not and allows their development. It is also necessary to consider the way in which the problem to which they provide answers is constructed. The choice of data deemed relevant depends on the complexity of this preliminary problem. They may be objective, but this does not guarantee the intention behind the choice that led to their selection. Because this book is intended to retrace the steps of medical electricity applied to the brain, choices had to be made to address the problem. In this book, the focus has been on the problems of mental activity, its cerebralization and the possibility of providing treatment and improvement.

Furthermore, medical electricity is taken, as a technique, in an imaginary of progress, in a Promethean context where humans control others and their environment. While this development of electrical techniques in the mental sphere is the result of cerebral reductionism, we could also evoke the concept of *extensionism* which ranges from the microscopic levels, that is the properties of neural networks, to the macroscopic levels which correspond here to psychological properties. Although this movement has implied the rejection of animal spirits and still implies that of dualism and the search for a single principle to understand organic mechanics in correlation with the principles of the living, including mental activity; it is a question of refocusing research within the potentialities of matter in all its forms. This scheme itself is part of this history and joins the questions about mental functioning and how to intervene in it:

> By positing that neuroscience will be able to reveal the physical mechanisms at the service of psychological functions, I assume that it is the brain that performs these functions and that human capacities are, in fact, the capacities of the human brain. [CHU in FAU 08, p. 329, author's translation]

The physicalist framework of the history of medical electricity refers to the fact that the data and measurements are aimed at the objectification of brain structures and their consequences on the subject. It is important to see that they remain the subject of multiple interpretations linked to the societal, medical and scientific context in which they take place. The interweaving of all these parameters is constitutive of an electrical imaginary that emerged and evolved during the 19th Century in two forms: on the one hand, electrophysiology which appeared as an instrument for exploring the limits between life and death; on the other hand, external electricity was available in many devices aiming at treating and improving the individual. Thus, can we correlate the history of medical electricity applied to the brain with the theme of human augmentation? There are a few elements pointing in the direction of a positive response. We would like to conclude this study on the theme of a dilution of the boundaries between therapeutic medicine and improving medicine. Beyond correlating it with technical and medical progress, it seems relevant to analyze this theme in relation to the medical and symbolic impact of electricity:

> It has been argued that if medicine strictly speaking is the set of practices that lead from the pathological to the normal, it is currently opening up to a set of practices that lead from normal to improved or, better still, from ordinary to modified or transformed. [HOT 15, p. 194, author's translation]

Not only does the history of medical electricity allow us to approach the history of anthropotechnics, insofar as it marks the entry of medicine into medical engineering, but it also questions the theme of the improvement of the subject. What is an improvement if not something that aims at an easier accomplishment of one or more actions? Medicine and treatment can already appear as improvement techniques. The pathological dimension must be removed in order to shift the questioning towards a correlation between the medical techniques of electrical stimulation and individual benefit. A questioning that began and renewed itself, as early as the 18th Century, has contributed to nourishing a medical reflection on improvement. The question precedes the techniques used to answer it. Cognitive enhancement is a growing field that begins with inventions that have entered our daily lives, such as glasses. Invasive cognitive enhancement measures complement these devices, internalizing their purpose into brain structures:

> Here, cyber-neuronal implants are designed to allow the individual with the device to access information online through devices built into his or her body or brain. [HOT 15, p. 194, author's translation]

Thus, at the end of this study, the issues of improving cognition through brain implants or the links between humans and techniques have been highlighted.

Appendix 1

Historical Points of Reference

1745–1765	The first applications of electricity to ailments of the body
1770–1800	Electrification of mental and convulsive diseases: madness as a therapeutic horizon
1800–1840	Electrophysiological explorations of the boundaries between life and death: a dualist medicine
1840–1900	Electrical stimulation of the brain and the notion of control: towards a holistic medicine
1900–1950	The development of electrophysiological explorations and electrotherapy
1980–2010	Deep brain stimulation: a new frontier in the field of psychiatry?

Table A1.1. *The five periods that mark the path from electricity to brain science*

Events	Works
1730: Gray included the human body in the list of conductive bodies.	Gray, E.: “Dernière lettre de M. Etienne Gray de la Société royale, à M. Granville Wheler, écuyer, de la Société royale, sur les révolutions que de petits corps suspendus font, par, l’électricité, d’occident en orient autour d’autres corps plus considérables, de la même manière que les planètes tournant autour du soleil”, *Transactions, philosophiques de la Société royale de Londres.* 1738, vol. 1736, pp. 54–55.
1743: Krüger took over the conductive body experiment for therapeutic purposes.	

1744: Kratzenstein applied an electric treatment for women with paralysis of the little finger.	Kratsenstein, C.-T.: *Lettres d'un physicien sur l'avantage dont l'électricité peut être pour l'art de guérir,* Halle, 1746.
1745: Invention of the Leyden jar, the first electric capacitor designed to store static electricity charges.	Musschenbroek van, P.: *Cours de physique expérimentale et mathématique,* translated by M. Sigaud de La Fond, Paris, Briasson, vol. 3, 1769.
1746: Appearance of the first electrostatic friction machines, designed to produce static electricity for medical use.	Nollet, J.-A.: *Essai sur l'électricité des corps,* Paris, Frères Guerin, 1746.
1748: Jallabert, from Geneva, marked the turning point of physics becoming medicalized.	Jallabert, J.: *Expériences sur l'électricité, avec quelques conjectures sur la cause de ses effets,* Geneva, Barillot & fils, 1748.
1749: François de Sauvages de la Croix was more successful in cases of hemiplegia.	A thesis was inspired by this: Deshays, J.-É.: *De hemiplegia per electricitalem curanda,* Montpellier, J. Martel, 1749.
1770: Ledru aimed at madness as a therapeutic horizon.	Ledru, N.-P.: *Rapport de MM. Cosnier, Maloet, Darcet Philip, Le Preux, Desessartz et Paulet, docteurs-régens de la faculté de médecine de Paris, sur les avantages reconnus de la nouvelle méthode d'administrer l'électricité dans les maladies nerveuses, particulièrement dans l'épilepsie et dans la catalepsie ; par Nic. Ph. Ledru, connu sous le nom de Comus. Ce rapport est précédé de l'aperçu du système de l'auteur sur l'agent qu'il emploie, et des avantages qu'il en a tirés,* Paris, Imprimerie Philippe-Denys Pierre, 1783.
1772: Abbé Sans put forward nervous diseases as preferred therapeutic targets.	Sans, J.: *Guérison de la paralysie par l'électricité: ou cette expérience physique employée avec succès dans le traitement de cette maladie regardée jusques à présent comme incurable*, Paris, chez Cailleau, 1772.
1775, Antoine van Haen multiplied therapeutic targets: chorea and menstrual disturbances of electrical treatment.	Haen van, A.: Litteris Remondiniadis, Venetiis, 1775.
1776: Sigaud-Lafond studied the influence of electricity in various diseases.	Sigaud de La Fond, J.-A.: *Traité de l'électricité, dans lequel on expose, et on démontre par expérience, toutes les découvertes électriques faites jusqu'à ce jour, pour servir de suite aux Leçons de physique du même auteur,* Paris, Laporte, 1776.

1779: Mauduit de la Varenne published his memoirs on the different ways of applying electricity.	Mauduyt de la Varenne, P.-J.-C.: *Mémoire sur le traitement électrique appliqué à 82 malades,* Paris, Imprimerie Royale, 1779.
1780: Mazars de Cazelles promoted therapies using static electricity.	Masars de Cazeles, F.: *Mémoire sur l'électricité médicale et l'histoire du traitement de vingt-et-un malades traités,* Paris, Mequignon l'aîné, 1780.
1791: Luigi Galvani's discoveries on animal electricity.	Galvani, L.: *De viribus electricitatis in motu musculari commentarius,* Ex typographia Instituti Scientiarum, Bologna,1791.
1800: Alessandro Volta discovered the battery.	Volta, A.: "On the electricity excited by the mere contact of conducting substances of different kinds", *Philosophical Transactions*, London, 90, part. 2, pp. 403–431, 1832.
1803: Aldini promoted Galvani's work throughout Europe while using the Voltaic pile.	Aldini, G.: *Précis des expériences galvaniques faites récemment à Londres et à Calais... suivi d'un extrait* d'*autres expériences,* Paris, P. Didot aîné, 1803.
1823: Johann Schweigger invented the galvanometer.	
1831: Michael Faraday updated electromagnetic induction. Induction currents compete with direct currents in medicine.	Faraday, M.: *Experimental Researches in Electricity*, (1839), Richard Taylor and William Francis, London, 1855.
1896: D'Arsonval's therapy with high-frequency currents.	Arsonval d', A.: "Action physiologique et thérapeutique des courants à haute fréquence", *Société internationale des électriciens,* Paris, Gauthier-Villars et fils, 1896.

Table A1.2. *18th–19th Centuries – chronological points of reference: when physics became medical*

Experimental facts	Implications
1803: Aldini carried out, notably in the presence of Pinel, the first transcranial stimulations. He announced positive effects for the treatment of melancholy.	Electricity, applied to the skull, is able to modify mood and emotional behavior.
1840: Thomas Laycock advocated the application of galvanization to restore contact between the consciousness and the body.	The links between body and mind can be regulated by means of medical electrical techniques.

1846: Carpenter stated the theory that man is still subject to primary instincts that bring him closer to his animality.	Localized electrification can prevent primitive instincts, in an evolutionary context, from expressing themselves.
1848: From Galvani to Bois-Reymond's experiments.	Naturalization of "animal spirits". Electrical stimuli can initiate and modify vital processes.
1870: Fritsch and Hitzig practised electrical stimulation of the brain in anaesthetized dogs and discovered that it evoked localized movements of the body and limbs.	The brain is excitable. Electrical stimulation of the cerebral cortex can produce movement.
1932: Hess, by practicing diencephalic stimulation in unanesthetized cats, pointed out that it evoked motor effects and well-organized emotional reactions.	Motor and emotional manifestations can be evoked by electrical stimulation of the brain in conscious animals.
1954: Works of Delgado. In isolated animals, learning, conditioning, instrumental responses, pain and pleasure were either evoked or inhibited by electrical stimulation of the brain in rats, cats and monkeys.	Psychological phenomena can be controlled by electrical stimulation of specific areas of the brain.
1955: Delgado, by experimenting on colonies of cats and monkeys, showed that aggression, domination, increase in and other social interactions were evoked, modified or inhibited by radiostimulation of specific brain areas.	Social behavior can be controlled by radiostimulation of specific areas of the brain.
1954: Development and refinement of the Penfield method. In patients, brain stimulation during surgery blocks the thinking process, inhibits speech and movement or, in other cases, induces pleasure, laughter, friendliness, verbal production, hostility, fear, hallucinations and memories.	Human mental functions can be influenced by electrical stimulation of specific areas of the brain.

Table A1.3. *Some dates in the context of brain control and the modulation of mental illness by electricity*

1874: Bartholow stimulated the human cerebral cortex.	Bartholow, R. *Experiments on the Functions of the Human Brain*, Br Med J., vol. 1 (700), p. 727, May 30, 1874.
1948: Pool implanted electric stimulators in humans.	Pool, J.L. "Psychosugery of older people", *J. Geriatr. Assoc.*, 1954
1950: Heath used stereotactic techniques to implant electrodes on humans to relieve schizophrenia or mood disorders, cyclothymia).	Heath, R.G.; Peacock, S.M.; Miller, W. "Induced paroxysmal electrical activity in man recorded simultaneously through subcortical and scalp electrodes", *Trans. Am. Neurol. Assoc.*, vol. 3, pp. 247–250, 1953. Heath, R.G.: *Studies in Schizophrenia*, Cambridge, Harvard University Press, 1954.
1960: Hassler described the effects of stimulation on tremors.	Hassler, R.; Riechert, T.; Mundinger, F. *et al.* "Physiological observations in stereotaxic opera tions in extrapyramidal motor disturbances", *Brain*, vol. 83, pp. 337–350, 1960.
1975: Bechtereva's work on the effects and links between ganglia and movement disorders.	Bechtereva, N.P.; Bondartchuk, A.N.; Smirnov, V.M.; Meliutcheva, L.A.; Shandurina, A.N. "Method of electrostimulation of the deep brain structures in the treatment of some chronic diseases", *Confin Neurol.*, vol. 37, pp. 136–140, 1975.
1979: Cooper's work on the thalamus and tremors.	Cooper, I.S.; Upton, A.R.M.; Amin, I. "Reversibility of chronic neurologic deficits. Some effects of electrical stimulation of the thalamus and internal capsule in man", *Appl Neurophysiol.*, vol. 43, pp. 244–258, 1980.
1987–1994: Benabib and deep brain stimulation in Parkinson's Disease.	Benabib, A. L.; Pollak, P.; Louveau, A.; Henri, S.; de Rougemont, J. "Combined (thalamotomy and stimulation) stereotactic surgery of the VIM thalamic nucleus for bilateral Parkinson disease", *Appl. Neurophysiol.*, vol. 50, pp. 344–346, 1987.
1999: Nuttin's experiments and research on stimulation and implantation of electrodes in internal capsules in obsessive-compulsive disorders.	Nuttin, B.; Cosyns, P.; Demeulemeester, H. *et al.* "Electrical stimulation in anterior limbs of internal capsules in patients with obsessive compulsive disorder", *Lancet*, vol. 354 (9189), p.1526, 1999.
1999: Vanderwalle's research on thalamus stimulation in Tourette's syndrome.	Visser-Vandewalle, V.; van der Linden, C.; Groenewegen, H.J. *et al.* "Stereotactic treatment of Gilles de la Tourette syndrome by high frequency stimulation of thalamus", *Lancet*, vol. 353, p.724, 1999.
2005: Mayberg's experiments for stimulation of the Brodmann area in the treatment of depression.	Mayberg; H., Lozano, A.M.; Voon, V.; McNeely, H.E.; Seminowicz; D.; Hamani, C., Schwalb, J.M., Kennedy, S.H. "Deep brain stimulation for treatment-resistant depression", *Neuron*, vol. 45, pp. 651–660, 2005.

Table A1.4. *Expanded history of deep brain stimulation*

Appendix 2

Comparison of Tables of Contents

Tables of contents compared to show that Nollet, like Franklin and Jallabert, approached the laws of electricity at the same time as its possible application to the human body

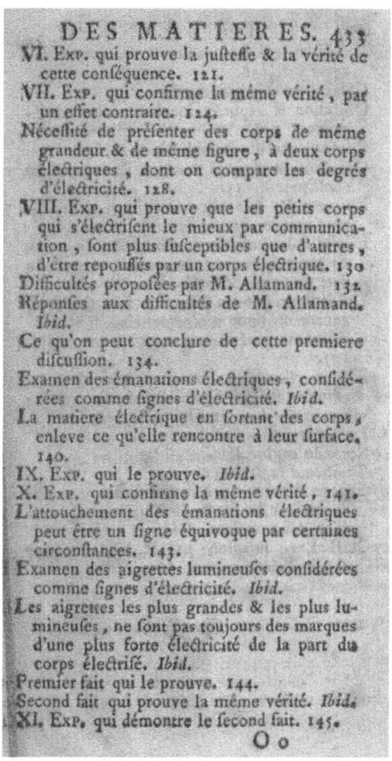

DES MATIERES. 433

Oo

Figure A2.1. *Nollet, J.-A.: Recherches sur les causes particulières [...], op. cit. 1748, p. 433*

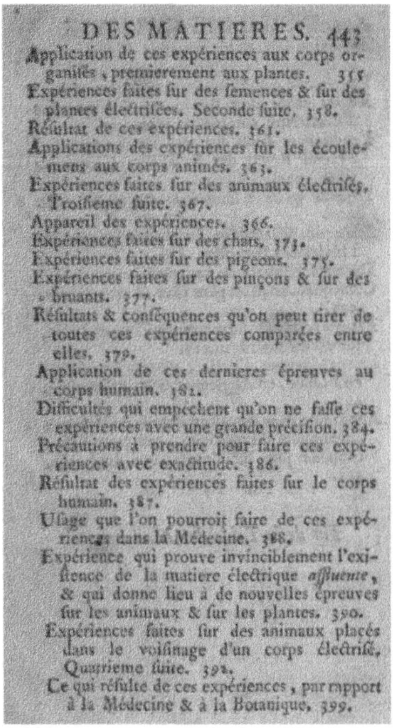

DES MATIERES. 443

Application de ces expériences aux corps organisés, premierement aux plantes. 355

Expériences faites sur des semences & sur des plantes électrisées. Seconde suite, 358.

Résultat de ces expériences. 361.

Applications des expériences sur les écoulemens aux corps animés. 363.

Expériences faites sur des animaux électrisés. Troisieme suite. 367.

Appareil des expériences. 366.

Expériences faites sur des chats. 373.

Expériences faites sur des pigeons. 375.

Expériences faites sur des pinçons & sur des bruants. 377.

Résultats & conséquences qu'on peut tirer de toutes ces expériences comparées entre elles. 379.

Application de ces dernieres épreuves au corps humain. 382.

Difficultés qui empechent qu'on ne fasse ces expériences avec une grande précision. 384.

Précautions à prendre pour faire ces expériences avec exactitude. 386.

Résultat des expériences faites sur le corps humain. 387.

Usage que l'on pourroit faire de ces expériences dans la Médecine. 388.

Expérience qui prouve invinciblement l'existence de la matiere électrique *affluente*, & qui donne lieu à de nouvelles épreuves sur les animaux & sur les plantes. 390.

Expériences faites sur des animaux placés dans le voisinage d'un corps électrisé. Quatrieme suite. 392.

Ce qui résulte de ces expériences, par rapport à la Médecine & à la Botanique. 399.

Figure A2.2. *Nollet (1748) devotes many chapters to the links between electricity and medicine, while exploring the mechanisms and principles of electric force. Nollet, J.-A.: Recherches sur les causes particulières [...], op. cit. 1748, p. 443.*

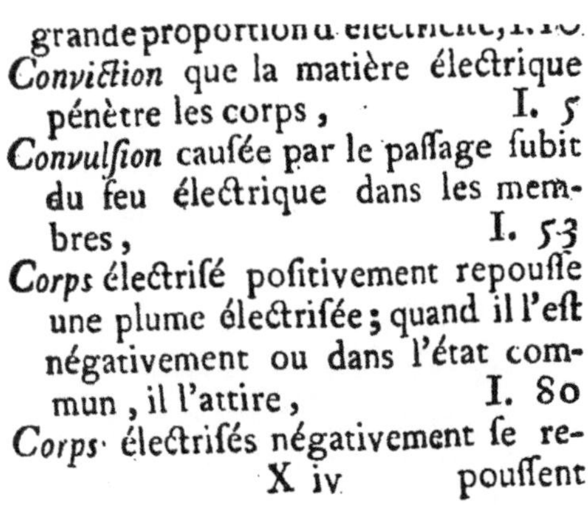

grande proportion d'électricité, I. 10

Conviction que la matière électrique pénètre les corps, I. 5

Convulsion causée par le passage subit du feu électrique dans les membres, I. 53

Corps électrisé positivement repousse une plume électrisée; quand il l'est négativement ou dans l'état commun, il l'attire, I. 80

Corps électrisés négativement se re-

X iv poussent

Figure A2.3. *Franklin, B.: Expériences et observations sur l'électricité faites à Philadelphie en Amérique, second edition revised, corrected and extended by M. D'Alibard, Paris, Durand, 1756, 2 vols, p. 327*

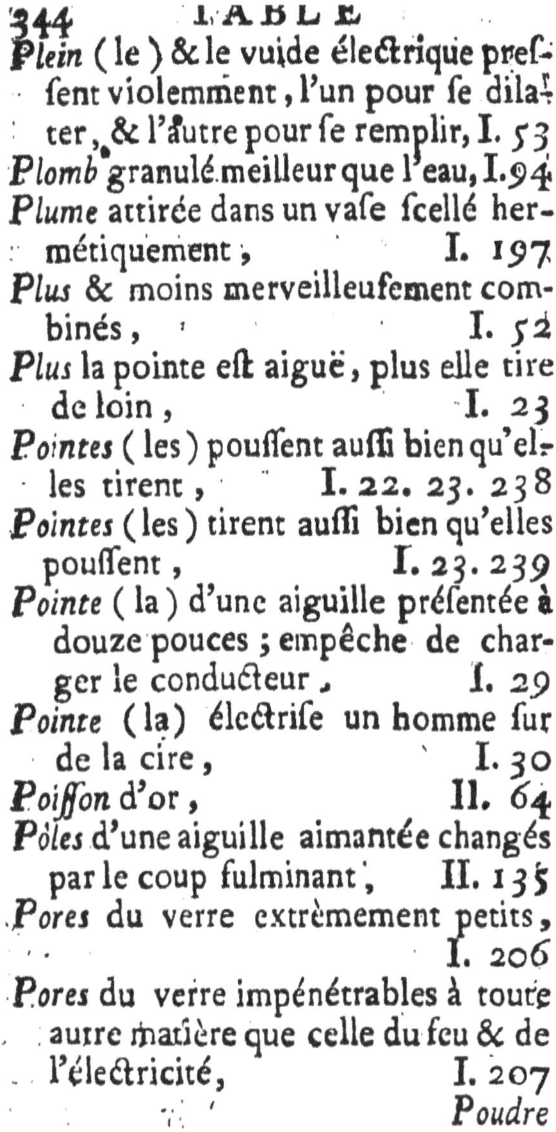

344 TABLE

Plein (le) & le vuide électrique preſ-ſent violemment, l'un pour ſe dila-ter, & l'autre pour ſe remplir, I. 53

Plomb granulé meilleur que l'eau, I. 94

Plume attirée dans un vaſe ſcellé her-métiquement, I. 197

Plus & moins merveilleuſement com-binés, I. 52

Plus la pointe eſt aiguë, plus elle tire de loin, I. 23

Pointes (les) pouſſent auſſi bien qu'el-les tirent, I. 22. 23. 238

Pointes (les) tirent auſſi bien qu'elles pouſſent, I. 23. 239

Pointe (la) d'une aiguille préſentée à douze pouces; empêche de char-ger le conducteur, I. 29

Pointe (la) électriſe un homme ſur de la cire, I. 30

Poiſſon d'or, II. 64

Pôles d'une aiguille aimantée changés par le coup fulminant, II. 135

Pores du verre extrèmement petits, I. 206

Pores du verre impénétrables à toute autre matière que celle du feu & de l'électricité, I. 207

Poudre

Figure A2.4. *Franklin (1756) discusses the penetration of bodies by electricity, the resulting convulsions as well as the properties of tips and glass as conductive objects. Franklin, B.: Expériences et observations sur l'électricité faites à Philadelphie en Amérique, second edition revised, corrected and extended by M. D'Alibard, Paris, Durand, 1756, 2 vols. p. 344*

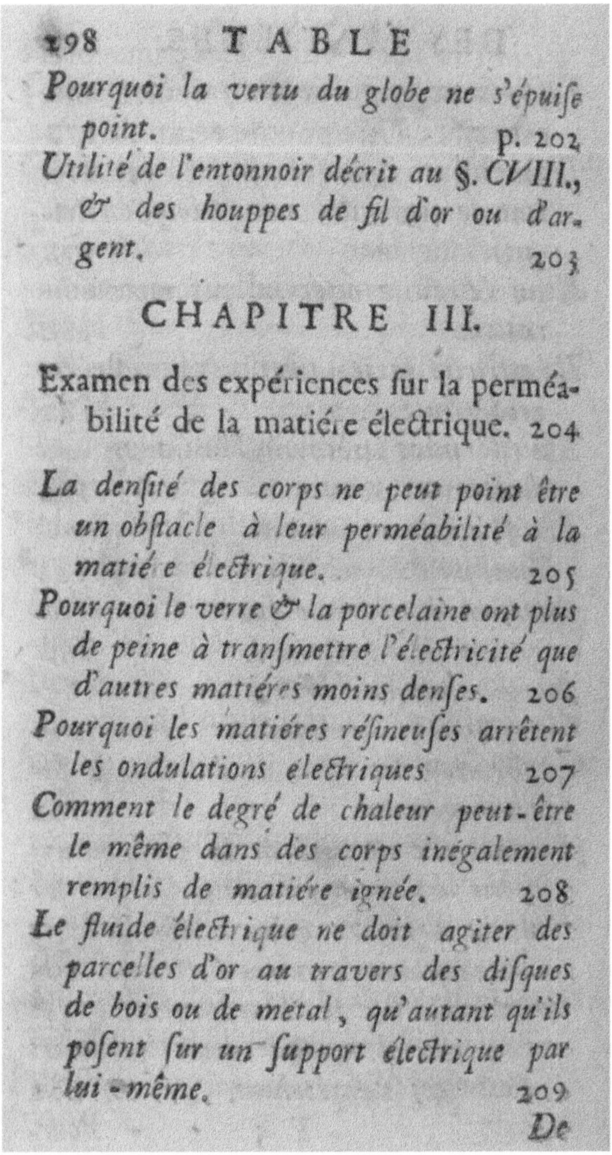
298 TABLE

Pourquoi la vertu du globe ne s'épuiſe point. p. 202

Utilité de l'entonnoir décrit au §. *CVIII., & des houppes de fil d'or ou d'argent.* 203

CHAPITRE III.

Examen des expériences ſur la perméabilité de la matiére électrique. 204

La denſité des corps ne peut point être un obſtacle à leur perméabilité à la matiére électrique. 205

Pourquoi le verre & la porcelaine ont plus de peine à tranſmettre l'électricité que d'autres matiéres moins denſes. 206

Pourquoi les matiéres réſineuſes arrêtent les ondulations electriques 207

Comment le degré de chaleur peut-être le même dans des corps inégalement remplis de matiére ignée. 208

Le fluide électrique ne doit agiter des parcelles d'or au travers des diſques de bois ou de metal, qu'autant qu'ils poſent ſur un ſupport électrique par lui-même. 209

De

Figure A2.5. *Jallabert, J.: Expériments sur l'électricité, avec quelques conjectures sur la cause de ses effets, Geneva, Barillot & fils, 1748, p. 298*

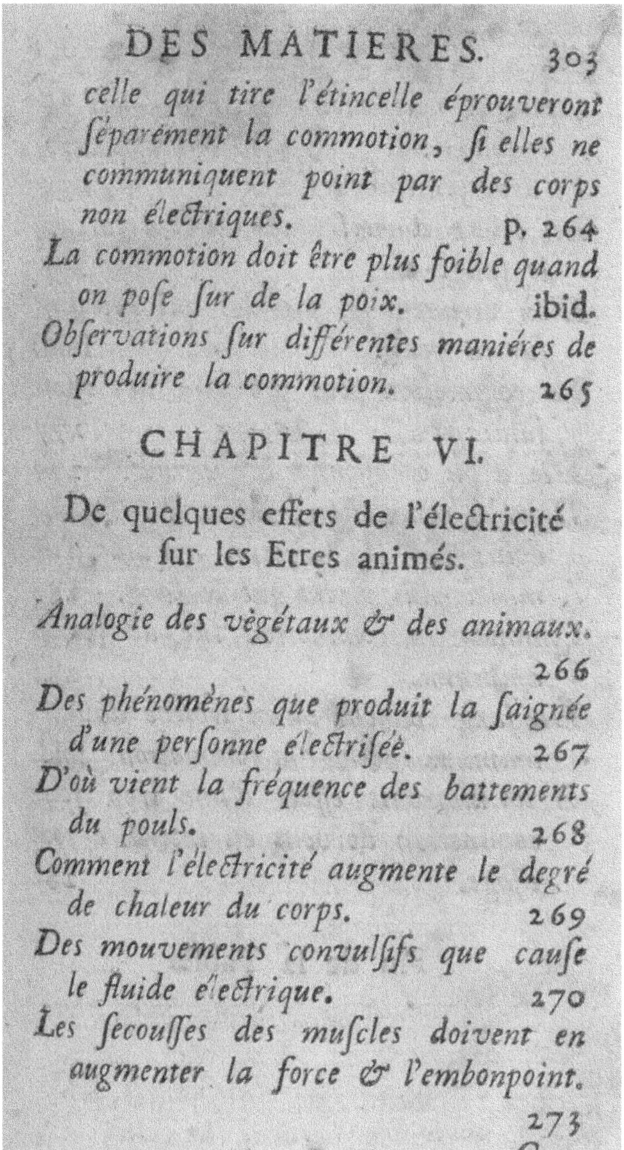

DES MATIERES. 303

CHAPITRE VI.

De quelques effets de l'électricité sur les Etres animés.

Com-

Figure A2.6. *Jallabert (1748) goes, in his treatise, from chapters on the laws of electricity to the effects of this force on human beings. Jallabert, J.: Expériences sur l'électricité, avec quelques conjectures sur la cause de ses effets, Geneva, Barillot & fils, 1748, p. 303*

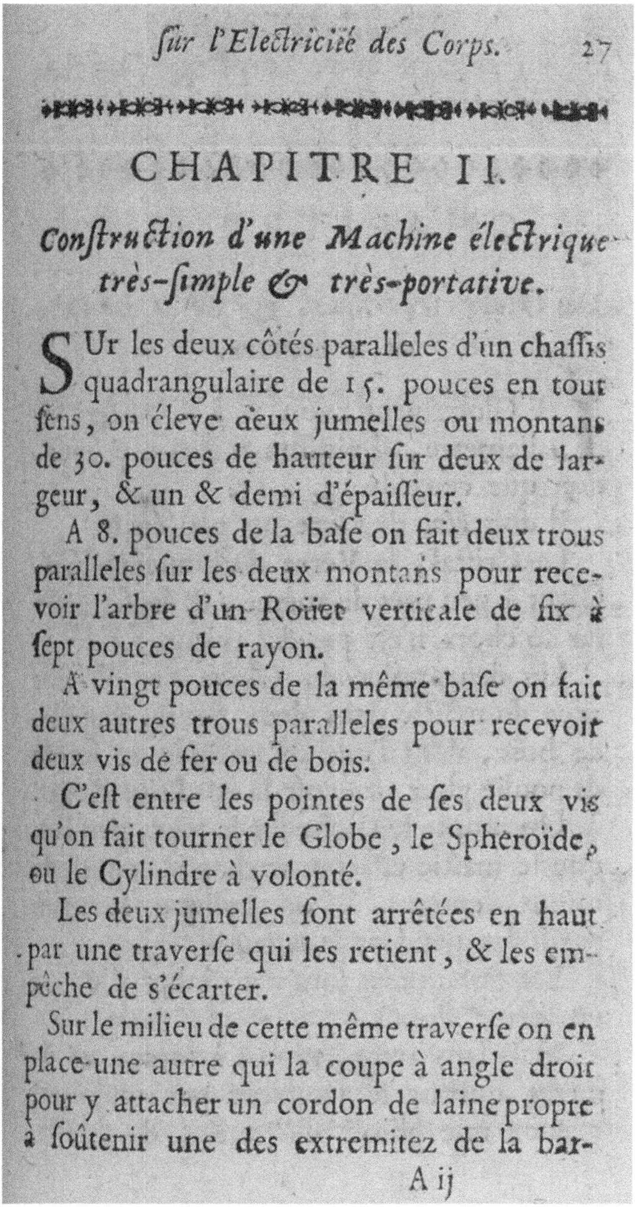

ſur l'Electricité des Corps. 27

CHAPITRE II.

Conſtruction d'une Machine électrique très-ſimple & très-portative.

SUr les deux côtés paralleles d'un chaſſis quadrangulaire de 15. pouces en tout ſens, on éleve deux jumelles ou montans de 30. pouces de hauteur ſur deux de largeur, & un & demi d'épaiſſeur.

A 8. pouces de la baſe on fait deux trous paralleles ſur les deux montans pour recevoir l'arbre d'un Roüet verticale de ſix à ſept pouces de rayon.

A vingt pouces de la même baſe on fait deux autres trous paralleles pour recevoir deux vis de fer ou de bois.

C'eſt entre les pointes de ſes deux vis qu'on fait tourner le Globe, le Spheroïde, ou le Cylindre à volonté.

Les deux jumelles ſont arrêtées en haut par une traverſe qui les retient, & les empêche de s'écarter.

Sur le milieu de cette même traverſe on en place une autre qui la coupe à angle droit pour y attacher un cordon de laine propre à ſoûtenir une des extremitez de la bar-

A ij

Figure A2.7. *Morin, J.: Nouvelle dissertation sur l'électricité des corps, Paris, Veuve Estienne et Fils, 1748, ch. II, p. 27*

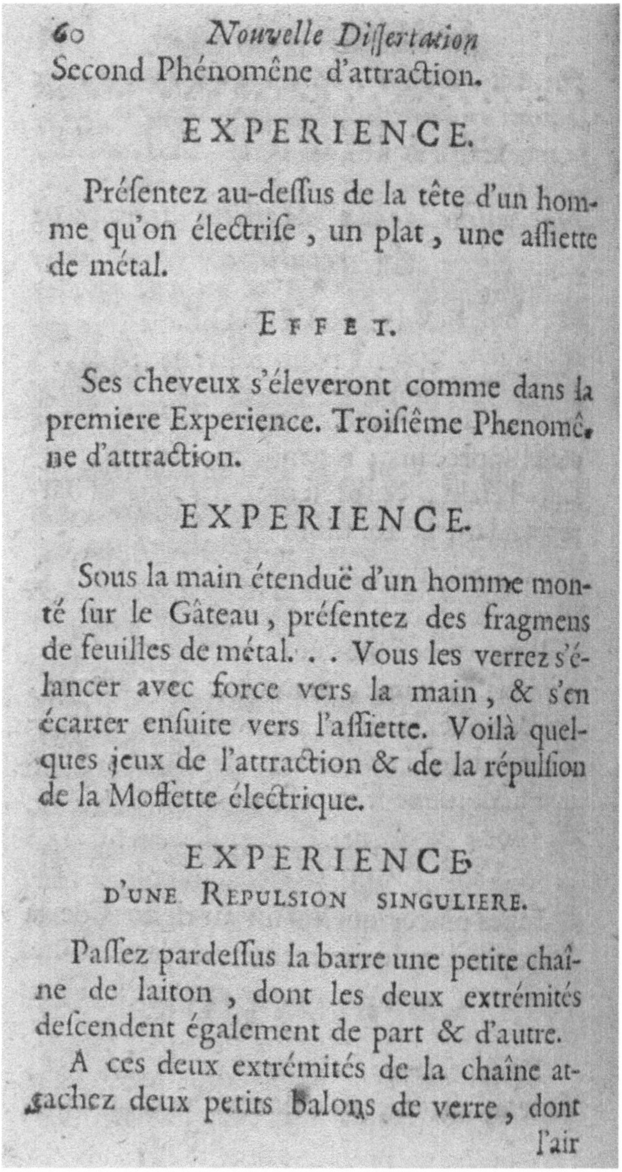

60 *Nouvelle Diſſertation*

Second Phénomêne d'attraction.

EXPERIENCE.

Préſentez au-deſſus de la tête d'un homme qu'on électriſe, un plat, une aſſiette de métal.

EFFET.

Ses cheveux s'éleveront comme dans la premiere Experience. Troiſiême Phenomêne d'attraction.

EXPERIENCE.

Sous la main étenduë d'un homme monté ſur le Gâteau, préſentez des fragmens de feuilles de métal. . . Vous les verrez s'élancer avec force vers la main, & s'en écarter enſuite vers l'aſſiette. Voilà quelques jeux de l'attraction & de la répulſion de la Moffette électrique.

EXPERIENCE

D'UNE REPULSION SINGULIERE.

Paſſez pardeſſus la barre une petite chaîne de laiton, dont les deux extrémités deſcendent également de part & d'autre. A ces deux extrémités de la chaîne attachez deux petits Balons de verre, dont l'air

Figure A2.8. *In Morin's treatise (1748), we can see that he indicates how to build a machine (p. 27) and then in the same treatise, he explores the effects of electricity on a male subject (p. 60). Morin, J.: Nouvelle dissertation sur l'électricité des corps, Paris, Veuve Estienne et Fils, 1748, ch. II, p. 60*

References

[ABE 05] ABELSON J.L., CURTIS G.C., SAGHER O. *et al.*, "Deep brain stimulation for refractory obsessive-compulsive disorder", *Biol. Psychiatry*, vol. 57, no. 5, pp. 510–516, 2005.

[ADA 84] ADAMS G., *An Essay on Electricity: In Which the Theory and Practice of That Useful Science, are Illustrated by a Variety of Experiments, Arranged in a Methodical Manner: To Which is Added, An Essay on Magnetism*, Tycho Brahe's Head, London, 1784.

[ADR 34] ADRIAN E.D., MATTHEWS B.H.C., "The Berger rhythm: Potential changes from the occipital lobes in man", *Brain*, vol. 57, no. 4, p. 373, 1934.

[ALB 15–16] ALBIGNAC H., "Le manicomio d'Aversa : une histoire tumultueuse d'une institution d'origine muratienne", in *Rencontres*, SIDEF, pp. 95–100, 2015–2016.

[ALD 94] ALDINI G., *De animali electricitate dissertationes duae*, Ex typographia Instituti Scientiarum Bologna, 1794.

[ALD 03a] ALDINI G., *An Account of the Galvanic Experiments Performed by John Aldini, on the Body of a Malefactor Executed at Newgate, Jan. 17, 1803: With a Short View of Some Experiments Which Will Be Described in the Author's New Work Now in the Press*, Cuthell and Martin, London, 1803.

[ALD 04b] ALDINI G., *Essai théorique et expérimental sur le galvanisme, avec une série d'expériences faites en présence des commissaires de l'Institut national de France, et en divers amphithéâtres de Londres*, Avec planches, Piranesi, Lucchesini, Paris, 1804.

[ALD 03c] ALDINI G., *Précis des expériences galvaniques faites récemment à Londres et à Calais... suivi d'un extrait d'autres expériences*, P. Didot aîné, Paris, 1803.

[ALD 03d] ALDINI G., *Traduction d'un commentaire de J. Aldini sur un mémoire de Galvani ayant pour titre : des Forces de l'électricité dans le mouvement musculaire*, in CASSIUS J.-J.-J., *Précis succinct des principaux phénomènes du galvanisme*, Delaplace et Goujon, Paris, 1803.

[ALT 59] ALTHAUS J., *A Treatise on Medical Electricity*, London, 1859.

[ALT 79] ALTHAUS J., "The functions of the brain", *Nineteenth Century,* vol. 6, pp. 1021–1032, 1879.

[ALT 99] ALTHAUS J., "Old age and rejuvenescence", *Lancet*, pp. 149–152, 1899.

[ALT 78] ALTICK R., *The Shows of London*, Belknap Press, Cambridge, Massachusetts, 1978.

[ALZ 05] ALZAGAA A.G., VARONB J., BASKETT P., "Charles Kite: The clinical epidemiology of sudden cardiac death and the origin of the early defibrillator", *Resuscitation*, vol. 64, pp. 7–12, 2005.

[AMO 08] AMORETTI C., *Della raddomanzia ossia della eletrometria animale, ricerche fisiche e storiche*, Giuseppe Marelli stampatore, Milan, 1808.

[AND 54] ANDRAUD M., *Théâtre scientifique; Électricité; Galvani, drame en cinq actes*, Guillaumin et Compagnie, Paris, 1854.

[AND 13] ANDRAULT R., "Définir le vitalisme. Les lectures de Claude Bernard", in DUCHESNEAU F., KUPIEC J.-J., MORANGE M., *Claude Bernard et la méthode de la physiologie*, Éditions Rue d'Ulm, Paris, 2013.

[AOU 04] AOUIZERATE B., CUNY E., MARTIN-GUEHL C. *et al.*, "Deep brain stimulation of the ventral caudate nucleus in the treatment of obsessive-compulsive disorder and major depression. Case report", *Journal of Neurosurgery*, vol. 101, pp. 482–486, 2004.

[AOU 05] AOUIZERATE B., CUNY E., MARTIN-GUEHL C. *et al.*, "Stimulation cérébrale profonde du striatum ventral dans le traitement du trouble obsessionnel-compulsif avec dépression majeure", *M/S : médecine sciences*, vol. 21, no. 10, pp. 787–893, 2005.

[AOU 09] AOUIZERATE B., CUNY E., BARDINET E. *et al.*, "Distinct striatal targets in treating obsessive-compulsive disorder and major depression", *Journal of Neurosurgery*, vol. 111, no. 4, pp. 775–779, 2009.

[ARL 64] ARLIDGE J.T., "Quarterly report on the progress of psychological medicine", *The Journal of Mental Science*, vol. IX, July, pp. 572–591, 1864.

[ARM 09] ARMINJON M., "L'homoncule de Penfield", https://www.espacestemps.net/articles/homoncule-penfield/, 2009.

[ARN 69] ARNDT R., "Die Electricität in der Psychiatrie", *Arch Psychiatr Nervenkr*, vol. 2, pp. 259–261, 1869.

[ARS 81–82a] D'ARSONVAL A., "Utilisation des forces naturelles. Avenir de l'électricité", *La Revue scientifique de la France et de l'étranger : revue des cours scientifiques*, pp. 370–372, 1881–1882.

[ARS 81–82b] D'ARSONVAL A., "Utilisation des forces naturelles par l'électricité", *La Revue scientifique de la France et de l'étranger : revue des cours scientifiques*, pp. 550–555, 1881–1882.

[ARS 81–82c] D'ARSONVAL A., "Discussion de la commission d'électrophysiologie", *La Revue scientifique de la France et de l'étranger : revue des cours scientifiques*, pp. 725–729, 1881–1882.

[ARS 96a] D'ARSONVAL A., "Action physiologique et thérapeutique des courants à haute fréquence", *Société internationale des électriciens*, Gauthier-Villars et fils, Paris, 1896.

[ARS 96b] D'ARSONVAL A., "Recherches sur la décharge électrique de la torpille", *J. Phys. Theor. Appl.*, vol. 5, no. 1, pp.149–154, 1896.

[ART 81] ARTHUIS A., *L'électricité statique et l'hystérie, mémoire précédé d'une Lettre à M. le professeur Charcot, par le docteur A. Arthuis*, Octave Doin, Paris, 1881.

[ASH 13] ASHCROFT F., *The Spark of Life: Electricity in the Human Body*, Penguin, London, 2013.

[AUE 08] AUENBRUGGER L., *Nouvelle méthode pour reconnaître les maladies internes de la poitrine par la percussion de cette cavité,* Migneret, Paris, 1808.

[AUM 55] D'AUMONT A., "Electricité médicinale", *Encyclopédie, ou dictionnaire raisonné des sciences, des arts et des métiers*, Volume 5, Briasson, David, Le Breton, Durand, Paris, 1755.

[AUZ 59] AUZOUY T., *Des Troubles fonctionnels de la peau et de l'action de l'électricité chez les aliénés*, Dard, Nancy, 1859.

[AUZ 67] AUZOUY T., *Compte-Rendu moral, administratif et médical, du service de l'asile*, Vignancour, Pau, 1867.

[AXE 83] AXENFELD A., *Traité des névroses* (2nd ed.), Baillière, Paris, 1883.

[BAB 03] BABINSKI J., "Guérison d'un cas de Mélancolie à la suite d'un accès provoqué de Vertige Voltaïque", *Revue neurologique,* vol. 10, pp. 525–531, 1903.

[BAB 09] BABINSKI J., *Démembrement de l'hystérie traditionnelle : pithiatisme,* Imprimerie de la semaine médicale, Paris, 1909.

[BAC 31–32] BACHELARD G., "Noumènes et microphysique", *Recherches philosophiques*, vol. I, pp. 55–65, 1931–1932.

[BAC 69] BACHELARD G., *Le nouvel esprit scientifique,* (1934), 10th ed., PUF, Paris, 1969.

[BAC 12] BACOPOULOS-VIAU A., "La danse des corps figés. Catalepsie et imaginaire médical au XIXe siècle", *Revue d'histoire du XIXe siècle*, vol. 44, pp. 164–184, 2012.

[BAG 00] BAGLIVI G., *De fibra motrice, et morbosa, nec non de experimentis, ac morbis salivae, bilis et sanguinis*, Perugia, 1700.

[BAI 40] BAILLARGER J., *Recherches sur la structure de la couche corticale des circonvolutions du cerveau*, Baillière, Paris, 1840.

[BAL 15] BALLY J., LÜSCHER C., BERNEY A. *et al.*, "Stimulation cérébrale profonde : nouvelles cibles et nouvelles indications", *Rev. Med. Suisse,* vol. 11, pp. 977–982, 2015.

[BAR 58] BARTHEZ P.-J., *Nouveaux éléments de la science de l'homme,* 3rd ed., Baillière, Paris, 1858.

[BAR 74] BARTHOLOW R., "Experiments on the functions of the human brain", *Br Med J.*, May 30, vol. 1, no. 700, p. 727, 1874.

[BAR 82] BARTHOLOW R., *Medical Electricity: A Practical Treatise on the Applications of Electricity to Medicine and Surgery*, 2nd ed., Henry C. Lea's Son & Co., Philadelphia, 1882.

[BAR 06] BARRAS V., PANESE F., "L'utopie médicale de la réanimation des corps (XIXe–XXIe siècles)", *Mouvements*, vol. 45–46, no. 3, pp. 36–42, 2006.

[BAR 18] BARBARA J.-G., "Localiser les enregistrements et les stimulations électriques du cerveau par la stéréotaxie", *Bulletin d'Epistémologie et d'histoire des sciences de la vie*, vol. 25, no. 2, pp. 169–186, 2018.

[BAY 25] BAYLE A.-L.-J., *Nouvelle doctrine des maladies mentales*, Gabon et compagnie, Paris, 1825.

[BEA 66] BEARD G.M., *Electricity as a Tonic*, New York, 1866.

[BEA 69] BEARD G.M., "Neurasthenia, or Nervous Exhaustion", *Boston Medical and Surgical Journal*, vol. 80, no. 3, pp. 217–221, 1869.

[BEC 53] BECCARIA G., *Dell'elettricismo artificiale e naturale libri due,* nella stampa di Filippo Antonio Campana, Turin, 1753.

[BEC 57] BECQUEREL L.A., *Traité des applications de l'électricité à la thérapeutique médicale et chirurgicale*, Baillière, Paris, 1857.

[BEC 75] BECHTEREVA N.P., BONDARTCHUK A.N., SMIRNOV V.M. *et al.*, "Method of electrostimulation of the deep brain structures in the treatment of some chronic diseases" *Confin Neurol.*, vol. 37, pp. 136–140, 1975.

[BEC 08] BECHTEL W., "L'épistémologie des données objectives", in FAUCHER L., POIRIER P., *Des neurosciences à la philosophie : Neurophilosophie et philosophie des neurosciences*, Syllepses, Paris, 2008.

[BED 99] BEDDOES TH., *Contributions to Physical and Medical Knowledge Principally from the West of England*, Biggs & Cottle, Bristol, 1799.

[BEJ 99] BEJJANI B.P., DAMIER P. *et al.*, "Transient acute depression induced by high frequency deep-brain stimulation", *New England Journal of Medicine*, vol. 340, pp. 1476–1480, 1999.

[BEL 91] BELTRAN A., *La Fée électricité*, Poche, Paris, 1991.

[BEL 14] BELKIN G.S., *Death Before Dying: History, Medicine, and Brain Death*, Oxford University Press, New York, 2014.

[BEN 87] BENABIB A.L., POLLAK P., LOUVEAU A. *et al.*, "Combined (thalamotomy and stimulation) stereotactic surgery of the VIM thalamic nucleus for bilateral Parkinson disease", *Appl. Neurophysiol.*, vol. 50, pp. 344–346, 1987.

[BEN 11] BENADHIRA R., BOUAZIZ N., SIDHOUMI D. *et al.*, "Stimulation cérébrale en psychiatrie : situation actuelle et perspectives", *L'information psychiatrique*, vol. 87, no. 9, pp. 705–713, 2011.

[BEN 17] BENABIB A.-L., *Les traitements de la maladie de Parkinson*, institut Diderot, Paris, 2017.

[BEN 84] BENGUIGUI I., *Théories électriques du XVIIIe siècle : Correspondance Nollet, Jallabert*, Georg, Geneva, 1984.

[BER 38] BERNOUILLI D., *Hydrodynamica, sive de Viribus et Motibus Fluidorum commentarii. Opus Academicum*, Dulsecker, Strasbourg, 1738.

[BER 80] BERTHOLON P., *De l'électricité du corps humain dans l'état de santé et de maladie ; ouvrage dans lequel on traite de l'électricité de l'atmosphère, de son influence, de ses effets sur l'économie animale*, Didot, Paris, 1780.

[BER 64–89] BERTIN E., "Mort (physiologie)", *Dictionnaire encyclopédique des sciences médicales*, vol. 9, pp. 517–579, 1864–1889.

[BER 68] BERNSTEIN J., "Ueber den zeitlichen Verlauf der negativen Schwankung des Nervenstroms", *Pflüger's Arch. f. d. ges. Physiol.*, vol. 1, pp. 173–207, 1868.

[BER 50] BERNARD C., *Lettres beaujolaises*, Cuvier, Villefranche-en-Beaujolais, 1950.

[BER 66] BERNARD C., *Introduction à l'étude de la médecine expérimentale,* Garnier Flammarion, Paris, 1966.

[BER 69] BERGER H., "On the electroencephalogram of man: The fourteen original reports on the human electroencephalogram", *EEG Clin Neurophysiol,* suppl. 28, pp. 1–350, 1969.

[BER 01] BERNARDI W., "La controverse sur l'électricité animale dans l'Italie du XVIIIe siècle : Galvani, Volta et... d'autres", *Revue d'histoire des sciences*, vol. 54-1, pp. 53–70, 2001.

[BER 05] BERTRAND R., CAROL A., PELEN J.-N., *Les narrations de la mort*, PUP, Aix-en-Provence, 2005.

[BER 07] BERTUCCI P., "Sparks in the dark: the attraction of electricity in the eighteenth century", *Endeavour*, vol. 31, no. 3, pp. 88–93, 2007.

[BIA 50] BIANCHINI G.F., *Recueil d'expériences faites à Venise sur la médecine électrique*, Le Mercier, Paris, 1750.

[BIA 77] BIANCHI M., "Recueil périodique d'observations de médecine", *Journal de physique, de chemie, d'histoire naturelle et des arts*, vol. 9, pp. 413–436, 1777.

[BIC 99] BICHAT M.F.X., *Traité des membranes en général et de diverses membranes en particulier*, Veuve Richard, Paris, (*A Treatise on the Membranes in General and On Different Membranes in Particular*, translated by HUSSON M., Cummings and Hilliard, Boston, 1813. Available at: https://archive.org/details/2543054R.nlm.nih.gov/page/n4/mode/2up), 1799.

[BIC 99–00] BICHAT M.F.X., *Recherches physiologiques sur la vie et la mort*, Brosson, Gabon et Compagnie, Paris, 1799–1800.

[BIC 39] BICHAT M.F.X., *Recherches sur la vie et la mort, suivies de la division la plus naturelle des phénomènes physiologiques, par Buisson, et des expériences sur le principe de la vie, par Legallois,* M. Gautret, Paris, 1839 (*Physiological Researches on Life and Death*), translated by GOLD F. and HAYWARD G., 2018. Available at: http://www.gutenberg.org/files/56440/56440-h/56440-h.htm), 1839.

[BIC 53] BICKFORD R.G., PETERSEN M.C., DODGE H.W. *et al.*, "Observations on depth stimulation of the human brain through implanted electrographic leads", *Proc. Staff. Meet Mayo. Clin.*, vol. 28, pp. 181–187, 1953.

[BLO 00] BLONDEL C., GOUZEVITCH I., DEBLOCK G. *et al.*, *Bibliographie francophone des ouvrages et articles relatifs à l'électricité et au magnétisme publiés avant 1820*, CNAM, Paris, 2000.

[BLO 10] BLONDEL CH., "La reconnaissance de l'électricité médicale et ses 'machines à guérir' par les scientifiques français (1880–1930)", *Annales historiques de l'électricité*, vol. 1, no. 8, pp. 37–51, 2010.

[BOR 08] BORCK C., "Recording the brain at work: The visible, the readable, and the invisible in electroencephalography", *Journal of the History of the Neurosciences,* vol. 17, pp. 367–379, 2008.

[BOR 18a] BORCK C., *Brainwaves: A Cultural History of Electroencephalography*, Routledge, London and New York, 2018.

[BOR 18b] BORCK C., "Machines, minds, electric brains: Stimulation/recording as practical framework and thought style", *Bulletin d'Epistémologie et d'histoire des sciences de la vie*, vol. 25, no. 2, op. 187–202, 2018.

[BOU 96] BOURDIN J.-C., *Les matérialistes au XVIIIe siècle*, Payot, Paris, 1996.

[BOS 57] BOS VAN DEN I.J., BIKKER L., *De vivis humani corpori solidis*, Luchtmans, Moens, Leiden, 1757.

[BOS 93] BOSSI L., "L'âme électrique", in CLAIR J., *L'âme au corps, Arts et sciences 1793–1993*, catalogue de l'exposition, Réunion des Musées Nationaux, Gallimard, Electa, Paris, 1993.

[BOS 03] BOSSI L., *Histoire naturelle de l'âme,* PUF - Science, histoire et société, Paris, 2003.

[BOZ 54] BOZE G.M., *L'électricité son origine et ses progrés : poème en deux livres,* translated by M. l'Abbé Joseph A. de C., chez les héritiers Lankisch, Leipzig, 1754.

[BRA 27] BRAVAIS L.F., Recherches sur les symptômes et le traitement de l'épilepsie hémiplégique, Thesis, Paris, 1827.

[BRA 80] BRAZIER M.A.B., "The emergence of electrophysiology as an aid to neurology", in AMINOFF, M.J.L. (ed.), *Electrodiagnosis in Clinical Neurology*, Churchill Livingstone Inc., London, 1980.

[BRA 88] BRAZIER M.A.B., *A History of Neurophysiology in the 19th Century*, Raven Press, New York, 1988.

[BRE 99] BRESADOLA M., PANCALDI G., Luigi Galvani international workshop. Proceedings, Bologna, 1999.

[BRE 08] BRESADOLA M., "Animal electricity at the end of the eighteenth century: The many facets of a great scientific controversy", *Journal of the History of the Neurosciences*, vol. 17, pp. 8–32, 2008.

[BRE 10] BRENNI P., "Les courants à hautes-fréquences apprivoisés à travers la darsonvalisation et les spectacles publics (1890–1930)", *Annales historiques de l'électricité*, vol. 1, no. 8, pp. 53–71, 2010.

[BRI 56] BRIERRE DE BOISMONT A., *Du suicide et de la folie suicide: considérés dans leurs rapports avec la statistique, la médecine et la philosophie,* Baillière, Paris, 1856.

[BRI 59a] BRIERRE DE BOISMONT A., "Études cliniques sur la paralysie générale", *Annales médico-psychologiques, Journal de l'aliénation mentale et de la médecine légale des aliénés*, L. Martinet, Paris, 1859.

[BRI 59b] BRIQUET P., *Traité clinique et thérapeutique de l'hystérie*, Baillière et fils, Paris, 1859.

[BRO 16] BROUSSAIS F.J.V., *Examen de la doctrine médicale généralement adoptée, et des systèmes modernes de nosologie*, Gabon, Paris, 1816.

[BRO 53] BROCHIN M., "Des ressources que fournit l'électricité au diagnostic différentiel des Paralysies", *Annales médico-psychologiques, Journal de l'aliénation mentale et de la médecine légale des aliénés,* Martinet, Paris, 1853.

[BRU 16] BRUNONI A., NITSCHE M., LOO C., *Transcranial Direct Current Stimulation in Neuropsychiatric Disorders*, Springer, Switzerland, 2016.

[BUE 10] BUELTZINGSLOEWEN VON I., "Un fol espoir thérapeutique ? L'introduction de l'électrochoc dans les hôpitaux psychiatriques français (1941–1945)", *Annales historiques de l'électricité*, vol. 1, no. 8, pp. 93–104, 2010.

[BUR] BURGER J.R., *Brain Theory From A Circuits And Systems Perspective,* Springer-Verlag, New York, 2013.

[CAI 19] CAIRE M., *Soigner les fous : histoire des traitements médicaux en psychiatrie,* Nouveau Monde, Paris, 2019.

[CAL 57] CALDANI L.M.A., *Sull'insensività, ed irritabilità di alcune parti degli animali,* Bologna, 1757.

[CAM 16] CAMPENOT R.B., *Animal Electricity: How We Learned That the Body and Brain Are Electric Machines*, Harvard University Press, London, 2016.

[CAM 18] CAMBIAGHI M., PARENT A., "From galvanization of human bodies to the Modern Prometheus", *Medicina Historica*, vol. 2, no. 1, pp. 27–37, 2018.

[CAN 78] CANGUILHEM G., *On the Normal and the Pathological*, translated by FAWCETT C.R., D. Reidel Publishing Company, Dordrecht. Available at: https://books.google.co.uk/books/about/On_the_Normal_and_the_Pathological.html?id=TJFfBgAAQBAJ&printsec=frontcover&source=kp_read_button&redir_esc=y#v=onepage&q=philosophy&f=false, 1978.

[CAN 16] CANAVERO S., REN X., KIM C.Y., "HEAVEN: The Frankenstein effect", *Surg. Neurol. Int.*, September 13. Available at: http://surgicalneurologyint.com/surgicalint-articles/heaven-the-frankenstein-effect/, 2016.

[CAR 46] CARPENTER W.B., "Mr Noble on the brain and its physiology", *British and Foreign Medical Review*, vol. 22, pp. 488–544, 1846.

[CAR 77] CARPENTER W.B., *Mesmerism, Spiritualism, &c.*, Appleton and company, New York, 1877.

[CAR 03] CARMENA J.M., LEBEDEV M.A., CRIST R.E. *et al.*, "Learning to control a brain – machine interface for reaching and grasping by primates", *PLoS Biology*, vol. 1, no. 2, pp. 193–208, 2003.

[CAR 09] CARLSON C., DEVINSKY O., "The excitable cerebral cortex Fritsch G, Hitzig E. Über die elektrische Erregbarkeit des Grosshirns. Arch Anat Physiol Wissen 1870; 37, 300-32", *Epilepsy & Behavior*, vol. 15, no. 2, pp. 131–132, 2009.

[CAR 15] CAROL A., "Une histoire médicale des critères de la mort", *Communications*, vol. 2, no. 97, pp. 45–55, 2015.

[CAS 03] CASSIUS J.-J.-J., *Précis succinct des principaux phénomènes du galvanisme*, Delaplace et Goujon, Paris, 1803.

[CAS 48] CASTIGLIONI A., *Storia della médicina,* Mondadori, Verona, 1948.

[CAS 16] CASCIATO S., RITACCIO A.L., "Ezio Sciamanna: The Italian contribution to the origin of cortical stimulation mapping in humans", *Epilepsy & Behavior*, vol. 62, pp.186–188, 2016.

[CAT 75] CATON R., "The electrical currents in the brain", *Brit Med J.*, vol. 2, p. 278, 1875.

[CAU 08] CAULLET-VEAUMOREL L., "Médecine", *Encyclopédie méthodique*, vol. 8, Agasse, Paris, 1808.

[CAU 37] CAUSTICK C., *Terrible Tractoration, and Other Poems,* 2nd ed., Samuel Colman, Boston, 1837.

[CAV 85] CAVALLO T., *Traité complet d'électricité, trad. de l'anglais, enrichie de ses nouvelles expériences*, Paris, 1785.

[CAV 79] CAVENDISH H., *The Electrical Researches of the Honourable Henry Cavendish, Written Between 1771 and 1781*, University Press, Cambridge, 1879.

[CAZ 81] CAZIN A., *Traité théorique et pratique des piles électriques : Mesure des constantes des piles. Unités électriques : Description et usage des différentes espèces de piles,* Gauthier-Villars, Paris, 1881.

[CER 50] CERLETTI U., HANDELSMAN J., DRESZER, R. *et al.*, *Thérapeutique biologique : Indications respectives des méthodes de choc*, Hermann, Paris, 1950.

[CER 06] CERLETTI U., PASSIONE R., *Scritti sull'elettroshock*, F. Angeli, Milano, 2006.

[CHA 78] CHARCOT J.M., "Treizième leçon, De l'Hystéro-épilepsie", in *Œuvres complètes,* Adrien Delahaye et Emile Lecrosnier Editeurs, Paris, 1878.

[CHA 82] CHARCOT J. M., "Phénomènes produits par l'application sur la voûte du crane du courant galvanique, pendant la période léthargique de l'hypnotisme chez les hystériques", *Prog Med.*, vol. 10, no. 20–21, pp. 63–64, 1882.

[CHA 89] CHARCOT J.M., "Leçons du mardi à la Salpêtrière", *Progrès Médical*, Emile Lecrosnier Editeurs, Paris,1889.

[CHA 17] CHARDIN C., *Précis d'électricité médicale exposant le principe et la méthode électro-cinésique vasculaire*, Chardin, Paris, 1917.

[CHA 60] CHARDACK W.M., GAGE A.A., GREATBATCH W., "A transistorized, self-contained, implantable pacemaker for the long-term correction of complete heart block", *Surgery,* vol. 48, pp. 643–654, 1960.

[CHE 15] CHERICI C., DUPONT J.-C., *L'exploration cérébrale, Histoire récente et nouveaux outils*, Hermann, Paris, 2015.

[CHE 16] CHERICI C., *Anatomophysiologie du cerveau et du cervelet chez Vincenzo Malacarne (1744–1816)*, Hermann, Paris, 2016.

[CHE 18] CHERICI C., "L'institut neurologique de Montréal: les travaux de Wilder Penfield (1891–1976)", *Bulletin d'Epistémologie et d'histoire des sciences de la vie*, vol. 25, no. 2, pp. 157–168, 2018.

[CHE 19] CHERICI C., "Stimulation cérébrale, électricité et psychiatrie : perspectives historiques" in DESMOULINS-CANSELIER S., GAILLLE M., MOUTAUD B., *La stimulation cérébrale profonde de l'innovation au soin*, Hermann, Paris, 2019.

[CHU 08] CHURCHLAND S.P., "Que peut nous enseigner la neurobiologie au sujet de la conscience ?" in FAUCHER L., POIRIER P., *Des neurosciences à la philosophie : Neurophilosophie et philosophie des neurosciences*, Syllepses, Paris, 2008.

[CIS 33a] DE CISTERNAY DU FAY C.F., *Premier mémoire sur l'électricité, Histoire de l'électricité*, Académie royale des sciences, 1733.

[CIS 33b] DE CISTERNAY DU FAY C.F., *Second mémoire sur l'électricité, Quels sont les Corps qui sont susceptibles d'Electricité,* Académie royale des sciences, 1733.

[CIS 33c] DE CISTERNAY DU FAY C.F., *Troisième mémoire sur l'électricité : Des Corps qui sont le plus vivement attirés par les matières électriques,* Académie royale des sciences, 1733.

[CIS 33d] DE CISTERNAY DU FAY C.F., *Quatrième mémoire sur l'électricité : De l'Attraction & Répulsion des Corps Electriques*, Académie royale des sciences, 1733.

[CIS 34a] de Cisternay du Fay C.F., *Cinquième mémoire sur l'électricité, Où l'on rend compte des nouvelles découvertes,* Académie royale des sciences, 1734.

[CIS 34b] de Cisternay du Fay C.F., *Sixième mémoire sur l'électricité, Où l'on examine quel rapport il y a entre l'Electricité, & la faculté de rendre de la Lumière*, Académie royale des sciences, 1734.

[CLO 98] Clower W.T., "The transition from animal spirits to animal electricity: A neuroscience paradigm shift", *J. Hist. Neurosci.*, vol. 7, pp. 201–218, 1998.

[COB 09] Coben R., Evans J.R., *Neurofeedback and Neuromodulation Techniques and Applications,* Elsevier, London, 2009.

[COL 53] Colden D., Remarks on Nollet's Letters, New York Historical Society. Available at: https://franklinpapers.org/framedVolumes.jsp?vol=5&page= 135a, 1753.

[CON 78] Conry Y., "Thomas Willis ou le premier discours rationaliste en pathologie mentale", *Revue d'histoire des sciences*, vol. 31, no. 3, pp. 193–231, 1978.

[COO 80] Cooper I.S., Upton A.R.M., Amin I., "Reversibility of chronic neurologic deficits: Some effects of electrical stimulation of the thalamus and internal capsule in man", *Appl Neurophysiol.*, vol. 43, pp. 244–258, 1980.

[COR 06] Corvisart J.-N., *Essai sur les maladies et les lésions organiques du cœur et des gros vaisseaux*, Migneret, Paris, 1806.

[COR 91] Corsi P., *The Enchanted Loom: Chapters in the History of Neuroscience*, Oxford University Press, New York, 1991.

[COU 81] Courant Th., *De l'électricité médicale et magnétique, ses applications curatives préconisées par les sommités médicales comme l'auxiliaire le plus puissant du traitement des maladies de la poitrine et des affections du système nerveux, musculaire et sanguin*, Imprimerie des Apprentis-orphelins, Paris-Auteuil, 1881.

[CRI 52] Crimotel V., *Électricité, galvanisme et magnétisme, appliqués aux maladies nerveuses et chroniques,* Baillière, Paris, 1852.

[CRI 38] Crichton-Browne J., *The Doctor Remembers*, Duckworth, London, 1938.

[DAI 98] Dai J., Swaab D.F., Buijs R.M., "Recovery of axonal transport in 'dead neurons'", *Lancet,* vol. 351, pp. 499–500, 1998.

[DAL 08] Dalton J., *A New System of Chemical Philosophy*, R. Bickerstaff, Manchester, UK, 1808.

[DAL 89] Danion L., "L'Électrothérapie", *Journal d'électricité médicale*, vol. II, 1889.

[DAN 92] Danion L., "L'Électrothérapie", *Journal d'électricité médicale*, vols 1–3, pp. 52–53, 1892.

[DAR 01] Darmon P., "Des suppliciés oubliés de la Grande Guerre : les pithiatiques", *Histoire, économie et société*, vol. 1, pp. 49–64, 2001.

[DAS 07] Daston L., Galison P., *Objectivity*, Zone books, New York, 2007.

[DEB 08] DEBRU A., "Les enseignements de la torpille dans la médecine antique", in *Le médecin initié par l'animal. Animaux et médecine dans l'Antiquité grecque et latine*, Proceedings of the international conference held at the Maison de l'Orient and Méditerranée-Jean Pouilloux, October 26 and 27, pp. 39–47, Lyon, 2008.

[DEC 53] DECOURT E., *Nouvelle application de l'électricité : Guérison des douleurs au moyen de douches galvaniques*, Théolier Ainé, Saint-Etienne, 1853.

[DEL 43] DELMAS-MARSALET P., *L'électrochoc en thérapeutique et la dissolution-reconstruction*, Baillère, Paris, 1943.

[DEL 52] DELGADO J., HAMLIN H., CHAPMAN W.P., "Technique of intracranial electrode implacement for recording and stimulation and its possible therapeutic value in psychotic patients", *Confin. Neurol.*, vol. 12, pp. 315–319, 1952.

[DEL 69] DELGADO J., *Physical Control of the Mind: Toward a Psychocivilized Society*, Harper and Row, New York, 1969.

[DEL 06] DELON M., "Électriser, un mot d'ordre au siècle des Lumières", *Revue des Sciences Humaines,* vol. 1, no. 281, p. 40, 2006.

[DEL 16] DELHERM L., DAUSSEL H., "Les névrites chez les blessés de guerre et les agents physiques", pp. 256–257, in BERGOGNIE J. (ed.), *Archives d'électricité médicale et de physiothérapie*, tome XXVI, J. Hamel, Bordeaux, 1916.

[DEM 13] DEMAZEUX S., *Qu'est-ce que le DSM ? Genèse et transformations de la bible américaine de la psychiatrie,* Les Éditions d'Ithaque, Paris, 2013.

[DEN 12] DENYS D., FEENSTRA M., SCHUURMAN R., *Deep Brain Stimulation: A New Frontier in Psychiatry*, Springer, Berlin, 2012.

[DEP 08] DEPROUW S., Un héritage des Bonaparte : Le prix du galvanisme (1802–1815) et le prix Volta (1852–1888) : L'État et l'encouragement à la recherche sur l'électricité, Thesis, La Sorbonne, Paris. Available at: http://theses.enc.sorbonne.fr/2008/deprouw, 2008.

[DES 49] DESHAYS J.-É., *De hemiplegia per electricitalem curanda,* J. Martel, Montpellier, 1749.

[DES 40] DESPINE A., *De l'emploi du magnétisme animal et des eaux minérales dans le traitement des maladies nerveuses : suivi d'une observation très curieuse de guérison de névropathie*, Ballière, Paris, 1840.

[DES 51] DESCHAMPS M.-H., *Du Signe certain de la mort, nouvelle épreuve pour éviter d'être enterré vivant,* Masson, Paris, 1851.

[DIC 15] *Dictionnaire des sciences médicales*, (eau-emo), vol. 11, pp. 262–313, 1815.

[DID 51] DIDEROT D., *Le rêve de d'Alembert*, (1769), M. Didier, Paris, 1951.

[DID 12] DIDEROT D., *D'Alembert's Dream,* translated by JOHNSTON I. Available at: http://www.blc.arizona.edu/courses/schaffer/249/Before %20 Darwin%20-%20New/Diderot/Diderot,%20D'Alembert's%20Dream.htm, 2012.

[DIE 05] DIEDERICH N.J., KALTEIS K., STAMENKOVIC M. *et al.*, "Efficient internal pallidal stimulation in Gilles de la Tourette syndrome: A case report", *Mov. Disord.*, vol. 20, no. 11, pp. 1496–1499, 2005.

[DOP 44] DOPPELMAYER J.G., *Neu-entdeckte Phaenomena von bewunderns-würdigen Würckungen der Natur*, gedruckt by Johann Joseph Fleischmann, Nürnberg, 1744.

[DOR 46] DORSMAN C., CROMMELIN C.A., *Invention of the Leiden Jar*, Communication 97 from National Museum of History of Science of Leiden, pp. 276–280, 1946.

[DOR 56] DOREN VAN C., *Benjamin Franklin*, Aubier, Paris, 1956.

[DUB 33] DUBOIS (D'AMIENS) E.-F., *Histoire philosophique de l'hypochondrie et de l'hystérie*, Deville Cavellin, Paris, 1833.

[DUB 48–49] DU-BOIS REYMOND E.H., *Untersuchungen über thierische Elektricität*, von G. Reimer, Berlin, 1848–1849.

[DUC 62] DUCHENNE DE BOULOGNE G-B., *Mécanisme de la physionomie humaine, ou Analyse électrophysiologique de l'expression des passions*, Jules Renouard, Paris, 1862.

[DUC 72] DUCHENNE DE BOULOGNE G-B., *De l'électrisation localisée et de son application à la physiologie, à la pathologie et à la thérapeutique*, 3rd ed., Baillière, Paris, 1872.

[DUG 18] DUGUET J.-J., *Traité des scrupules, de leurs causes, de leurs espèces, de leurs suites dangereuses, de leurs remèdes généraux et particuliers*, J. Estienne, Paris, 1718.

[DUP 99a] DUPONT J.-C., *Histoire de la neurotransmission,* Presses Universitaires de France, Paris, 1999.

[DUP 99b] DUPONT J.-C., "Instrumentation et neurosciences : histoire de l'électrode", *Bull. Hist. Epistém. Sc. Vie*, vol. 6, no. 2, pp. 243–270, 1999.

[DUP 01a] DUPONT J.-C., "L'activité réflexe, ou le cerveau entre neurochimie et électrophysiologie", *Les Cahiers Alfred Binet*, vol. 12, no. 667, pp. 51–58, 2001.

[DUP 01b] DUPONT J.-C., "L'électricité : de l'essence de la vie à l'outil de diagnostic", *Pratiques*, vol. 12, pp. 90–93, 2001.

[DUP 08] DUPONT J.-C., "Charcot, à la conquête du cerveau", *Les génies de la science*, vol. 37, 2008.

[DUP 16] DUPONT J.-C., NAASSILA M., "Une brève histoire de l'addiction", *Alcoologie et addictologie*, vol. 38, no. 2, pp. 93–102, 2016.

[DUP 18] DUPONT J.-C., "Pathologies et modèles cérébraux. Le cas de l'épilepsie", *Bulletin d'Epistémologie et d'histoire des sciences de la vie*, vol. 25, no. 2, pp. 137–156, 2018.

[DUR 35] DURUP G., FESSARD A., "L'électroencéphalogramme de l'homme : observations psycho-physiologiques relatives à l'actions des stimulus visuels et auditifs", *Ann. Psychol.*, vol. 36, pp. 1–32, 1935.

[ELL 94] ELLENBERGER H.F., *Histoire de la découverte de l'inconscient*, Fayard, Paris, 1994.

[ELL 14] ELLIOTT P., "Electricity and the brain: An historical evaluation", in KADOSH R. C., *The Stimulated Brain: Cognitive Enhancement Using Non-Invasive Brain Stimulation*, Elsevier, Amsterdam, 2014.

[ERB 84] ERB W., *Traité d'électrothérapie*, Delahaye, Lecrosnier, Paris, 1884.

[ESQ 38] ESQUIROL E., *Des maladies mentales : considérées sous les rapports médical, hygiénique et médico-legal*, Volume 2, J.B. Baillière, Paris, 1838.

[EXP 19] "Exposé de quelques expériences faites sur le corps d'un supplicié immédiatement après son exécution : subi d'observations physiologiques et pratiques", *Bibliothèque universelle des sciences, belles-lettres, et arts*, vol. 10, pp. 128–136, 177–184, 1819.

[FAR 55] FARADAY M., *Experimental Researches in Electricity*, (1839), Richard Taylor and William Francis, London, 1855.

[FAU 08] FAUCHER L., POIRIER P., *Des neurosciences à la philosophie : Neurophilosophie et philosophie des neurosciences*, Syllepses, Paris, 2008.

[FAU 12] FAURE S., "Panorama des approches par stimulation cérébrale en neuropsychologie", *Revue de neuropsychologie*, vol. 2, no. 4, pp. 84–89, 2012.

[FAU 18] FAUCHER L., "Stimulation cérébrale profonde et augmentation morale", *Bulletin d'Epistémologie et d'histoire des sciences de la vie*, vol. 25, no. 2, pp. 221–238, 2018.

[FEI 16] FEINDEL W., LEBLANC R., *The Wounded Brain Healed: The Golden Age of the Montreal Neurological Institute, 1934–1984*, McGill-Queen's University Press, Canada, 2016.

[FER 74a] FERRIER D., *Recherches expérimentales sur la physiologie et la pathologie cérébrales*, Adrien Delahaye, Aux bureaux du progrès médical, Paris, 1874.

[FER 74b] FERRIER D., "Experiments on the brain of monkeys", *Phil Trans*, vol. 165, pp. 433–488, 1874.

[FER 79] FERRIER D., *De la localisation des maladies cérébrales*, Baillière, Paris, 1879.

[FER 91] FERRIER D., *Leçons sur les localisations cérébrales*, Lecrosnier et Babé, Paris, 1891.

[FIG 67] FIGUIER L., *Les Merveilles de la science*, Volume 2, Furne, Jouvet et cie, Paris, 1867.

[FIN 30] FINKLER W., "Die elektrische Schrift des Gehirns, Die Zickzackkurve der Menschenseele. - Bahnbrechende Versuche eines deutschen Psychiaters", *Neues Wiener Journal*, July 4, p. 7, 1930.

[FIN 94] FINGER S., *Origins of Neuroscience*, Oxford University Press, New York, 1994.

[FIN 99] FINGER S., *Minds Behind the Brain: A History of the Pioneers and Their Discoveries*, Oxford University Press, 1999.

[FIN 11] FINGER S., PICCOLINO M., *The Shocking History of Electric Fishes: From Ancient Epochs to the Birth of Modern Neurophysiology*, Oxford University Press, New York, 2011.

[FIN 12a] FINGER S., SMITH C.U.M., FRIXIONE E. *et al.*, *The Animal Spirit, Doctrine and the Origins of Neurophysiology,* Oxford University Press, New York, 2012.

[FIN 12b] FINN M.A., The West Riding Lunatic Asylum and the making of the modern brain sciences in the nineteenth century, Thesis, University of Leeds, 2012.

[FIN 15] FINKELSTEIN G., "Mechanical neuroscience: Emil du Bois-Reymond's innovations in theory and practice", *Frontiers in Systems Neuroscience*, vol. 9, no. 133, pp. 1–5, 2015.

[FLO 24] FLOURENS P., *Recherches expérimentales sur les propriétés et les fonctions du système nerveux dans les animaux vertébrés*, Crevot, Paris, 1824.

[FLO 99] FLOURENS A., *De l'Opothérapie*, Imprimerie G. Gounouilhou, Bordeaux, 1899.

[FON 75] FONTANA F., *Ricerche filosofiche sopra la fisica animale,* Cambiagi, Florence, 1775.

[FON 84] FONTANA F., *Opuscules physiques et chimiques*, translated by GIBELIN M. Nyon, Paris, 1784.

[FOR 63] FORTUNATI BIANCHINI M.J., *Recueil sur l'électricité médicale dans lequel on a rassemblé les principales pièces publiées par divers sçavans, sur les moyens de guérir en électrifiant les malades,* tome 2, Vincent and Didot Jeune, Paris, 1763.

[FOU 06] FOUCAULT M., *History of Madness*, translated by MURPHY J. and KHALFA J., Taylor & Francis Group, London, 2006.

[FRA 56] FRANKLIN B., *Expériences et observations sur l'électricité faites à Philadelphie en Amérique*, 2nd ed., revised, corrected and extended by D'ALIBARD M., Durand, Paris, 1756.

[FRA 87] FRANÇOIS-FRANCK CH.-É., *Cours du Collège de France. 1884–1885. Leçons sur les fonctions motrices du cerveau*, Octave Doin, Paris, 1887.

[FRA 18] FRANK L., *The Pleasure Shock, the Rise of Deep Brain Stimulation and its Forgotten Inventor,* Dutton, New York, 2018.

[FRE 50] FREUD S., *Ma vie et la psychanalyse*, (1925), Gallimard, Paris, 1950.

[FRE 73] FREUD S., "Nouvelles remarques sur les psychonévroses de défense", in *Névrose, psychose et perversion*, 1896, Presses Universitaires de France, Paris, 1973.

[FRE 84] FREUD S., "Rapport d'expert sur le traitement électrique des névroses de guerre" (1955), in *Résultats, idées, problèmes I*, Presses Universitaires de France, Paris, 1984.

[FRE 10] FREUD S., *L'homme aux rats : un cas de névrose obsessionnelle ; suivi de Nouvelles remarques sur les psychonévroses de défense*, translated by SKALLI C.C., TRIOL J., Payot & Rivages, Paris, 2010.

[FRE 13] FREUD S., *Névrose et psychose*, translated by CASANOVA N., NEUBURGER R., Payot & Rivages, Paris, 2013.

[FRE 14] FRENZA DE L., "La morte e l'elettricità. Esperienze di elettrofisiologia tra XVIII e XIX secolo", in DE CEGLIA F.P., *Storia della definizione di morte*, Franco Angeli, Milan, 2014.

[FRI 70] FRITSCH G., HITZIG E., "Über die elektrische Erregbarkeit des Grosshirns", *Archiv für Anatomie, Physiologie und wissenschaftliche Medicin*, pp. 300–332, 1870.

[GAL 91] GALVANI L., *De viribus electricitatis in motu musculari commentarius*, Ex typographia Instituti Scientiarum, Bologna,1791.

[GAL 94] GALVANI L., *Dell'uso e dell'attività dell'arco conduttore nelle contractioni dei muscoli*, S. Tommaso d'Aquino, Bologna, 1794.

[GAL 09] GALL F.J., SPURZHEIM G., Recherche sur le système nerveux en général et sur celui du cerveau en particulier, Mémoire présenté à l'Institut de France en 1808, Schoell et Nicolle, Paris, 1809.

[GAL 11] GALL F.J., SPURZHEIM G., *Des dispositions innées de l'âme et de l'esprit*, Schoell, Paris, 1811.

[GAL 41] GALVANI L., *Memorie sull'elettricità animale al celebre abate Lazzaro Spallanzani*, 1797, in GALVANI L. *Opere edite ed inedite ed. S. Gherardi*, Tip. di E. dall' Olmo, Bologna, 1841.

[GAL 69] GALL F.J., "Lettre ouverte au baron M.J.F de Retzer", in *Questions philosophiques, sociales et politiques, traitées d'après les principes de la physiologie du cerveau*, translated by FOSSATI J.L.A., Amyot, Paris, 1869.

[GAL 53] GALVANI L., *Commentary on the Effect of Electricity on Muscular Motion*, translated by GREEN R.M., Elizabeth Licht, Cambridge, Massachusetts, 1953.

[GAN 94] GANOT A., *Traité élémentaire de physique*, Librairie Hachette et Cie, Paris, 1894.

[GAR 73] GARDANE J.-J., *Gazette de la santé*, Thursday August 19, Ruault, Paris, 1773.

[GAR 93] GARDEY D., "Histoire de l'électricité en France – 1881–1918" *Flux*, vol. 12, pp. 55–57, 1993.

[GAR 13] GARDNER J., "A history of deep brain stimulation: Technological innovation and the role of clinical assessment tools", *Social Studies of Science*, vol. 43, no. 5, pp. 707–728, 2013.

[GAS 49] GASTAUT H., "L'activité électrique cérébrale en relation avec les grands problèmes psychologiques", *L'Année psychologique*, vol. 51, pp. 61–88, 1949.

[GAS 51a] GASTAUT H., "The electrical activity of the brain", *Annu. Rev. Physiol.*, vol. 13, pp. 297–326, 1951.

[GAS 51b] GASTAUT H., CORNIL L., CORRIOL J., "Evaluation of the degree of consciousness during epileptic petit mal paroxysms", *Rev. Neurol.*, Feb., vol. 84, no. 2, pp. 149–151, Paris, 1951.

[GAS 53] GASTAUT H., FERRER S., CASTELLS C., "Action de la diétliylamide de l'acide d-lysergique (LSD 25) sur les fonctions psychiques et l'électroencéphalogramme", *Confin. Neurol.*, pp. 102–120, 1953.

[GAS 54] GASTAUT H., "The brain stem and cerebral-electrogenesis in relation to consciousness", in ADRIAN E.D., BREMER F., JASPER H. (eds), *Brain Mechanisms and Consciousness,* Blackwell, Oxford, 1954.

[GAU 75] GAUTIER-DAGOTY J.-F., *Exposition anatomique des organes des sens, jointe à la névrologie entière du corps humain, et conjectures sur l'électricité animale et le siège de l'âme*, Demonville, Paris, 1775.

[GAU 97] GAUTHIER G., LARAT. J., *Revue international d'électrothérapie et de radiothérapie*, vols 2–3, pp. 241–266, 1897.

[GAZ 13] GAZZANIGA M.S., KALDY P., *Le libre arbitre et la science du cerveau*, Odile Jacob, Paris, 2013.

[GEI 02] GEIGER CH.-F., *Dissertation sur le galvanisme et son application*, Guilleminet, Paris, 1802.

[GEN 16] GENTILI M., "Syndrome post-traumatique lors de la première guerre mondiale et littérature", *Histoire des sciences médicales*, vol. 1, no. 2, pp. 171–174, 2016.

[GER 91] GERHARD R., "La méthode hallérienne en physiologie", *Dix-Huitième Siècle,* vol. 23, pp. 75–84, 1991.

[GIA 08] GIANNINI G., *De la nature des fièvres et de la meilleure méthode de les traiter*, Léopold Collin, Paris, 1808.

[GIB 41] GIBBS F.A., GIBBS E.L., *Atlas of Electroencephalography*, Lew A. Cummings, Cambridge, Massachusetts, 1941.

[GLO 69] GLOOR P., "Hans Berger on the electroencephalogram: The fourteen original reports on the human electroencephalogram", *Electrenceph. Clin. Neurophysiol. Suppl.*, vol. 28, 1969.

[GOD 20] GODWIN W., *Of Population: An Enquiry Concerning the Power of Increase in the Numbers of Mankind, Being an Answer to Mr. Malthus's Essay on that Subject*, Longman, Hurst, Rees, Orme and Brown, London, 1820.

[GOU 02] GOULET D., "La vigueur retrouvée : la promesse des ceintures électrique", *Cap-aux-Diamants*, vol. 69, pp. 33–36, 2002.

[GRA 31–32] GRAY S., "A letter to Cromwell Mortimer, containing several experiments concerning Electricity", *Philosophical Transactions of the Royal Society*, vol. 37, pp. 18–44, 1731–1732.

[GRA 36] GRAY E., "Détail de plusieurs expériences sur l'électricité, dictée le 14 février 1736, par M. Etienne Gray, de la Société royale, la veille de sa mort, à M. Cromwel Mortimer de la Société royale, pour être communiquées à la Société", *Transactions philosophiques de la Société royale de Londres*, vol. 1736, pp. 281–284, 1736.

[GRA 38a] GRAY E., "Dernière lettre de M. Etienne Gray de la Société royale, à M. Granville Wheler, écuyer, de la Société royale, sur les révolutions que de petits corps suspendus font, par, l'électricité, d'occident en orient autour d'autres corps plus considérables, de la même manière que les planètes tournant autour du soleil", *Transactions, philosophiques de la Société royale de Londres*, vol. 1736, pp. 54–55, 1738.

[GRA 38b] GRAY E., "Lettre de M. Etienne Gray, de la Société royale, à M. Cromwel Mortimer, docteur en médecine et secrétaire de la Société royale, contenant des expériences et des observations sur la lumière produite par la communication de l'électricité aux animaux et aux corps animés, et sur quelques effets très surprenant de cette lumière", *Transactions philosophiques de la Société royale de Londres*, vol. 1735, pp. 22–34, 1738.

[GRA 38c] GRAY E., "Lettre de M. Etienne Gray, de la Société royale, au docteur Mortimer, docteur en médecine et secrétaire de la Société royale, contenant des expériences sur l'électricité", *Transactions philosophiques de la Société royale de Londres*, vol. 1735, pp. 174–178, 1738.

[GRA 41a] GRAY E., "Lettre de Etienne Gray à M. Cromwel Mortimer, Secrétaire de la Société royale contenant de nouvelles expériences sur l'électricité", *Transactions philosophiques de la Société royale de Londres*, vol. 1732, pp. 226–238, 1741.

[GRA 41b] GRAY E., "Lettre de M. Etienne Gray à M. Cromwel Mortimer, docteur en médecine, secrétaire de la Société royale, contenant diverses expériences sur l'électricité", *Transactions philosophiques de la Société royale de Londres*, vol. 1731, pp. 23–55, 1741.

[GRA 41c] GRAY E., "Lettre de M. Gray à M. Mortimer, Secrétaire de la Société royale, sur l'électricité de l'eau, de Charter-House, le 20 janvier 1732", *Transactions philosophiques de la Société royale de Londres*, vol. 1732, pp. 7–12, 1741.

[GRA 41d] GRAY E., "Lettre de M. Gray à M. Mortimer, Secrétaire de la Société royale contenant la suite de ses expériences sur l'électricité, de Charter-House, le 7 juin 1732", *Transactions philosophiques de la Société royale de Londres*, vol. 1732, pp. 78–87, 1741.

[GRA 90] GRAY E., "Art. II. Effets de l'étincelle électrique sur des animaux et parties d'animaux", *Abrégé des transactions philosophiques de la Société royale de Londres*. vol. 2, pp. 313–314, 1790.

[GRA 01] GRAPENGIESSER J.C., "Recherches sur le galvanisme, et sur son, usage dans le traitement de certaines maladies", in BREWER DELAROCHE C.C. (ed.), *Essai sur le galvanisme*, extract of the VIIIth volume of the Bibliothèque germanique médico-chirurgicale, Volume 8, Croullebois, Fuchs, Paris, 1801.

[GRE 02] GREEN MUSSELMAN E., "The governor and the telegraph: Mental management in British natural philosophy" in RHYS MORUS I. (ed.), *Bodies/Machines*, Berg, Oxford, 2002.

[GRE 03] GREENBERG B.D., REZAI A.R., "Mechanisms and the current state of deep brain stimulation in neuropsychiatry", *CNS Spectrums*, vol. 8, no. 7, pp. 522–526, 2003.

[GRO 09] GRONCHI-PERRIN A., VINGERHOETS F., "Effets cognitivo-comportementaux de la stimulation cérébrale profonde dans la maladie de Parkinson", *Revue de neuropsychologie*, vol. 1, pp. 59–63, 2009.

[GUC 05] GUCHET X., *Les sens de l'évolution technique*, Léo Scheer, Paris, 2005.

[GUE 09] GUEDRON M., "L'imaginaire des fluides au XVIIIe siècle", *Sociétés & Représentations*, vol. 2, no. 28, p. 173–186, 2009.

[GUI 54] GUITARD M.I., *Histoire de l'électricité médicale, comprenant l'étude des instruments et appareils, le résumé des auteurs, un choix d'observations*, Masson, Paris, 1854.

[GUI 77] GUICHARD A., *Historique de l'électricité animale, leçon faite à la séance de rentrée, le 6 novembre 1877*, Lachèse, Belleuvre et Dolbeau, Angers, 1877.

[HAC 98] HACKING I., *L'âme réécrite : étude sur la personnalité multiple et les sciences de la mémoire*, Les empêcheurs de tourner en rond, Paris, 1998.

[HAL 55] VON HALLER A., *Dissertation sur les parties irritables et sensibles des animaux*, translated by TISSOT M., Marc-Michel Bousquet et comp., Lausanne, 1755.

[HAL 92] VON HALLER A., *Mémoire sur la sensibilité, mémoire sur l'irritabilité : réponse à quelques objections*, Masson, Paris, 1892.

[HAL 07] HALLIDAY S., *Science and Technology in the Age of Hawthorne, Melville, Twain and James: Thinking and Writing Electricity*, Palgrave, New York, 2007.

[HAM 14] HAMANI C., PILITSIS J., RUGHANI A.I. *et al.*, "Deep brain stimulation for obsessive-compulsive disorder: Systematic review and evidence-based guideline sponsored by the American Society for Stereotactic and Functional Neurosurgery and the Congress of Neurological Surgeons (CNS) and endorsed by the CNS and American Association of Neurological Surgeons", *Neurosurgery*, vol. 75, pp. 327–333, 2014.

[HAN 92] HANSEN B., "American physicians' 'discovery' of homosexuals, 1880–1900: A new diagnosis in a changing society", in ROSENBERG C. and GOLDEN J. (eds), *Framing Disease: Studies in Cultural History*, Rutgers University Press, New Brunswick, 1992.

[HAR 10] HARIZ M.I., BLOMSTEDT P., ZRINZO L., "Deep brain stimulation between 1947 and 1987: The untold story"; *Neurosurg Focus*, vol. 29, pp. 1–11, 2010.

[HAS 60] HASSLER R., RIECHERT T., MUNDINGER F. *et al.*, "Physiological observations in stereotaxic operations in extrapyramidal motor disturbances", *Brain*, vol. 83, pp. 337–350, 1960.

[HEA 53] HEATH R.G., PEACOCK S.M., MILLER W., "Induced paroxysmal electrical activity in man recorded simultaneously through subcortical and scalp electrodes", *Trans. Am. Neurol. Assoc.*, vol. 3, pp. 247–250, 1953.

[HEA 54a] HEATH R.G., *Studies in Schizophrenia*, Harvard University Press, Cambridge, 1954.

[HEA 54b] HEATH R.G., GALLANT D.M., "Activity of the human brain during emotional thought", in *The Role of Pleasure in Behavior*, Hoeber Medical Division, Harper & Row, New York, 1954.

[HEA 57] HEATH R.G., MONROE R.R., MICKLE W.A. *et al.*, "Correlation of rhinencephalic electrograms with behavior", *EEG Clin. Neurophysiol.*, vol. 9, pp. 623–643, 1957.

[HEA 58] HEATH R.G., "Correlation of electrical recordings from cortical and subcortical regions of the brain with abnormal behavior in human subjects", *Confin. Neurol.*, vol. 18, pp. 305–315, 1958.

[HEA 80] HEATH R.G., LLEWELLYN R.C., ROUCHELL A.M., "The cerebellar pacemaker for intractable behavioral disorders and epilepsy: Follow-up report", *Biol. Psychiatry*, vol. 15, no. 2, pp. 243–256, 1980.

[HEA 96] HEATH R.G., *Exploring the Mind–Brain Relationship*, Moran Printing, Inc., Baton Rouge, Los Angeles, 1996.

[HEL 59] HELVETIUS C.-A., *De l'esprit*, Durand [Bassompierre], Paris, 1759.

[HEL 50] HELMHOLTZ H., "Note sur la vitesse de propagation de l'agent nerveux dans les nerfs rachidiens", *C. R. Acad. Sci.*, vol. 30, pp. 204–206, 1850.

[HEL 51] HELMHOLTZ H., "Deuxième note sur la vitesse de propagation de l'agent nerveux", *C. R. Acad. Sci.*, 32, pp. 262–265, 1851.

[HIP 03] HIPPOCRATES, *De la maladie sacrée*, volume II, part 3, chapter 20, Les Belles Lettres, Paris, 2003.

[HOD 52] HODGKIN A.L., HUXLEY A.F., "A quantitative description of membrane current and its application to conduction and excitation in nerve", *J. Physiol.*, vol. 117, pp. 500–544, 1952.

[HOE 65] HOEFER F., *La chimie enseignée par la biographie de ses fondateurs : R. Boyle, Lavoisier, Priestley, Scheele, Davy, etc.*, Hachette, Paris, 1865.

[HOL 70] D'HOLBACH P.T., *Système de la nature ou des loix du monde physique & du monde moral*, Marc-Michel Rey, Amsterdam, 1770.

[HOR 84] HORSLEY V., "Case of occipital encephalocele in which a correct diagnosis was obtained by means of the induced current", *Brain*, vol. 7, pp. 228–243, 1884.

[HOR 87] HORSLEY V., "Remarks on ten consecutive cases of operations upon the brain and cranial cavity to illustrate the details and safety of the method used", *Br Med J.*, vol. 1, pp. 863–865, 1887.

[HOR 90] HORSLEY V., "Remarks on the surgery of the central nervous system", *Br. Med. J.*, vol. 2, pp. 1286–1292, 1890.

[HOT 15] HOTTOIS G., MISSA J.-N., PERBAL L., *Encyclopédie du transhumanisme et du posthumanisme, L'humain et ses prefixes*, Vrin, Paris, 2015.

[HUM 99] VON HUMBOLDT A., *Expériences sur le galvanisme et en général sur l'irritation des fibres musculaires et nerveuses*, translated by JADELOT J.F.N, Didot jeune, J. -F. Fuchs, Paris, 1799.

[HUR 16] HURREN E.T., *Dissecting the Criminal Corpse: Staging Post-Execution Punishment in Early Modern England*, Palgrave Macmillan, London, 2016.

[IMB 06] IMBERT M., *Traité du cerveau*, Odile Jacob, Paris, 2006.

[ING 85] INGENHOUSZ J., *Nouvelles expériences et Observations sur divers objets de physique*, chez P. Théophile Barrois le jeune, Paris, 1785.

[JAC 84] JACKSON J.H., "The Croonian Lectures", *Royal College of Physicians*, pp. 660–663, 1884.

[JAC 90] JACKSON J. H., "The Lumleian Lectures", *The British Medical Journal*, pp. 820–827, 1890.

[JAC 21] JACKSON J.H., "Sur l'évolution et la dissolution du système nerveux", *Archives suisses de neurologie et de psychiatrie*, 1884, vol. VIII, 1921.

[JAC 31–32] JACKSON J.H., *Selected Writings of John Hughlings Jackson*, Volume 2, Hodder & Stoughton, London, 1931–1932.

[JAL 48] JALLABERT J., *Expériences sur l'électricité, avec quelques conjectures sur la cause de ses effets*, Barillot & fils, Geneva, 1748.

[JAN 08–11] JANET P., RAYMOND F., *Les obsessions et la psychasthénie*, Volume 2, Alcan, Paris, 1908–1911.

[JAS 90] JASPERS K., *Strindberg et Van Gogh, Swedenborg-Hölderlin : étude psychiatrique comparative*, Editions de minuit, Paris, 1990.

[JEN 98] JENIKE M.A., "Neurosurgical treatment of obsessive-compulsive disorder", *Br. J. Psychiatry*, vol. 35 (suppl), pp. 79–90, 1998.

[JOB 58] JOBERT DE LAMBALLE A.J., *Des appareils électriques : des poissons électriques*, Baillière, Mallet-Bachelier, Paris, 1858.

[JON 17] JONES K.T., PETERSON D.J., BLACKER K.J. *et al.*, "Frontoparietal neurostimulation modulates working memory training benefits and oscillatory synchronization", *Brain Res.*, vol. 1967, pp. 1–29, 2017.

[KEE 79] KEEL O., *La généalogie de l'histopathologie, une révision déchirante : Philippe Pinel, lecteur discret de J.C Smyth,* Vrin, Paris, 1979.

[KEE 01] KEEL O., *L'avènement de la médecine clinique moderne en Europe : 1750–1815 : politiques, institutions et savoirs*, Presse de l'Université de Montréal, Montreal, 2001.

[KEV 85] KEVORKIAN J., "A brief history of experimentation on condemned and executed humans", *Journal of the National Medical Association*, vol. 77, no. 3, pp. 215–226, 1985.

[KIL 66] KILLAM K.F., KILLAM E.K., NAQUET R., "Study of responses evoked by intermittent light stimulation in monkeys presenting paroxysmal responses to this type of stimulation", *Revue Neurologique*, vol. 115, pp. 422–423, 1966.

[KIM 99] KIM E.E., TOOLE B.A., "Ada and the first computer: The collaboration between Ada, countess of Lovelace, and computer pioneer Charles Babbage resulted in a landmark publication that described how to program the world's first computer", *Scientific American*, pp. 76–81, May 1999.

[KIR 12] KIRKLAND K., *Neuroelectricity: The Past, Present, and Future of Brain Stimulation,* Neuroniverse, 2012.

[KIT 88] KITE C., *An Essay on the Recovery of the Apparently Dead,* C. Dilly in the poultry, London, 1788.

[KRA 46] KRATSENSTEIN C.-T., *Lettres d'un physicien sur l'avantage dont l'électricité peut être pour l'art de guérir*, Halle, 1746.

[KRA 12] KRAUSE F., *Chirurgie du cerveau et de la moelle épinière*, translated by BOURGUET, F., Gittler, Paris, 1912.

[KRA 34] KRAUSE F., *Surgical Operation of the Head*, Allied Book Company, New York, 1934.

[KRE 38] KREEZER G., "The electro-encephalogram and its use in psychology", *The American Journal of Psychology*, vol. 51, no. 4, pp. 737–759, 1938.

[LAB 28] LA BEAUME M., *Du Galvanisme appliqué à la médecine et de son efficacité dans le traitement des affections nerveuses de l'asthme, des paralysies, des douleurs rhumatismales, des maladies chroniques en général et particulièrement des maladies chroniques de l'estomac, des intestins, du foie, etc.*, Selligue, Béchet, Paris, 1828.

[LAM 65] LA METTRIE J.O., *L'homme machine*, Frederic Henry, Paris, 1865.

[LAP 38] LAPICQUE L., *La chronaxie et ses applications physiologiques*, Hermann, Paris, 1938.

[LAP 43] LAPICQUE L., *La machine nerveuse,* Flammarion, Paris, 1943.

[LAP 70] LAPLASSOTTE F., "Quelques étapes de la physiologie du cerveau du XVIIe au début du XIXe siècle", *Annales : Economies, sociétés, civilisations*, vol. 3, pp. 599–613, 1970.

[LAY 40a] LAYCOCK T., *An Essay on Hysteria: Being an Analysis of Its Irregular and Aggravated Forms: Including Hysterical Hemorrhage, and Hysterical Ischuria: With Numerous Illustrative Cases,* Haswell, Barrington and Haswell, Philadelphia, 1840.

[LAY 40b] LAYCOCK T., *A Treatise on the Nervous Disorders of Women: Comprising an Inquiry into the Nature, Causes, and Treatment of Spinal and Hysterical Disorders*, Longman, Orme, Brown, Green, and Longmans, London, 1840.

[LAY 50] LAYCOCK T., "On the treatment of cerebral hysteria and of moral insanity in women by electro-galvanism", *The Medical Times*, vol. 22, pp. 57–58, 1850.

[LAY 63] LAYCOCK T., "On the Naming and Classification of Mental Diseases and Defects", *The Journal of Mental Science,* vol. 46., no. IX, pp. 153–170, 1863.

[LEB 68] LEBRETON P.-A., *Des Différentes variétés de la paralysie hystérique*, Delahaye, Paris, 1868.

[LED 83] LEDRU N.-P., Rapport de MM. Cosnier, Maloet, Darcet Philip, Le Preux, Desessartz et Paulet, docteurs-régens de la faculté de médecine de Paris, sur les avantages reconnus de la nouvelle méthode d'administrer l'électricité dans les maladies nerveuses, particulièrement dans l'épilepsie et dans la catalepsie : par Nic. Ph. Ledru, connu sous le nom de Comus. Ce rapport est précédé de l'aperçu du système de l'auteur sur l'agent qu'il emploie, et des avantages qu'il en a tirés, Imprimerie Philippe-Denys Pierre, Paris, 1783.

[LEV 14] LEVEQUE M., "Le renouveau d'une discipline controversée : la psychochirurgie", *PSN*, vol. 2, no. 12, pp. 19–35, 2014.

[LEF 03] LEFF A., "Thomas Laycock and the romantic genesis of the cerebral reflex", *ACNR*, vol. 3, no. 1, pp. 26–27, 2003.

[LEJ 18] LE JEUNE K., "Mise en perspective historique des liens entre épilepsie, neurologie et psychiatrie, période XIXe-XXIe siècles", *Bulletin d'Epistémologie et d'histoire des sciences de la vie*, vol. 25, no. 2, pp. 121–136, 2018.

[LEN 86] LENOIR T., "Models and instruments in the development of electrophysiology, 1845–1912", *Historical Studies in the Physical and Biological Sciences*, vol. 17, no. 1, pp. 1–54, 1986.

[LEW 19] LEWIS A., *The Brontës and the Idea of the Human: Science, Ethics, and the Victorian Imagination*, Cambridge University Press, United Kingdom, 2019.

[LLE 17] LLEDO P.-M., *Le cerveau, la machine et l'humain*, Odile Jacob, Paris, 2017.

[LIP 13] LIPSMAN N., GLANNON W., "Brain, mind and machine: What are the implications of deep brain stimulation for perceptions of personal identity, agency and free will?", *Bioethics*, vol. 27, no. 9, pp. 465–470, 2013.

[LIT 12] LITTLEFIELD M.M., JOHNSON M.J., *The Neuroscientific Turn: Transdiplinarity in the Age of the Brain,* University of the Michigan Press, USA, 2012.

[LOB 59] LOBB H., *On the Curative Treatment of Paralysis and Neuralgia, and Other Affections of the Nervous System With the Aid of Galvanism*, Baillière, London, 1859.

[LOE 99] LOEB L., "Consumerism and commercial electrotherapy: The medical battery company in nineteenth-century London", *Journal of Victorian Culture*, vol. 4, no. 2, pp. 252–275, 1999.

[LOU 47] LOUIS A., *Observations sur l'électricité, Où l'on tâche d'expliquer son mécanisme et les Effets sur l'économie animale ; avec des remarques sur son usage*, Osmont & Delaguette, Paris, 1747.

[LOZ 08] LOZANO A.M., MAYBERG H.S. *et al.*, "Subcallosal cingulate gyrus deep brain stimulation for treatment-resistant depression", *Biological Psychiatry*, vol. 64, no. 6, pp. 461–467, 2008.

[LUC 17] LUCAS K., *The Conduction of Nervous Impulse*, Longmans Green, London, 1917.

[MAC 31] MACERONE F., "An account of some remarkable electrical phenomena seen in the Mediterranean, with some physiological deductions", *Mechanics' Magazine*, vol. 15, pp. 93–100, 1831.

[MAG 16] MAGENDIE F., *Précis élémentaire de physiologie*, Méquignon-Marvis, Paris, 1816.

[MAL 76] MALACARNE V., *Nuova esposizione della vera struttura del cerveletto umano*, G. Briolo, Turin, 1776.

[MAL 80] MALACARNE V., *Encefalotomia nuova universale*, G. Briolo, Turin, 1780.

[MAL 92] MALACARNE V., "Lettre n. 1929–19130, datée du 29 août 1792 et adressée à Charles Dénina, Fonds Malacarne", *Académie des sciences de Turin*, 1792.

[MAL 98] MALACARNE V., "Dell'esistenza e dell'influenza di diversi sistemi nell'economia animale : Divisione generale de'sistemi, e meravigliosa estensione del sistema cutaneo" in BRERA L.V. (ed.), I commentari medici, Callazzi, Pavia, 1798.

[MAL 99] MALACARNE V., "Dell'esistenza di diversi altri sistemi nell'economia animale, dedotte specialmente da nuove osservazioni patologiche" in BRERA L.V. (ed.), *I commentari medici,* Callazzi, Pavia, 1799.

[MAL 03] MALACARNE V., *Qu'elles sont les influences sympathiques qu'exercent réciproquement, les uns sur les autres, les divers systèmes et organes de l'économie vivante* ?, Mémoire de la Société Médicale d'Emulation de Paris, Volume 5, Veuve Richard, Paris, 1803.

[MAL 08] MALACARNE V., "Conoscendo dalla organizzazione del cervelletto in ispezie, e forse anche da più attento esame del cervello e dalla midolla spinale che queste viscere formano qualche cosa di somigliante alla colonna galvanica del Volta, cercare per mezzi di sperimenti se fossero mai la sorgente principale del fluido galvanico animator materiale della macchina animale", *Giornale della società di incoraggiamento delle scienze e delle arti di Milano*, vol. 4, pp. 122–130, 1808.

[MAN 52] MANGIN ABBE DE, GUER J.A., *Histoire générale et particulière de l'électricité, ou ce qu'en ont dit de curieux et d'amusant, d'utile et d'intéressant, e réjouissant et de badin, quelques physiciens d'Europe*, Rollin, Paris, 1752.

[MAN 63] MANGIN A., *Le feu du ciel. Histoire de l'électricité et de ses principales applications: idées des anciens, premières observations, machine électrique, bouteille de leyde, paratonnerre, phénomènes naturels,* Mame et cie, Tours, 1863.

[MAN 72] MANNELLI M.A., "Cenni sui contribuiti italiani alla neurofisiologia, Da Leonardo da Vinci a Camillo Golgi", *Episteme, rivista critica di storia delle scienze medicine e biologiche, Grafiche Signorelli*, vol. 1, pp. 196–232, January–March 1972.

[MAR 73] MAREY E.-J., *La machine animale, locomotion terrestre et aérienne*, G. Baillière, Paris, 1873.

[MAR 73] MARRIGUES A., *Suite de la guérison de la paralysie par l'électricité, d'après la méthode de l'abbé Sans*, André-Charles Cailleau, Paris, 1773.

[MAR 78] MAREY E.-J., *La méthode graphique dans les sciences expérimentales et principalement en physiologie et en médecine*, Masson, Paris, 1878.

[MAR 87] MARKS I., *Fears, Phobias, and Rituals: Panic, Anxiety, and Their Disorders*, Oxford University Press, USA, 1987.

[MAR 88] MARKS I., *Cure and Care of Neuroses,* Amer Psychiatric. Pub. Inc., Washington, DC, 1988.

[MAR 18] MARCIL Y., "Entre voyage savant et campagne médicale : le séjour en Italie de La Condamine (1755–1756)", *Diciottesimo Secolo*, anno III, pp. 23–46, 2018.

[MAS 80] MASARS DE CAZELES F., *Mémoire sur l'électricité médicale et l'histoire du traitement de vingt-et-un malades traités*, Mequignon l'aîné, Paris, 1780.

[MAT 40] MATTEUCCI C., *Essai sur les phénomènes électriques*, Carilliau, Goeury et Dalmont, Paris, 1840.

[MAT 42] MATTEUCCI C., "Deuxième mémoire sur le courant électrique propre de la grenouille et sur celui des animaux à sang chaud", *Ann. Chim. Phys.*, vol. 6, no. 3, pp. 301–339, 1842.

[MAT 44] MATTEUCCI C., *Traité des phénomènes électro-physiologiques des animaux*, Fortin, Masson et Cie, Paris, 1844.

[MAU 79] MAUDUYT DE LA VARENNE P.-J.-C., *Mémoire sur le traitement électrique appliqué à 82 malades*, Imprimerie Royale, Paris, 1779.

[MAU 81] MAUDUYT DE LA VARENNE P.-J.-C., *Avis sur l'électricité médicale*, l'imprimerie de Paris, imprimeur ordinaire du Roi, de la Société Royale de Médecine, Paris, 1781.

[MAU 84] MAUDUYT DE LA VARENNE P.-J.-C., *Mémoire sur les différentes manières d'administrer l'électricité, et observations sur les effets qu'elles ont produits*, Imprimerie Royale, Paris, 1784.

[MAY 03] MAYBERG H., "Positron emission tomography imaging in depression: A neural systems perspective", *Neuroimaging Clin. N. Am.*, vol. 13, pp. 805–815, 2003.

[MAY 05] MAYBERG H., LOZANO A.M., VOON V. *et al.*, "Deep brain stimulation for treatment-resistant depression", *Neuron*, vol. 45, pp. 651–660, 2005.

[MAY 09] MAYBERG H., "Targeted electrode-based modulation of neural circuits for depression", *J. Clin. Invest.*, vol. 119, pp. 717–725, 2009.

[MAZ 90] MAZZOLINI R., "Schémas et modèles de la machine pensante (1662–1762)", in CORSI P., *La fabrique de la pensée*, Electa, Milan, 1990.

[MER 08] MERMILLOD M., GALLAND F., MONDILLON L. *et al.*, "Troubles psychiatriques et stimulation cérébrale profonde : perspectives de recherche clinique et fondamentale", in S. CAMPANELLA *et al.* (eds), *Psychopathologie et neurosciences,* pp. 229–261, Brussels, 2008.

[MIC 13] MICOULAUD-FRANCHI J.-A., QUILES C., VION-DURY J., "Eléments pour une histoire de l'électricité et du cerveau en psychiatrie. Naissance et développement de la stimulation et de l'enregistrement électrique en neurophysiologie", *Annales Médico-Psychologiques,* vol. 171, pp. 318–328, 2013.

[MIL 48] MILLINGEN J.G., *The Passions or, Mind and Matter*, J. and D.A. Darling, London, 1848.

[MIL 91] MILANESI C., "La mort-instant et la mort-processus dans la médecine de la seconde moitié du siècle", *Dix-huitième Siècle*, vol. 23, pp. 171–190, 1991.

[MIN 09] MINOT J.-F., "Histoire du bistouri électrique chirurgical en France", *La Lettre des Amis du Musée de l'AP-HP*, vol. 14, pp. 42–45, June 2009.

[MIO 13] MIOCINOVIC S., SOMAYAJULA S., CHITNIS S., "History, applications, and mechanisms of deep brain stimulation", *Neurol.*, vol. 70, no. 2, pp. 163–171, 2013.

[MIT 72] MITCHELL S.W., *Injuries of Nerves and Their Consequences,* J.P. Lippencott, Philadelphia, 1872.

[MOD 10] MODOLO J., EDWARDS R., CAMPAGNAUD J., "Past, present and future of brain stimulation", *Math. Model. Nat. Phenom.*, vol. 5, no. 2, pp. 185–207, 2010.

[MON 36] MONIZ E., *Tentatives opératoires dans le traitement de certaines psychoses*, Masson, Paris, 1936.

[MON 90] MONTI M.T., *Congettura ed esperienza nella fisiologia di Haller. La riforma dell'anatomia animata e il sistema della generazione*, Olschki, Florence, 1990.

[MOR 48] MORIN J., *Nouvelle dissertation sur l'électricité des corps*, Veuve Estienne et Fils, Paris, 1748.

[MOR 61] MORGAGNI G.-B., *De sedibus et causis morborum*, Volume 2, Venise, 1761.

[MOR 37–39] MORGAGNI G.-B., *Recherche anatomiques sur le siège et les causes des maladies*, translated by DESTOUET D. Paris, 1837–1839.

[MOR 49] MORUZZI G., MAGOUN H.W., "Brain stem reticular formation and activation of the EEG", *Electroencephalography and Clinical Neurophysiology*, vol. 1, no. 4, pp. 455–473, 1949.

[MOR 11] MORLACCHI P., NELSON R.R., "How medical practice evolves: Learning to treat failing hearts with an implantable device", *Research Policy*, vol. 40, no. 4, pp. 511–525, 2011.

[MOS 25] MOST J. F., *La Guérison de l'épilepsie par un nouveau procédé, puissant, efficace, et peu coûteux : appuyée par de nombreux exemples*, Rey et Gravier, Paris, 1825.

[MOU 08] MOUTAUD B., "C'est un problème neurologique ou psychiatrique ? Anthropologie de la neuro-stimulation intracrânienne appliquée à des troubles psychiatriques", *Les Cahiers du Centre Georges Canguilhem*, vol. 1, no. 2, pp. 175–214, 2008.

[MOU 14] MOUTAUD B., CHAMAK B., *Neurosciences et société : enjeux des savoirs et pratiques sur le cerveau*, Armand Colin, Paris, 2014.

[MOU 18a] MOUTAUD B., HAXIAIRE C., FARNARIER C., *L'innovation en santé : technologies, organisations, changements*, Presses universitaires de Rennes, Rennes, 2018.

[MOU 18b] MOUTAUD B., ARANZAZU A., "À la croisée de tant de mondes. Quelle(s) histoire(s) pour la stimulation cérébrale profonde ?", *Bulletin d'Epistémologie et d'histoire des sciences de la vie*, vol. 25, no. 2, pp. 203–220, 2018.

[MUL 59] MULLAN S., "Illusions of comparative interpretation and emotion production by electric discharge and by electric stimulation in the cerebral cortex", *American Medical Association Archives of Neurology and Psychiatry*, vol. 81, no. 3, pp. 269–284, 1959.

[MUK 08] MUKHIDA K., BISHOP M., HONG M. *et al.*, "Neurosurgical strategies for Gilles de la Tourette's syndrome", *Neuropsychiatric Disease and Treatment*, vol. 4, no. 6, pp. 1111–1128, 2008.

[MUN 53a] MUNDY-CASTLE A.C., "The clinical significance of electroencephalography", *South African Medical Journal*, vol. 27, no. 15, pp. 298–310, 1953.

[MUN 53b] MUNDY-CASTLE A.C., "Electrical responses of the brain in relation to behavior", *British Journal of Psychology*, vol. 44, pp. 318–329, 1953.

[MUN 54] MUNDY-CASTLE A.C., HURST L.A., BEERSTECHER D.M., "The electroencephalogram in manic-depressive psychosis", *J. Ment. Sci.*, Jan., vol. 100, no. 418, pp. 220–240, 1954.

[MUS 69] MUSSCHENBROEK VAN P., *Cours de physique expérimentale et mathématique*, translated by SIGAUD DE LA FOND M., Briasson, Paris, 1769.

[NAU 03] NAUCHE J., Journal du galvanisme, de vaccine, etc., F. Buisson, Paris, 1803.

[NEU 81] NEUBURGER M., *The Historical Development of Experimental Brain and Spinal Cord Physiology before Flourens,* CLARKE E., (ed.), The Johns Hopkins University Press, Baltimore, 1981.

[NIC 86] NICOLAS M., "Expériences faites sur quelques animaux tombés dans l'asphixie occasionnée par les vapeurs du charbon", *Observations sur la physique, sur l'histoire naturelle et sur les arts*, vol. 1, no. 29, p. 231, 1786.

[NOB 28a] NOBILI L., "Sur la nature des courants électriques", *Bibliothèque universelle des sciences, belles lettres et arts*, pp. 118–144, 180–184, 1828.

[NOB 28b] NOBILI L., "Comparaison entre les deux galvanomètres les plus sensibles, la grenouille et le multiplicateur à deux aiguilles, suivie de quelques résultats nouveaux", *Annales de chimie et de physique*, pp. 225–245, 1828.

[NOL 38] NOLLET J.-A., *Programme ou idée générale d'un cours de physique expérimentale, avec un catalogue raisonné des instrumens qui servent aux expériences*, P.G. Le Mercier, Paris, 1738.

[NOL 46] NOLLET J.-A., *Essai sur l'électricité des corps*, Frères Guerin, Paris, 1746.

[NOL 48] NOLLET J.-A., *Recherches sur les causes particulières des phénomènes électriques et sur les effets nuisibles ou avantageux qu'on peut en attendre*, Frères Guerin, Paris, 1748.

[NOL 53–57] NOLLET J.-A., *Lettres sur l'électricité. Dans lesquelles on examine les dernières découvertes qui ont été faites sur cette matière, les conséquences que l'on peut en tirer*, Hippolythe-Louis Guerin & Louis-François Delatour, Paris, vol. 3, 1753–1757.

[NUN 53] NUNN T.W., *Inflammation of the Breast, and Milk Abscess,* William Brinton, London, 1853.

[NUN 86] NUNN T.W., *A Page in the History of Ovariotomy in London*, Macdonald, London, 1886.

[NUT 99] NUTTIN B., COSYNS P., DEMEULEMEESTER H. *et al.*, "Electrical stimulation in anterior limbs of internal capsules in patients with obsessive compulsive disorder", *The Lancet*, vol. 354, no. 9189, p. 1526, 1999.

[NYE 90] NYE D.E., *Electrifying America: Social Meanings of a New Technology*, MIT Press, Cambridge, 1990.

[NYS 02] NYSTEN P.-H., *Nouvelles expériences galvaniques : faites sur les organes musculaires de l'homme et des animaux à sang rouge : dans lesquelles, en classant ces divers organes sous le rapport de la durée de leur excitabilité galvanique, on prouve que le cœur est celui qui conserve le plus longtemps cette propriété*, Levrault, Paris, 1802.

[ØRS 54] ØRSTED H.-C., *Der Geist in der Natur,* Carl B. Lorck, Leipzig, 1854.

[PAR 41] PARCHAPPE DE VINAY M., *Traité théorique et pratique de la folie : observations particulières et documens nécroscopiques*, Béchet jeune et Labé, Paris, 1841.

[PAL 47] PALLAS E., *De l'influence de l'électricité atmosphérique et terrestre sur l'organisme, et de l'effet de l'isolement électrique considéré comme moyen curatif et préservatif d'un grand nombre de maladies*, Masson, Paris, 1847.

[PAR 58] PARENTE G., *Origini e vicende ecclesiastiche della città di Aversa*, Cardamone, Naples, 1858.

[PAR 04] PARENT A., "Giovanni Aldini: From animal electricity to human brain stimulation", *Can. J. Neurol. Sci.*, vol. 31, no. 4, pp. 576–584, 2004.

[PAU 10] PAULA GRUITHUISEN VON F., *Anthropologie oder von der Natur des menschlichen Lebens und Denkens : für angehende Philosophen und Aerzte*, Lentner, Munich, 1810.

[PAY 65] PAYEN J., "Les origines de la bobine d'induction", *Revue d'histoire des sciences et de leurs applications*, vol. 18, no. 3, pp. 315–318, 1965.

[PEN 35] PENFIELD W., EVANS J., "The frontal lobe in man: A clinical study of maximum removals", *Brain*, vol. 58, pp. 115–133, 1935.

[PEN 37] PENFIELD W., BOLDREY E., "Motor and sensory representation in the cerebral cortex of man as studied by electrical stimulation", *Brain*, vol. 60, no. 4, pp. 389–443, 1937.

[PEN 40] PENFIELD W., HEBB D.O., "Human behavior after extensive bilateral removal from the frontal lobes", *Archives of Neurology and Psychiatry*, vol. 44, pp. 421–438, 1940.

[PEN 50] PENFIELD W., RASMUSSEN T., *The Cerebral Cortex of Man*, The Macmillan Company, New York, 1950.

[PEN 52] PENFIELD W., JASPER H., FEINDEL W., "Localization of epileptic discharge in temporal lobe automatism", *Transactions of the American Neurological Association*, vol. 77, pp. 14–17, 1952.

[PEN 54] PENFIELD W., JASPER H., *Epilepsy and the Functional Anatomy of the Brain*, Little, Brown & Co, Boston, 1954.

[PEN 75] PENFIELD W., *Mystery of the Mind: A Critical Study of Consciousness and the Human Brain*, Princeton University Press, New Jersey, 1975.

[PEN 77] PENFIELD W., *No Man Alone: A Surgeon's Life,* Little, Brown & Co, Boston, 1977.

[PEN 59] PENFIELD W., "The interpretive cortex: The stream of consciousness in the human brain can be electrically reactivated", Science, vol. 129, pp. 1719–1725, 1959.

[PER 91] PERA M., *The Ambiguous Frog: The Galvani–Volta Controversy on Animal Electricity,* Princeton University Press, Princeton, 1991.

[PET 87] PETETIN J.-H.-D., *Mémoire sur la découverte des phénomènes que présentent la catalepsie et le somnambulisme, symptômes de l'affection hystérique essentielle*, Lyon, 1787.

[PET 02–03] PETETIN J.-H.-D., *Théorie du galvanisme, ses rapports avec le nouveau mécanisme de l'électricité,* Brunot, Reymann, Paris, Lyon, 1802–1803.

[PHI 83] PHILIP A., Rapport sur les avantages reconnus de la nouvelle méthode d'administrer l'électricité dans les maladies nerveuses particulièrement dans l'épilepsie et la catalepsie par M. Ledru, Ph.-Den. Pierres, Paris, 1783.

[PIC 98] PICCOLINO M., "Animal electricity and the birth of electrophysiology: The legacy of Luigi Galvani", *Brain Research Bulletin*, vol. 46, no. 5, pp. 381–407, 1998.

[PIC 03] PICCOLINO M., BRESADOLA M., *Rane, torpedini e scintille : Galvani, Volta e l'elettricità animale*, Bollati Boringhieri, Turin, 2003.

[PIC 10] PICAUD R., *Les machines à électricité de l'abbé Nollet*, Univ. Européenne, Paris, 2010.

[PIC 13] PICCOLINO M., BRESADOLA M., *Shocking Frogs: Galvani, Volta, and the Electric Origins of Neuroscience*, Oxford University Press, New York, 2013.

[PID 10] PIDOUX V., "Expérimentation et clinique électroencéphalographiques entre physiologie, neurologie et psychiatrie (Suisse, 1935–1965)", *Revue d'histoire des sciences,* vol. 2, no. 63, pp. 439–472, 2010.

[PIG 39] PIGEAIRE J., *Puissance de l'électricité animale, ou du magnétisme vital et de ses rapports avec la physique, la physiologie et la médecine*, Ballière, Paris, 1839.

[PIN 07] PINEL P., *Nosographie philosophique*, Brosson, Paris, 1807.

[PIV 47] PIVATI J.F., *Lettere della elletricita medica*, Venezia, 1747.

[POE 44] POE E.A., "The premature burial", *The Philadelphia Dollar Newspaper*, July, 1844.

[POE 45] POE E.A., "Some words with a mummy", *American Review*, April, 1845.

[POE 48] POE E.A., *Eureka : A Prose Poem, or An Essay on the Material and Spiritual Universe*, Wiley & Putnam, New York, 1848.

[POO 54] POOL J.L., "Psychosugery of older people", *J. Geriatr. Assoc.*, vol. 2, pp. 456–465, 1954.

[PRA 13] PRAVDICH-NEMINSKY V., "Ein Versuch der Registrierung der elektrischen Gehirnerscheinungen", *Zentralblatt für Physiologie*, vol. 27, pp. 951–960, 1913.

[PRE 71] PREMUDA L., *Metodo e conoscenza da Ippocrate ai nostri giorni*, CEDAM, Padua, 1971.

[PRI 67] PRIESTLEY J., *The History and Present State of Electricity, with Original Experiments*, J. Dodsley, London, 1767.

[PRI 71] PRIESTLEY J., *Histoire de l'électricité*, Herissant, Paris, 1771.

[PRO 09] PROUST J., "Le contrôle de soi : vers un homme nouveau ?", *Le Débat,* vol. 5, no. 157, pp. 124–143, 2009.

[QUE 12] QUETEL C., *Histoire de la folie : De l'Antiquité à nos jours,* Tallandier, Paris, 2012.

[RAM 09–11] RAMÓN Y CAJAL S., *Histologie du système nerveux de l'homme et des vertébrés*, Maloine, Paris, 1909–1911.

[RAM 96] RAMÓN Y CAJAL S., *Recollections of My Life*, translated by HORNE CRAIGIE E., MIT Press, Cambridge, Available at: books.google.co.uk/books/about/Recollections_of_My_Life.html?id=fzVQZekeA4kC&printsec=frontcover&source=kp_read_button&redir_esc=y#v=onepage&q&f=false, 1996.

[RAS 10] RASMUSSEN A., "L'électrothérapie en guerre : pratiques et débats en France (1914–1920)", *Annales historiques de l'électricité*, vol. 1, no. 8, pp. 73–91, 2010.

[REI 07a] REIL J.C., "Fragmente über die Bildung des kleinen Gehirnes im Menschen", *Archiv für die Physiologie*, vol. 8, pp. 1–58, 1807.

[REI 07b] REIL J.C., "Üeber die Eigenschaften des Ganglien systems und sein Verhältnis zum cerebral en Syste", *Archiv für die Physiologie*, vol. 7, no. 2, pp. 189–254, 1807.

[REM 58] REMUSAT DE P., "Physiologie expérimentale. – Du Système nerveux et de l'application médicale de l'Électricité", *Revue des deux mondes*, 2nd period, tome 18, pp. 407–435, 1858.

[REM 18] REMONDIERE R., "Évolution historique de la publicité et vulgarisation de l'électrothérapie au XIXe et au XXe siècles", *Kinésithérapie, la Revue*, vol. 18, no. 202, pp. 29–35, 2018.

[RHY 98] RHYS MORUS I., *Frankenstein's Children: Electricity, Exhibition, and Experiment in Early-Nineteenth-Century London*, Princeton University Press, Princeton, 1998.

[RHY 99] RHYS MORUS I., "The measure of man: Technologizing the Victorian body", *History of Science*, vol. 37, pp. 249–282, 1999.

[RHY 02] RHYS MORUS I., *Bodies/Machines*, Berg, Oxford, 2002.

[RHY 04] RHYS MORUS I., *Michael Faraday and the Electrical Century*, Icon, London, 2004.

[RHY 09a] RHYS MORUS I., *When Physics Became King*, University of Chicago Press, Chicago, 2009.

[RHY 09b] RHYS MORUS I., "Radicals, romantics and electrical showmen: Placing galvanism at the end of the English Enlightenment", *Notes and Records of the Royal Society of London*, vol. 63, 3, pp. 263–275, 2009.

[RHY 10] RHYS MORUS I., "Worlds of wonder: Sensation and the Victorian scientific performance", *Isis*, vol. 101, no. 4, pp. 806–816, 2010.

[RHY 11] RHYS MORUS I., *Shocking Bodies: Life, Death & Electricity in Victorian England*, History Press, Cheltenham, 2011.

[RIC 85] RICHER P., *Études cliniques sur la grande hystérie ou hystéro-épilepsie*, 2nd Edition, revised and explanded, Delahaye and Lecrosnier, Paris, 1885.

[RIC 07] RICHET C., LANGLOIS P., LAPICQUE L., *Dictionnaire de physiologie*, Alcan, Paris, 1907.

[ROL 09] ROLANDO L., *Saggio sopra la vera struttura del cervello dell'uomo e degli animali e sopra il sistema nervoso*, Stamperia da S.S.R.M Privilegiata, Sassari, 1809.

[ROL 22] ROLANDO L., *Inductions physiologiques et pathologiques sur les différentes espèces d'excitabilité et d'excitement, sur l'irritation, et sur les puissances excitantes, débilitantes et irritantes*, Caille et Ravier, Paris, 1822.

[RON 43] RONDEPIERRE J., "Vié J : 'Essai de traitement de l'épilepsie par l'électrochoc'", *Paris Médical*, no. 51, p. VII, 1943.

[ROU 91] ROUSSEAU G., *Discours marchand et discours scientifique : la représentation publicitaire de la ceinture électrique au début du siècle*, pp. 187–213, Les Presses de l'Université Laval, Quebec, 1991.

[RUT 16] RUTKA R., DENIS A., VERCUEIL L. *et al.*, "Crises psychogènes non épileptiques : état des connaissances et apports de l'évaluation des traitements émotionnels", *Neurosciences affectives et santé mentale*, vol. 41, no. 1, pp. 122–139, 2016.

[SAM 08] SAMUELI J.-J., BOUDENOT J.-C., "L'invention de la pile électrochimique par Volta", *Bibnum* [Online]. Available at: http://journals.openedition.org/bibnum/719, 2008.

[SAN 72] SANS J., *Guérison de la paralysie par l'électricité : ou cette expérience physique employée avec succès dans le traitement de cette maladie regardée jusques à présent comme incurable*, Cailleau, Paris, 1772.

[SAR 25] SARLANDIERE J.-B., *Mémoires sur l'électropuncture, considérée comme moyen nouveau de traiter efficacement la goutte, les rhumatismes et les affections nerveuses, et sur l'emploi du moxa japonais en France : suivis d'un traité de l'acupuncture et du moxa*, Delaunay, Paris, 1825.

[SAU 49] SAUL L.J., DAVIS H., DAVIS P.A., "Psychologic correlations with the electroencephalogram", *Psychosomatic Medicine*, vol. 11, no. 6, pp. 361–376, 1949.

[SAY 15] SAYER E., "The organization of electro-therapy in military hospitals", *Proceedings of the Royal Society of Medicine*, vol. IX, no. I, pp. 39–46, 1915.

[SCH 48] SCHWEIGGER J.F.C., *Über das Elektron der Alten*, Greifswald, 1848.

[SCH 41] SCHWAB R.S., "The clinical application of electroencephalography", *Medical Clinics of North America*, vol. 25, no. 5, pp. 1477–1489, 1941.

[SCH 08] SCHWALB J.M., HAMANI C., "The history and future of deep brain stimulation", *Neurotherapeutics: The Journal of the American Society for Experimental NeuroTherapeutics*, vol. 5, pp. 3–13, 2008.

[SCI 82] SCIAMANNA E., "Fenomeni prodotti dall'applicazione della corrente elettrica sulla dura madre e modificazione del polso cerebrale. Ricerche sperimentali sull'uomo", *Atti della R. Accademia dei Lincei, Memorie della classe di scienze Fisiche, Matematiche e Naturali*, vol. 13, pp. 25–42, 1882.

[SCI 02] SCIALOM D., *Associations névroso-organiques : hystérie et neurasthénie*, Firmin, Montane et Sicardi, Montpellier, 1902.

[SEG 01] SEGALA M., "Électricité animale, magnétisme animal, galvanisme universel : à la recherche de l'identité entre l'homme et la nature", *Revue d'histoire des sciences*, vol. 54, no. 1, pp. 71–84, 2001.

[SEQ 15] SEQUEIRA H., D'HONDT F., "Magnétoencéphalographie et exploration de la dynamique cérébrale : Du cerveau à l'esprit", in CHERICI C. and DUPONT J.-C. (eds), *L'exploration cérébrale, Histoire récente et nouveaux outils*, pp. 131–150, Hermann, Paris, 2015.

[SHA 07] SHAHED J., POYSKY J., KENNEY C. *et al.*, "GPi deep brain stimulation for Tourette syndrome improves tics and psychiatric comorbidities", *Neurology*, vol. 68, no. 2, pp. 159–160, 2007.

[SHE 18] SHELLEY M., *Frankenstein; or, The Modern Prometheus*, Lackington, Allen & Company, London, 1818.

[SHE 49] SHERRINGTON C., *Man and His Nature: Broadcast Talks in Religion and Philosophy,* CUP Archive, Cambridge, England, 1949.

[SHE 61] SHEER D., *Electrical Stimulation of the Brain: An Interdisciplinary Survey of Neurobehavioral Sciences*, University of Texas Press, Austin, 1961.

[SIE 30] SIEBER F.-G., *Découverte sur la guérison de la paralysie, de l'émiplégie et de l'apoplexie*, Imprimerie Henry, Paris, 1830.

[SIG 76] SIGAUD DE LA FOND J.-A., *Traité de l'électricité, dans lequel on expose, et on démontre par expérience, toutes les découvertes électriques faites jusqu'à ce jour, pour servir de suite aux Leçons de physique du même auteur*, Laporte, Paris, 1776.

[SIR 11] SIRONI V.A., "Origin and evolution of deep brain stimulation", *Frontiers in Integrative Neuroscience,* vol. 5, no. 42, pp. 1–5, August 2011.

[SMI 14] SMITH C.U.M., WHITAKER H., *Brain, Mind and Consciousness in the History of Neuroscience History, Philosophy and Theory of the Life Sciences*, Springer, The Netherlands, 2014.

[SNE 95] SNETLAGE L., *Nouveau Dictionnaire Français*, Jean Chrétien Dieterich Libraire, Göttingen, 1795.

[SOE 13] SOEIRO D., "On artificial and animal electricity: Alessandro Volta vs. Luigi Galvani", *Journal of Philosophy of Life*, vol. 3, no. 3, pp. 212–237, 2013.

[SPI 47] SPIEGEL E.A., WYCIS H.T., MARKS M. *et al.*, "Stereotaxic apparatus for operations on the human brain", *Science,* vol. 106, pp. 349–350, 1947.

[SRI 55] SRIBONIUS L., *Compositiones medicae. Joannes Rhodius recensuit, notis illustrauit, lexicon scriboniaum adiecit*, Patavii, P. Frambotti, 1655.

[STE 86] STEVENSON R., *The Strange Case of Dr. Jekyll and Mr. Hyde and Other Stories*, Longsman, Green and co., London, 1886.

[STO 40] STOKES W., BELL J., *Lectures on the Theory and Practice of Physic,* Haswell, Barrington and Haswell, Philadelphia, 1840.

[STO 76] STOCK J.T., "James Cumming: A pioneer in electrical instrumentation", *Journal of Chemical Education*, vol. 53, no. 1, pp. 29–31, 1976.

[TAB 18] TABB K., "Struck, as it were, with madness", in CHERICI C., DUPONT J.C., WOLFE T.C. (eds), *Physique de l'esprit : Empirisme, médecine et cerveau (XVIIe-XIXe siècles)*, Hermann, Paris, 2018.

[TAL 07] TALAIRACH J., *Souvenirs des études stéréotaxiques du cerveau humain*, John Libbey Eurotext, Paris, 2007.

[TAL 09] TALAN J., *Deep Brain Stimulation, A New Treatment Shows Promise in the Most Difficult Cases,* Dana press, New York and Washington, 2009.

[TAN 02] TAN T.-C., BLACK P.M., "Sir Victor Horsley (1857–1916): Pioneer of Neurological Surgery", *Neurosurgery*, vol. 50, pp. 607–612, 2002.

[TAR 08] TARSY D., VITEK J.L., STARR P.A. *et al.*, *Deep Brain Stimulation in Neurological and Psychiatric Disorders*, Humana press, Boston, 2008.

[TEI 59] TEILLEUX L., *De l'application de l'électricité au traitement de l'aliénation mentale*, Masson, Paris, 1859.

[TEM 71] TEMKIN O., *The Falling Sickness*, 2nd ed., Johns Hopkins University Press, Baltimore, 1971.

[TEM 04] TEMEL Y., VISSER-VANDEWALLE V., "Surgery in Tourette syndrome", *Mov. Disord.*, vol. 19, pp. 3–14, 2004.

[TER 99] TERRY J., *An American Obsession: Science, Medicine and Homosexuality in Modern Society*, University of Chicago Press, Chicago, 1999.

[TES 40] TESTE A., *Manuel pratique de magnétisme animal, exposition méthodique des procédés employés pour produire les phénomènes magnétiques et leur application à l'étude et au traitement des maladies*, Paris, 1840.

[TES 00] TESLA N., "The Problem of Increasing Human Energy", *The Century Magazine*, vol. XXXVIII, pp. 175–211, 1900.

[TES 20] TESLA N., *Mechanical Therapy*. Available at: https://teslauniverse.com/nikola-tesla/articles/mechanical-therapy, 2020.

[THE 93] THELWALL J., *An Essay Towards a Definition of Animal Vitality*, T. Rickaby, London, 1793.

[THE 02] THELWALL J., *Selections, and Original Articles, Read and Recited, in Illustration of Mr. Thelwall's Lectures on the Science and Practice of Elocution*, R. Hurst, Wakefield, 1802.

[TIB 77] TIBBITS H., *Handbook of Medical and Surgical Electricity*, 2nd ed., J. A. Churchill, London, 1877.

[TIG 77] TIGGES W., "Behandlung der Psychosen mit Elektricität", *Allg. Zeitsch. für Psychiatrie*, vol. 34, p. 731 onwards, 1877.

[TIL 15] TILLOCH A., "Experiments of professor Aldini on galvanism", *Philosophical Magazine and Journal*, vol. XLV, pp. 220–225, 1815.

[TIL 53] TILT E.J., *On Diseases of Women and Ovarian Inflammation*, Churchill, London, 1853.

[TOR 55] TORLAIS J., "Un prestidigitateur célèbre, chef de service d'électrothérapie au XVIIIe siècle, Ledru dit Comus, 1731–1807", *Histoire de la médecine*, vol. 5, pp. 13–25, 1955.

[TOR 56] TORLAIS J., "Une grande controverse scientifique au XVIIIe siècle. L'abbé Nollet et Benjamin Franklin", *Revue d'histoire des sciences et de leurs applications*, vol. 9, no. 4, pp. 339–349, 1956.

[TOR 63] TORLAIS J., "Qui a inventé la bouteille de Leyde ?", in *Revue d'histoire des sciences et de leurs applications*, vol. 16, no. 3, pp. 211–219, 1963.

[TOS 55] TOSETTI U., *Sull'insensibilità di alcune parti degli animali, lettera...al signor Dottore G. Valdambrini*, Rome, 1755.

[TRO 88] TROOSTWYK VAN P.A., KRAYENHOFF C.R.T., *De l'Application de l'électricité à la physique et à la médecine*, Changuion, Amsterdam, 1788.

[TRO 55] TROUSSEAU A., *De l'épilepsie, leçons cliniques faites à l'Hôtel-Dieu*, Aux principales Librairies Médicales, Paris, 1855.

[TYN 76] TYNDALL, J., *Lessons in Electricity*, 2nd ed., Longmans, Green, and Co., 1876.

[TYN 85] TYNDALL J., *Leçons sur l'électricité*, 2nd ed., Gauthier-Villars, Paris, 1885.

[UNI 54-69] *L'union magnétique: journal de la Société philanthropico-magnétique de Paris,* n. 1-348, Paris, 1854-1869.

[URE 19] URE A., "An account of some experiments made on the body of a criminal immediately after execution, with physiological and practical observations", *Quarterly Journal of Science*, vol. 6, pp. 283–294, 1819.

[VAS 02] VASSALLI-EANDI A.M., *Expériences et observations sur le fluide de l'électro-moteur de Volta*, Turin, 1802.

[VEN 65] VENEL G.-F., "Médicament" in *Encyclopédie ou dictionnaire raisonné des sciences et des arts, par une société des gens de lettres,* Samuel Fauche, Neuchâtel, 1765.

[VEN 82] VENEL G.F., *Précis de matière médicale*, Volume 2, Richard, Caille et Ravier, Paris, 1782.

[VID 05] VIDAL F., "Le sujet cérébral : une esquisse historique et conceptuelle", *PSN. Psychiatrie, sciences humaines, neurosciences*, vol. 3, no. 11, pp. 37–48, 2005.

[VIG 82] VIGOUROUX P., *De l'électricité statique et de son emploi en thérapeutique*, Baillière, Paris, 1882.

[VIN 16] VINCENT C., "Le traitement des phénomènes hystériques par la 'rééducation intensive'", *Archives d'électricité médicale*, pp. 405–416, 1916.

[VIS 99] VISSER-VANDEWALLE V., VAN DER LINDEN C., GROENEWEGEN H.J. *et al.*, "Stereotactic treatment of Gilles de la Tourette syndrome by high frequency stimulation of thalamus", *Lancet,* vol. 353, p.724, 1999.

[VOL 90] VOLTA A., "On the electricity excited by the mere contact of conducting substances of different kinds", *Philosophical Transactions of the Royal Society*, vol. 2, pp. 403–431, 1790.

[VOL 93] VOLTA A., "Account of some discoveries made by Mr. Galvani, of Bologna; with experiments and observations on them. In two letters from Mr. Alexander Volta, F.R.S. Professor of Natural Philosophy in the University of Pavia, to Mr. Tiberius Cavallo, F.R.S.", in *Philosophical Transactions of the Royal Society of London,* Royal Society, London, 1793.

[VOL 02] VOLTA A., "De l'électricité dite Galvanique", *Annales de Chimie*, vol. 40, 1802.

[VOL 14] VOLTA A., *L'identita del fluido electtrico col cosi detto fluido galvanico vittoriosamente dimostrata con nuove esperienze ed osservazione*, Memoria comunicata al Signore Pietro Configliachi, Pavie, 1814.

[WAL 74] WALSH J., "Experiments on the torpedo or the electric ray at La Rochelle and isle de Ré in June–July 1772", *Observations sur la Physique*, vol. IV, pp. 205–217, 1774.

[WAL 34] WALTER G., *Thought and Brain: A Cambridge Experiment*, The Spectator, London, 1934.

[WAL 05] WALLACE J., *D. H. Lawrence, Science and the Posthuman*, Palgrave, Basingstoke, 2005.

[WAR 02] WARWICK K., *I, Cyborg*, University of Illinois Press, New York, 2002.

[WAR 12] WARWICK K., *Artificial Intelligence: The Basics*, Routledge, London and New York, 2012.

[WIC 14] WICKENS A.P., *A History of the Brain: From Stone Age Surgery to Modern Neuroscience,* Psychology Press, 2014.

[WIE 88] WIETHE TH., *Formulaire de la Faculté de médecine de Vienne donnant les prescriptions thérapeutiques utilisées par les professeurs Albert, Bamberger, Benedikt*, translation by VOGT, E., Reinwald, Paris, 1888.

[YAT 10] YATMAN M., *Galvanism, Proved to be a Regular Assistant Branch of Medicine: Also, "Interesting Inquiries Concerning this Influence, With Regard to Living Actions" [...],* G. Hayden, London, 1810.

[ZAN 10] ZANETTI F., "L'électricité du corps humain chez l'abbé Bertholon et quelques contemporains", *Annales historiques de l'électricité*, vol. 8, pp. 9–20, 2010.

[ZAN 17] ZANETTI F., *L'Électricité médicale dans la France des Lumières*, Voltaire Foundation, Oxford, 2017.

[ZIM 14] ZIMMER C., *Et l'âme devint chair*, Zones sensibles, Le Kremlin-Bicêtre, 2014.

[ZOL 85] ZOLA E., *The Beast in Man*, translated by BROWN A., Grafton, London, 1985.

Index of Names

C

D

E, F

G

N

O, P

R

S

T

U, V

W, Y, Z

Index of Terms

I, L

M

N

O, P

R, S, V

Printed and bound by CPI Group (UK) Ltd, Croydon, CR0 4YY

15/09/2026

14962358-0001